Access and Mediation

Age of Access?
Grundfragen der
Informationsgesellschaft

Edited by
André Schüller-Zwierlein

Editorial Board
Herbert Burkert (St. Gallen)
Klaus Ceynowa (München)
Heinrich Hußmann (München)
Michael Jäckel (Trier)
Rainer Kuhlen (Konstanz)
Frank Marcinkowski (Düsseldorf)
Rudi Schmiede (Darmstadt)
Richard Stang (Stuttgart)

Volume 11

Access and Mediation

Transdisciplinary Perspectives on Attention

Edited by
Maren Wehrle, Diego D'Angelo,
and Elizaveta Solomonova

DE GRUYTER
SAUR

ISBN 978-3-11-135313-5
e-ISBN (PDF) 978-3-11-064724-2
e-ISBN (EPUB) 978-3-11-064305-3
ISSN 2195-0210

Library of Congress Control Number: 2021947259

Bibliographic information published by the Deutsche Nationalbibliothek
The Deutsche Nationalbibliothek lists this publication in the Deutsche Nationalbibliografie;
detailed bibliographic data are available on the Internet at http://dnb.dnb.de.

© 2023 Walter de Gruyter GmbH, Berlin/Boston
This volume is text- and page-identical with the hardback published in 2022.
Cover image: Stockbyte/Retrofile RF/Getty Images
Typesetting: Integra Software Services Pvt. Ltd.
Printing and binding: CPI books GmbH, Leck

www.degruyter.com

Contents

Maren Wehrle
Introduction. Access and Mediation: Attention Beyond Selectivity —— 1

Part 1: Attention and Access

Diego D'Angelo
Chapter 1
Introspection and Access: Some Conceptual Remarks on Attention and the Sense of Agency —— 23

Susanne Schmetkamp
Chapter 2
Aesthetic Attention and Change of Perspectives —— 43

Felipe León
Chapter 3
Attention in Joint Attention: From Selection to Prioritization —— 65

Miguel Segundo-Ortin and Glenda Satne
Chapter 4
Sharing Attention, Sharing Affordances: From Dyadic Interaction to Collective Information —— 91

Natalie Depraz
Chapter 5
Attention as Vigilant Openness —— 113

Yuko Ishihara and Olaf Witkowski
Chapter 6
Different Ways of Attending to Experience: Formalizing the Phenomenological Epoché to Translate Between Science and Philosophy —— 135

Part 2: **Attention and Mediation**

Elizaveta Solomonova and Michelle Carr
Chapter 7
The Role of Attention and Intention in Dreams —— 163

Luis R. Sandoval and Betzamel López
Chapter 8
Improving Attention in Psychosis With Digital Tools —— 189

Bas de Boer
Chapter 9
Attending to Your Lifestyle: Self-Tracking Technologies and Relevance —— 217

Galit Wellner
Chapter 10
Attention and Technology: From Focusing to Multiple Attentions —— 239

Cor van der Weele
Chapter 11
How Can Attention Seeking Be Good? From Strategic Ignorance to Self-Experiments —— 259

Lauren Hayes and Juan M. Loaiza
Chapter 12
Exploring Attention Through Technologically-Mediated Musical Improvisation: An Enactive-Ecological Perspective —— 279

Maren Wehrle
Introduction. Access and Mediation: Attention Beyond Selectivity

Basically, everyone seems to know what attention is. It is a phenomenon that we encounter every day in our own experience, where it appears as a subjective performance in the form of a deliberately executed concentration. In psychology as well as in philosophy, attention is mostly regarded as an expression of the selective character of perception and thought, one that allows us to occupy ourselves with certain things usually for a particular purpose, while allowing other possible contents of consciousness to fade into the background and away from our focus. In this respect, the definition of attention by the psychologist William James has not lost its meaning: "Everyone knows what attention is. It is the taking possession of the mind, in clear and vivid form, of one out of what seems several simultaneously possible objects or trains of thought. [. . .] It implies withdrawal from some things in order to deal effectively with others [. . .]." (James 1890, 403–404).

The failure to concentrate or to employ selective attention strategies would seem to prevent our ability to execute a task, to work or to perform acts of thinking effectively; a lack of attention is seen as a distraction, as mind-wandering, or it can be defined in its more pathological forms as Attention Deficit disorder. Since the industrial revolution, from the mechanization of labor up to the recent digitalization of work and of our private environments, such 'lacks' of attention have increasingly become a problem for science, education and for the economy to solve. Starting with the sensory and informational demands of the technological workplace, such as the cockpits of British pilots in World War II, early cognitive psychology began its research on attention with the aim to respond to informational overloads. Such research assumes a limitation of the human processing system, in which only a few pieces of information can be processed properly per given time frame. In this sense, attention was understood as a kind of filter mechanism, one that 'decides' which information is permitted access to a deeper level of processing and thereby may find its way into our consciousness and memory. In the history of experimental research in cognitive psychol-

Acknowledgments: The editors would like to thank Karoline Mertens and especially Sören Fiedler at the Institute for Philosophy of the University of Würzburg for their help in editing the manuscript.

https://doi.org/10.1515/9783110647242-001

ogy from 1950 onwards, we continue to see attention conceptualized mostly as a selective cognitive or neuronal mechanism in the form of an auditory filter (Broadbent 1958), a visual spotlight (Posner et al. 1980), an attentional set (Folk et al. 1992) or saliency map (Jonides and Yantis 1988), and which is either spatially, task or goal oriented, or driven by salient stimuli. Even though, since James's insight in the 19th Century, we have acknowledged that we have the capacity to vividly possess objects or thoughts with our mind, how we might (better) control, capture or guide this capacity for 'attention', and ultimately avoid distractions, seems to remain a problem.

Today, one is not only confronted with an increasing amount of, and accelerated, visual stimuli and linguistic information within one's workplace, but, with the advent of ICT's (Internet Communication Technologies) and social media, the stimuli and information distinctions between the public and private domains are increasingly blurred. In the age of the internet, where social media sites, companies and influencers fight for our time, clicks, and feedback, attention is indeed the most valuable currency. The pressing question, here, seems to concern whether we are in fact in control of our own attention; do we possess the modern stream of information or does the world of infinite content and click-bait rabbit-holes possess us instead? That said, attention is about more than control over our own capacity for it: after all, we need and do in turn seek inter-personal recognition and hence we strive to capture, and perhaps even manipulate, the attention of others.

1 No One Knows What Attention Is

Although attention seems more important and existentially needed than ever before, do we really know what we seek, lack or want to control about it? What do we even mean by 'attention'?

If we take a closer look at the empirical research on attention in the last few decades, we might come to doubt the self-evidence of James' statement about what attention is. Instead, we are confronted with the tension between two methodological perspectives. First, from a personalistic perspective, attention is described as a human behavior, as is demonstrated in James' formulation: the 'taking possession of the mind' (op. cit.). Second, from a naturalistic perspective, attention is seen as a neuronal or functional mechanism. This latter perspective does depend on the former: while empirical cognitive psychology departs from the (mere) descriptive phenomenon or problem of attention, its empirical methods continue to rely on the observation and on the interpretation of the attentional behavior of subjects. Despite this, the definition of attention is now formulated merely in a naturalistic way, that is, as a system "responsible for maximizing the efficient utilization of our limited capacities to process, store, and retrieve information" (Hommel 2019, 2290).

The shift from the description of a behavioral phenomenon or an experience of concentration, called 'attention', to the assumption that attention is nothing but a mechanism that causes or explains the respective behavior in question reflects another, less heuristic, shift from reflecting on attention as something, a phenomenon, that should itself be explained (the *explanandum*) to the cause or explanation of attentional behavior (the *explanans*). When, and how, did this happen? And, indeed, is the naturalistic definition able to explain what attention is or, even, why it is relevant to understand it? Even if we put aside the utilitarian wording and usage of economic and technical metaphors of this definition – which employs adjectives such as 'maximizing', 'efficient', 'utilization', 'limited capacities', 'processing', etc. – we are still left with the problem of circular reasoning, where an observed behavior is aligned with an underlying (set of) mechanism(s) that in turn causes and explains this behavior.

However, in an obvious discrepancy to this simple and general definition, when opening a psychological textbook or, simply, researching attention online, one cannot help but be confronted by an overwhelming variety of attentions or its sub-functions. Here, we find attention defined as *vigilance* (general status of high awareness), as *feature integration* (ability to selectively integrate information belonging to one event within and across sensory modalities; cf. Treisman and Gelade 1980) or as *selective attention* (ability to ignore misleading information or an irrelevant spatial location). Furthermore, it is differentiated between *spatial attention* (ability to prioritize processing of events from a particular location); *focused attention* (selection of external events for further internal processing); *involuntary attention* (processing of irrelevant information); attention as *visual search* (where one systematically searches for a target object or event); *divided attention* (ability to perform multiple tasks at one time which is enhanced through training and habit formation); *selective attention for action* (where attention is the ability to control spatial parameters of eye movements to select features or objects relevant for a task at hand); *goal-centered attention* (where attention enables us to prioritize one goal over another); *object-centered attention* (where attention allows us to prioritize one object, memory item, thought or conscious representation over others); and, last but not least *sustained attention* (the consolidation of information for later use, and/or the concentration in anticipation of a possible event over some time) (cf. Hommel 2019, 2290; Eysenck and Keyne 2000; Styles 2006).

In light of this variety of forms or subfunctions of attention, it would seem highly unlikely that one might be able to identify only one set of either functionally or neuronally-defined mechanisms. Moreover, all of the above identified forms of attention are directly related to the empirical research designed to test specific and isolated attentional tasks or subfunctions rather than the processes themselves. For example, attention is often studied via visual search tasks, in which subjects are asked to identify a target with respective features, such as a simple picture of a geometrical figure, like a colored circle or square, with the influence of distractors or location cues; attentional success is based on the time it takes to find the search target. The choice of tasks, therefore, represents how individual researchers operationalize 'attention' or 'attentional behavior'. Debates in the history of attention research are thus necessarily related to specific experimental designs and paradigms that measure specific tasks, which the researchers then assume to represent 'attention' or attentional behavior as such.

Since the 1950s the research debates in cognitive psychology centered around, among other things, the following questions: is attention voluntary or involuntary? Conscious or automatic? Stimulus driven or, rather, task-driven? Controlled either exogenously or endogenously? Where does the supposed selection occur: at an early stage of sensory processing (visual features)? Or, at later stages (identified objects, linguistic meanings)? Does attention refer merely to input control or also to behavioral output (action control)? Such questions imply opposing or binary solutions. Indeed, viewed from a philosophical, especially phenomenological or enactive perspective (cf. D'Angelo 2020), these debates appear at times artificial as they tend to lose the connection to attention as a dynamic real-life phenomenon: within these highly specialized and isolated debates, some of which span over decades, one can seldom find an attempt to understand what attention is in a more comprehensive or integrative way.

The analytical psychological approach to attention, which invokes the neat separation of different functions like attention, intention, decision-making, perception, and relates them to separate functional and or physiological systems, seems to make the task of understanding attention increasingly difficult. If one wants to neatly differentiate between what psychology understands as 'intention' (i.e., the willingly carried out action or focusing behavior) and 'attention', then one must render the latter concept a mere selection of sensory input that is not in any way related to action (i.e., output). But, would this definition have anything in common then with the phenomenon we started out explaining? Why should a human subject or organism choose this stimuli, object or event over a variety of others if it is not in any way related to what one needs, wants or has to do, understood in a practical, aesthetic or epistemological sense?

Why are some stimuli, objects or events more relevant, preferred or salient in the first place? These terms are by definition relative, especially with regard to the (experienced) environment, that is, the objective context or spatial horizon, or with regard to the organism in question. Something emerges as salient in contrast to other stimuli or objects, like a red tomato within a green field but not within a field of red tulips. Moreover, what is salient for living beings, like humans, is not only relative to the natural, spatial or featural context, but to the respective bodily needs, conditions and skills in relation to their environment. In this sense, not every object or environment affords the same attention or action for all organisms. For cultural beings, like humans, saliency is also relative to what one has experienced and learnt before (what one is used to), what one individually, socially or culturally prefers or finds relevant in general or with regard to a current task, project or goal.

Even a brief consideration of attention research in cognitive psychology brings us to the conclusion that, contra James, "no one knows what attention is" (Hommel et al. 2019). Given this situation, Hommel et al. argue for an alternative approach: instead of the common analytical approach – that starts with a complex phenomenon or concept and tries to subdivide it into simpler parts – they suggest to take a *synthetic* approach. By synthetic, they mean starting from a simple function or mechanism that has been shown to be behaviorally relevant. In this context, Hommel et al. turn away from the search for brain regions or functional mechanism that cause or explain attention and instead try to identify the development of relevant neuronal circuits in the evolution of early vertebrates that have enabled "animals to select and control interactions with their environments to achieve goals and avoid negative outcomes" (Hommel et al. 2019, 2297). They argue that what one actually finds when investigating neuronal circuits is not a separate mechanism but rather a set of interacting processes that ultimately create the phenomenon of selectivity. These processes interact not because they belong to the same dedicated system but because "the human brain and body evolved this way and selectivity was a necessary feature to achieve efficient behavior" (ibid., 2298).

The critique provided by Hommel, among others, aligns well with the conceptual critique coming from philosophy, which understands attention not as a separate mechanism or cognitive mode but rather as a structuring quality or unity within different modes of consciousness, cognition or sensory experience (Mole 2010; Watzl 2017). In their special emphasis on evolution, behavior and action, the proposed synthetic approach furthermore connects well with ecological (Gibson 1979), embodied (Varela et al. 1991; Gallagher 2005; Thompson 2007) and enactive approaches (Noë and O'Reagan 2001; Noë 2004) in cognitive science, philosophy of mind and phenomenology (cf. Husserl 1973, 1989; Merleau-Ponty

2012; Doyon and Wehrle 2020). Independent of internal variations and conceptual differences, all of these approaches agree that perception and action cannot be investigated separately and that cognition in general must be understood as embodied, enactive, embedded and extended.

2 Attention Beyond Selectivity

At a closer look, however, this synthetic and more comprehensive approach seems to suffer from a reductive tendency or un-reflected presupposition. Attention is conceptualized here primarily as selectivity, where selectivity is understood as means to an end of an efficient action or outcome. Although selection is surely behaviorally relevant and can be interpreted as basis or precursor for what we describe as attention, it does not have to follow that (a) the function of attention is merely or always selective or that (b) selection necessarily relates to goal-oriented behavior.

Firstly, if one defines selection in a merely exclusive way, one assumes that selection (and thus attention) is an all-or-nothing phenomenon, where stimuli are either selected (processed) or not. Furthermore, the reason why an animal or human selects some stimuli over another is exclusively understood in relation to efficiency or avoidance of negative outcomes. This seems in wording very similar to the 'conventional' definition of attention, which the synthetic approach also criticized. However, the need for a selection of stimuli or objects must not be equated with an exhaustive or definitive exclusion of other stimuli. Rather, selection could be one aspect of an overall structuring and integrating function. As such, one has to take not only the spatial but also the temporal dimension of the respective subject-environment relations into account. Attentional structuring therefore means that an organism temporarily engages more vividly with some stimuli than others. This does not imply, however, that other stimuli, features, aspects, objects or events are not processed or experienced at all or will not be relevant at a later stage or where previously focused.

Secondly, attention as a phenomenon might have evolved with regards to goal-directed actions, but it cannot – at least in its distinguished human form – be reduced to mere goal-oriented behavior. Attention does play a role in meditation, in play and creative behavior as well as in thinking, in imagining in aesthetic experience, in social interactions, and in empathy, but none of these essential human behaviors are, in the strict sense of the word, goal-oriented or merely aim at efficiency or at the avoidance of negative outcomes. One could even go so far as to argue that goal-oriented behavior is not primary for the evolution and development of most animals and humans, but rather is an open way of exploring and getting to know one's environment. This can be observed in the playful way some animals and human children discover their world and learn how to orientate themselves within it. Moreover, such a playful or explorative behavior in humans and other primates is not merely individual but also social; it is accompanied, supported and/or guided by caregivers and calls for imitation, interaction, and empathy. Most real-life selection or attention is thus not primarily individual or goal-directed in a strict sense, but indirectly or directly social. Attention is *indirectly* social in that one is implicitly guided not only by the natural but also by the social relevance and affordances of one's environment; it is *directly* social in the form of joint or shared attention which is the condition for any form of human cooperation and communication, and as such also 'behaviorally relevant' for survival.

To illustrate why it is important to look beyond (mere) selectivity when investigating attention as a dynamic phenomenon, a closer look at the concluding statement of Hommel et al.'s paper (2019) might be useful. The authors state: "everything an individual does throughout their life (distant and recent past) creates, reinforces, and shapes selection. Turning to the left makes us ignore stimuli on the right, picking one apple makes us overlook the others, saying one word prevents us from uttering any other. And each of the different selections results in all ranges of rewards, from positive gains to negative losses. Selection and reward are thus inherent ingredients of all our lives and the way we lead them" (Hommel et al. 2019, 2298; cf. also Allport 1987).

But does turning to the left really make us 'ignore' or 'overlook' the stimuli to our right? Rather, would it not be more appropriate to say that they are still conscious in the background, implicitly retained as that which we have previously perceived (i.e., before turning to the right)? These stimuli are in our minds but their presence may not necessarily be explicit; regardless, they impact what we expect to see and how we see new stimuli, including that which is to the left of us. Turning to the left enters the prior stimuli on the right into the temporal horizon of perception or stream of consciousness. This suggests that stimuli on the left (towards which we have now turned) are perceived in light of, and integrated with, what we have previously seen (Husserl 1991).

Picking one apple makes this apple and not all other apples available for closer inspection or ready to eat – that is true. But does this mean that we overlook all other apples? Again, the other apples are still there as the spatial horizon or thematic field (cf. Husserl 1983, 195ff., Husserl 2001, 42ff.; Gurwitsch 1964), in which the apple we now pick is spatially but also thematically embedded. The other apples are not currently picked by us, but are nonetheless experienced in the spatial horizon or periphery. They are ready to be picked by us (at a later time) or by other animals. The fact that we can pick *one* apple already implies that there are more apples from which we can choose. On the side of the organism, it implies that we have a need, interest or appetite for apples; and, even more importantly, that we are able to pick this apple such that we have the bodily condition (free moving arms and legs) and acquired the habitual skill (of how best to pick apples) (cf. Merleau-Ponty 2012). Picking apples thus presupposes an 'I-can' on the side of the organism (Husserl 1989, 159; Merleau-Ponty 2012, 139), which translates to a specific affordance with regard to apples (as something that affords us to be picked, cf. Gibson 1979).

Of course, it is also not possible to utter more than one word at a time, but this one word, like a tone in a melody, can only be uttered and in turn heard as part of a temporal context of other words (e.g., the words uttered/heard before or uttered/heard after). Moreover, a word would not have the meaning of a 'word' (and not mere noise) if it were not for other words. For a word to be uttered, it must be integrated into the meaningful context of other words, which is to say in the context of a language as a whole. The subject in turn must know and understand this language, and be physically and mentally able to utter words that are understandable to others.

What hopefully becomes clear in this short illustration is that selection is never an all-or-nothing, nor indeed an 'either/or' phenomenon, but one always embedded in an overall attentional behavior. Selection is not an exclusion or erasure of other not-selected objects, but rather is one aspect of a more general change of structure (foreground and background, before-after). This structuring enables, on the one hand, a closer, more detailed engagement with some of the available stimuli or objects, while embedding and integrating them within a coherent and meaningful experience on the other. The selected stimuli are thereby part of an overall spatial and temporal horizon of experience. Without these other objects and contexts the selected stimuli would be neither meaningful nor relevant for the respective organism.

"Everyone knows what attention is" – James' seemingly self-evident definition of attention points to the fact that for our experience and action not all possible objects, events or aspects of the world are equally important or present (more or less vivid, detailed, thematic) to us at a given time and within a given spatial context. Attention thus, in principle, represents a subjective differentiation or structuring of that which is objectively perceivable or noticeable – or, in the language of cognitive psychology, processable or computable. This specific engagement thereby goes beyond the scope of mere perception, it is more and less than just perception. Attention is more specific than perception in that it refers only to the singled-out feature(s), space, events, task(s) and divides the general perceptual field into a foreground and background, theme and thematic field (Gurwitsch 1964), focus and horizon. It goes beyond the scope of mere perception because its objects are present in a more thematic, vivid or detailed way, but also because attention cannot be reduced to the level of sensory perception, but is characteristic of all kinds of cognition as well as action. One could say, attention as engagement is an action, that is, we *turn* our attention towards something either mentally or literally when moving our eyes and body, and perspective (Wehrle 2013; Wehrle and Breyer 2015).

In this latter sense, attention is not merely a *static possession of a selected item*, the *result of a differentiation of a given visual field* into parts or stimuli we focus on and those we merely vaguely notice in the periphery. Approached dynamically, attention is a turning towards or away from something: a change of perspective; a striving and aiming for something. In this sense, attention determines *what we come to see*, that is to say, it opens up future horizons of perception and action. While static attention is the function of (selective) structuring of the given (cf. Gurwitsch 1964; Mole 2010; Watzl 2017), dynamic attention in its intertwinement of (habitual or current) subjective interest and external affection is the motor of every future perception and action (cf. Husserl 1973, 2004; Wehrle 2010, 2015).

3 Attention as Access and Mediation

Already in 1945, Maurice Merleau-Ponty criticized the opposing definitions of rationalism and empiricism and their binary conceptualizations of attention. Attention, according to Merleau-Ponty, is neither rational or empirical, top-down or bottom-up, initiated or controlled by the mind or the world, but rather it characterizes our specific, situated and explorative relation to the world. As he puts it in *The Phenomenology of Perception*:

> How could one real object among all objects be able to arouse an act of attention, given that consciousness already possesses them all? What was lacking for empiricism was an internal connection between the object and the act it triggers. What intellectualism lacks is the contingency of the opportunities for thought. Consciousness is too poor in the first case and too rich in the second for any phenomenon to be able to *solicit* it. Empiricism does not see that we need to know what we are looking for, otherwise we would not go looking for it; intellectualism does not see that we need to be ignorant of what we are looking for, or again we would not go looking for it. They are in accord in that neither grasps consciousness *in the act of learning*, neither accounts for this "circumscribed ignorance," for this still "empty" though already determinate intention that is attention itself.
> (Merleau-Ponty 2012, 30).

Taking this phenomenological description as our source of inspiration and starting point, this volume endeavors to define *attention* by embarking on a more comprehensive approach and thereby moving beyond seeing it as a static act of selection of environmental aspects, however behaviorally relevant. Therefore, we – the editors and authors alike – will address attention as a general *possibility* and specific *ability* that is learnt, socially shaped, enabled, skilled and/or impaired, *to access* one'e environment in a differentiated, structured, integrative and meaningful way. This 'access', as we see it, occurs on *different levels*: from sensory reception to practical action, to focused perception, thinking and reflecting. Each level is characterized by different modes of self and world-awareness, that is, from implicit and operative to explicit and thematic awareness, and they can be descriptively distinguished by *different modes* of cognition and action (e.g., practical-bodily, focal, global, joint, shared, aesthetic or moral attention).

By drawing attention to a notion of *access*, we want to point to the dynamic nature of attention itself and hence emphasize its temporality and operative character; indeed, to access something is an *action* and, as such, it *takes time*. Our understanding of access therefore includes implicit and/or explicit relations of access to oneself as an actor, and is in this sense closely related to the concepts of agency, skill and/or habits. This point will emerge as salient in a number of chapters of this volume (cf. D'Angelo; Solomonova and Carr; Sandoval and López; Hayes and Loaiza; de Boer).

At the same time, *access* points to the fact that *there is more than one way to access*, disclose or address the world or information within it; one can never access all aspects of a given world or the information it contains simultaneously. Every *access* is thereby concretely *situated*, which is to say that it is dependent on natural, cultural, social conditions and contexts as well as on the experiential past, skills and current goals of the individual. The chapters of this volume will draw out this point (cf. León; Segundo-Ortin and Satne; Hayes and Loaiza; van der Weele). As such, we understand attention as an ability, current act or as the possibility of access to an organism's environment. This means that attention cannot be a mere spotlight or mere highlighting of something that one already knows, possesses or has in one's mind or consciousness. Accessing something means, by definition, the possibility of *exploring or finding something new* that is or was formerly unseen and/or unknown (at least by that particular subject). In turn, attention as access cannot be reduced to the objective implication of external stimuli, something that one only needs in order to register or to select independently of individual or cultural circumstances.

Accessing something is thereby *not limited to a voluntary or explicit form of intentionality* (or directedness) towards the world, ourselves or others. Every active access or selection relies on passive processes or on operative intentionality, as Husserl and Merleau-Ponty would define them. Before we explicitly turn towards something or single something out in perception, it first has to affect us passively, has to be given or made present to us in some way (cf. Depraz). The fact that we can thematically access things, that is, focus or inspect something in more detail, and thus pay attention to them, does not mean that the other aspects of our perception (i.e., what is not yet or not anymore in the center of our attention) are excluded, unconscious or unprocessed, nor does it mean that this renders our overall experience unchanged. Every *new* experience, *access* or sense of the world *re-frames, re-arranges the current and possible field of experience.*

Finally, approaching attention in terms of approach calls for a *transdisciplinary approach*. Attention is not only studied systematically, then, with regard to its different levels or modes, but it is also approached from the point of view of different disciplinary perspectives that themselves have approach to different insights and methodologies. With such a transdisciplinary approach, descriptive and explanatory, first-person and third-person, behavioral observation or subject reports, and objective measurement must seek to be combined in fruitful ways.

However, concretely-speaking, current acts of attention are always *mediated* – that is, shaped by one's material, biological, economic, cultural or social circumstances, that is, concrete situation. In this regard, the concrete environmental, that is, spatial, temporal and social, context of attentional behavior becomes crucial. Especially nowadays specific cases of attention are often mediated in the literal sense by technologies. In line with post-phenomenological approaches (cf. Ihde 1990; Verbeek 2005), we understand technologies not as neutral intermediaries, but as mediums that do actively shape how one experiences and understands themselves and the world. Print media, telephone connectivity, television and smartphones, Internet Communication Technologies, search engines, virtual reality, the Internet of Things, all of these technologies shape how and what we can access of the world. They mediate how the world becomes present to people who relate to them in turn, and thereby co-constitute what stands out as relevant for us. Moreover, they also shape how we (scientifically) conceive, understand and conceptualize phenomena such as attention (cf. Wellner). Digital and augmented technology as well as internet communication devices can therefore be depicted as extensions of our bodily, sensory and informational access to the world. Even as extensions, however, mediated access can both broaden and limit one's experiential horizon; it can open up new perspectives and possibilities, but also close-off other perspectives and forms of access, especially when it is too selective, individualized or restricted. Personalized algorithms, for example, may present you with custom-fit information, tailored to past, present and supposedly future individual preferences, which has its uses when looking for suitable products and booking a vacation, but becomes problematic if what you are looking for needs to be grounded in information that is depersonalized and more objective. While personalized algorithms are problematic because we are not aware of the ways in which information is selected for us, self-tracking technologies in turn make us overly attentive to our behavior and habits.

Attention is in this sense an expression of the situated and mediated as well as interested (and thus selective) and meaningful (coherent and integrative) ways in which we are affected, directed and can address our environment, ourselves and others. On the one hand, attention is needed to access, which is to say to notice, to order, to concentrate, to integrate and to make sense of information about the world and others. On the other hand, humans, persons or social groups need the attention of others in order to be seen and heard, which consists in our participating of the cultural, social and/or political mediating of this very world.

4 Contents

In this volume, we propose a transdisciplinary framework that views attention from a particular angle: as a means of approaching, thereby disclosing the world in a practical and meaningful way. Moreover, we seek to focus on how attention is and can be technologically *mediated*, that is, how attention is shaped by technology, and how in turn technology can help us understand attention through the ways it enables and/or limits subjects to access information and participate in a shared world. Under the umbrella of access and mediation, this volume brings together theoretical, applied and empirical approaches to the subject matter, including the history of philosophy; philosophy of mind; philosophy of aesthetics; phenomenology and post-phenomenology; enactive theory; embodied cognitive science and cognitive neuroscience; ecological psychology and psychiatry; media studies and social theory. All of the contributions emphasize the dynamic and social character of attention by focusing on the role of the body, motricity/action, agency, temporality, habit, affordances, interaction, sociality, culture and technologies.

In the first part of this volume – *Attention as Access* – the chapters investigate the relation of attention to agency, aesthetics, and to joint action and social norms. In this regard, the volume reflects on the different levels and modes of attention as well as on the role of specific forms of access and attention that are in play in either philosophical or scientific research.

In his contribution, *Diego D'Angelo* (philosopher) carefully poses the question and discusses to what extent we need to be attentive in order to feel like we are the authors or owners of our actions. Indeed, in addition to this, to what extent can we access our own mental states and intentions, and how are introspection and agency related? He argues that, although (some) attention to the things we are doing is necessary for a sense of agency, this does not imply introspection. Rather this experience related to attention refers to an implicit feeling of being bodily active that has to accompany every action we assign agency to.

Susanne Schmetkamp (philosopher) emphasizes the normative character of attention, which is seldomly addressed in empirical research on attention. Schmetkamp introduces, in this regard, a differentiation between a usual perceptional or practical operative attention and a specific form of aesthetic attention. She argues that aesthetic attention is an integral part of aesthetic experience. Aesthetic attention is not only characterized by its (somehow aesthetic) object, but it must be defined by its qualitative character; it is an intensified and unified experience that differs from mere interest. In this sense, it is not a task-related but an intrinsic form of attention, which is to say that it comes with a change of perspective: we literally see something differently when we attend to it aesthetically. Aesthetic attention is thus a specific style of disclosure or access that is related to mindfulness, which in turn also has ethical implications and impact.

That attention is a social rather than merely individual behavior becomes obvious when we look at research on joint and shared attention. The investigation of attentional employment involves more than one person and occurs in joint or triadic ways. Attention is as such necessarily a communicative and social phenomena, a fact which is mostly underrepresented in empirical attention research in cognitive psychology. In his chapter, *Felipe León* (philosopher) shows why approaches that understand attention as an organizing or shaping experience, rather than as a mere selective or filtering function, are better suited to explain the transition from solitary to joint attention. Thus, he argues for a mutual enlightenment between research on attention in cognitive psychology and that on joint attention in philosophy, evolutionary anthropology and developmental psychology.

Attention as a phenomenon of access can and must be shared, and is in turn motivated and shaped by shared (social) environmental affordances. *Miguel Segundo-Ortin* and *Glenda Satne* (philosophers) introduce a long-needed relational account of social affordances, which is distinguished from a representational account. They distinguish between affordances that are provided directly by dyadic or triangular relations of agents and collective affordances that operate indirectly and lie at the basis of socio-cultural forms of life. A mail box, for example, is only salient for subjects who have the intention to mail a letter. Moreover, a mail box only invites those who want to mail a letter and, presumably, who live in a community or society with a shared analogue writing culture and a postal system. Affordances thus only exist in relation to a shared socio-normative context, and yet such a context relies on direct social interaction at its basis.

In this regard, the question of the extent to which we must share attention and affordances in order to access and guarantee an objective world, a world that is valid for all, is again made urgent. And, in what sense is attention as access necessary for mutual empathy and understanding? *Natalie Depraz* (philosopher) understands attention in this regard as the basic capacity for a vigilant openness towards the world. The fact that we can focus or reflect on something specific or that we can be affected by it already presupposes a general state of sensitivity or responsiveness. This aspect of attention as an initial openness or tendency to explore and to being curious about the world has been long ignored in the history of the philosophy of attention. Nonetheless, such an openness or responsiveness is not only the condition of every specific access to information, but also of empathy and ethical behavior. In this sense openness or responsiveness can be intentionally and methodologically cultivated as practices of meditation or reflection. The phenomenological epoché introduced by Edmund Husserl, for example, is based on a shift of our attention from the perceived objects in the world to the very way this world (and object) is accessed by us, and motivates in turn an inquiry into the conditions of this access.

When we reflect on our experience, our attention shifts, for example, from the objects of our experience towards the experience of the objects. However, the way this experience is addressed either philosophically or scientifically is rather different. In their contribution, *Yuko Ishihara* (philosopher) and *Olaf Witkowski* (computational scientist) take the phenomenological method of 'bracketing' or epoché as their starting point in order to distinguish between three different ways of reflection on experience, the physical, psychological and the phenomenological-transcendental reflection. In developing a formal notation of these different ways of access, they provide a methodological framework which helps to clarify, compare, and translate philosophical and scientific accesses into each other and avoid logical misunderstandings. Moreover, the introduced matrix, which clarifies the conditions in which variables and operators require careful translation, helps to apply, mediate and implement philosophical differentiations into scientific research.

This leads over to the *second part* of our edited book – *Attention and Mediation* – which focuses on concrete, specific cases of attention (as in dreams, pathologies, musical performance, or social behavior) that often is mediated or investigated by way of technologies.

In the first chapter of this part, *Elizaveta Solomonova* and *Michelle Carr* (cognitive scientists), who both study the psychophysiology of sleep, investigate attention as a means of accessing and mediating the dream world. In their enactive approach to dreams, dreams are understood as embodied imagination. By way of this approach, they argue that attention is (a) a constitutive factor and (b) a trainable practice or skill that provides access and can mediate dream experience. Attention provides access to dreams via practices of recording and sharing dreams that enable, recall and enhance the richness of the dream experience; attention, in this respect, is a mediator of dreams in the sense that incubation, imagery rehearsal, and ultimately lucidity can be cultivated as cognitive skills enabling agency in the dream experience.

Luis Sandoval and *Betzamel López* (psychologists) also emphasize the role of attention as a trainable skill that can help patients to cope with psychosis. Schizophrenia, and other similar mental illnesses, are often accompanied by cognitive and attentional deficits. Sandoval and López thus combine interpersonal therapy with computerized neurocognitive treatment to improve cognitive abilities such that one might train attentional skills which then help patients to improve higher-order attention in schizophrenia. This treatment allows patients to quiet and 'encapsulate' their auditory hallucinations, which in turn expands their social skills and enables them to participate in social interactions again.

While in dream research and psychiatry, attention is revealed as a trainable skill that helps to establish or regain agency in limit cases, self-tracking technologies are explicitly designed to enhance agency and control of our bodies in everyday life. Contrary to personalized algorithms that are designed to capture and manipulate our attention, these health care technologies are designed with the purpose to help people lead a healthier lifestyle. In his contribution, *Bas de Boer* (philosopher) provides us with a careful analysis of how self-tracking technologies re-structure our attention in that they turn our body and habits into visible objects of relevance, indeed, objects that must be continuously attended to. Self-tracking devices are therefore not neutral technologies: they embody normative assumptions that pre-structure what it means to be healthy (e.g., taking at least 10,000 steps per day), and privilege certain ways of turning our habits into explicit 'projects'. Self-tracking devices thereby shape the attention of users by making some aspects of the world stand out at the expense of others.

Technologies do structure and shape what we see of the world and how we approach it. This is also true when we look at how different technological developments have influenced and framed our concepts and research on attention itself. In her chapter, *Galit Wellner* (philosopher) provides us with a genealogy of attention through the lens of major shifts in media development, from print media, telephone, television to cellphones and the internet. She argues that the concept of attention and the development of these technologies has been co-constitutive. The implication of this is that the dominant concept of attention as selective, which results from the dichotomy perceived between focus *versus* distraction, is in part attributable to the technologies available at the time. However, at least with the rise of the use of the mobile phone, which has made multi-tasking an everyday phenomenon, one has to admit that attention in the plural is possible or even that such multi-attentions are the norm rather than the exception.

What becomes obvious in this case is that the way we address and understand attention is mediated by technologies, but also by cultural and social pre-assumptions. Being distracted is not only considered epistemologically, evolutionary or economically problematic (as less efficient or a disadvantage for survival) but also socially problematic. Being distracted means that we are not able to concentrate on what is relevant (e.g., work) or to provide something or someone with the attention they deserve or demand. Being focused or paying attention is in this sense evaluated as socially and even sometimes morally good, whereas being distracted or seeking attention (as opposed to giving) is assumed to be a negative personal character trait. For this reason, *Cor van der Weele* (philosopher and biologist) asks why is it that seeking attention has such a bad moral reputation? This moral bias towards attention leads to a strategic ignorance in research when it comes to the investigation of the behavior of seeking attention. However, if we look at attention as a social phenomenon it becomes clear that we not only give but also hope to receive attention. Cor van der Weele appeals to a more sympathetic approach to attention-seeking. In an experiment she conducted with her own students, she was able to show that an explicit reflection on one's attention-seeking behavior can lead to more social understanding and acceptance.

Last but not least, *Lauren Hayes* (sound artist and researcher) and *Juan Loaiza* (philosopher) investigate attention in technologically-mediated musical improvisation. Digital technologies enable new participatory forms of musicking that in turn provides researchers with the possibility to investigate attention in action. In this respect, Hayes and Loaiza argue for an enactive and ecological psychological approach that can account for the dynamic, temporal and interactive dimension of attention. They investigate how attention unfolds in different temporalities, and how it depends on sensorimotor as well as social histories of past experiences (or input). In the context of musical improvisation that involves processes of sense-making and co-creation, the necessity of a sort of attention that is embodied and participatory becomes evident.

References

Allport, D. A. (1987): "Selection for Action: Some Behavioral and Neurophysiological Considerations of Attention and Action". In: H. Heuer; H. F. Sanders (eds): *Perspectives on Perception and Action* (395–419). Hillsdale: Lawrence Erlbaum.

Bekkering, H.; Neggers, S.F.W. (2002): "Visual Search Is Modulated by Action Intentions". *Psychological Science* 13, 370–374.

Broadbent, D (1958): *Perception and Communication*. London: Pergamon Press.

Comoli, E.; Das Neves Favaro, P.; Vautrelle, N.; Leriche, M.; Overton, P. G.; Redgrave, P. (2012). "Segregated Anatomical Input to Subregions of the Rodent Superior Colliculus associated with Approach and Defense". *Frontiers in Neuroanatomy* 6, 9.

Craighero, L.; Fadiga, L.; Rizzolatti, G.; Umiltà, C. A. (1999). "Action for Perception: A Motor-Visual Attentional Effect". *Journal of Experimental Psychology: Human Perception and Performance* 25, 1673–1692.

D'Angelo, D. (2020). "The Phenomenology of Embodied Attention". *Phenomenology and the Cognitive Sciences* 19, 961–978.

Doyon, M.; Wehrle, M. (2020). "Body". In: D. De Santis; B. C. Hopkins; C. Majolino (eds.): *The Routledge Handbook of Phenomenology and Phenomenological Philosophy* (Chapter 9). New York: Routledge.

Eysenck, M.W.; Keane, M.T. (2000). *Cognitive Psychology: A Student's Handbook*. 4th edition. Philadelphia: Psychology Press.

Fagioli, S.; Hommel, B.; Schubotz, R. I. (2007). "Intentional Control of Attention: Action Planning Primes Action Related Stimulus Dimensions". *Psychological Research* 71, 22–29.

Folk, C.L.; Remington, R.W.; Johnston J.C. (1992). "Involuntary Covert Orienting is Contingent on Attentional Control Settings". *Journal Experimental Psychology: Human Perception & Performance* 18, 1030–1044.

Gallagher, S. (2005). *How the Body Shapes the Mind*. Oxford: Oxford University Press.

Gibson, J.J. (1979). *The Ecological Approach to Visual Perception*. Boston, MA: Houghton Mifflin Harcourt.

Grossberg, S. (1973). "Contour Enhancement, Short Term Memory, and Constancies in Reverberating Neural Networks". *Studies in Applied Mathematics* 52, 213–257.

Gurwitsch, Aron (1964). *The field of Consciousness. Theme, Thematic Field, and Margin*. In: *Phaneomenologica. Collected Works III*, ed. by R. M. Zaner. Dordrecht: Springer.

Herrero, L.; Rodriguez, F.; Salas, C.; Torres, B. (1998). "Tail and Eye Movements Evoked by Electrical Microstimulation of the Optic Tectum in Goldfish". *Experimental Brain Research* 120, 291–305.

Hommel, B.; Chapman, C.S.; Cisek, Neyedli, H.F.; Song, J.-H.; Welsh, T.M. (2019). "No One Knows what Attention Is". *Attention, Perception & Psychophysics* 81, 2288–2303.

Hommel, B. (2010). "Grounding Attention in Action Control: The Intentional Control of Selection". In: B.J. Bruya (ed.): *Effortless Attention: A New Perspective in the Cognitive Science of Attention and Action* (121–140). Cambridge, MA: MIT Press.

Husserl, E. (2004). *Wahrnehmung und Aufmerksamkeit. Texte aus dem Nachlass (1893–1912)*. In: *Husserliana XXXVIII*, ed. by T. Vongehr/R. Giuliani. Dordrecht: Springer.

Husserl, E. (2001). *Analyses concerning Passive and Active Synthesis. Lecture on Transcendental Logic*. In: Collected Works LX, trans. by A. Steinbock. Dordrecht: Kluwer Academic Publishers.

Husserl, E. (1991). *On the Phenomenology of the Consciousness of Internal Time (1893–1917)*. In: Collected Works IV, trans. by J. B. Brough. Dordrecht: Kluwer Academic Publishers.

Husserl, E. (1989). *Ideas pertaining to a Pure Phenomenology and to a Phenomenological Philosophy. Second Book. Studies in the Phenomenology of Constitution*. In: Collected Works III, trans. by R. Rojcewicz/A. Schuwer. Dordrecht: Kluwer Academic Publishers.

Husserl, E. (1983). *Ideas pertaining to a Pure Phenomenology and to a Phenomenological Philosophy. First Book. General Introduction to a Pure Phenomenology*. In: Collected Works II, trans. by F. Kersten. Dordrecht: Kluwer Academic Publishers.

Husserl, E. (1973). *Experience and Judgement. Investigations into a Genealogy of Logic*. New York: Routledge and Keagan Paul.

Ihde, D. (1990): *Technology and the Lifeworld: From Garden to Earth*. Indianapolis: Indiana University Press.

James, W. (1890). *The Principles of Psychology*. Volume 1. New York: Henry Holt & Company.

Jonides, J.; Yantis, S. (1988). "Uniqueness of Abrupt Visual Onset in Capturing Attention". *Perception & Psychophysics* 43, 346–354.

Kahneman, D. (1973). *Attention and Effort*. Englewood Cliffs, NJ: Prentice-Hall.

Luck, S. J.; Gaspelin, N.; Folk, C.L.; Remington, R.W.; Theeuwes, J. (2021): "Progress toward Resolving the Attentional Capture Debate". *Visual Cognition* 29:1, 1–21.

Merleau-Ponty, M. (2012 [1945]). *The Phenomenology of Perception*, trans. by D. E. Landes. New York: Routledge.

Mole, C. (2010). *Attention is Cognitive Unison. An Essay in Philosophical Psychology*. Oxford: Oxford University Press.

Mysore, S. P.; Knudsen, E. I. (2011). "The Role of a Midbrain Network in Competitive Stimulus Selection". *Current Opinion in Neurobiology* 21, 653–660.

Navon, D. (1984). "Resources – a Theoretical Soup Stone?". *Psychological Review* 91, 216–234.

Noë, A.; O'Reagan, K. (2001). "A Sensorimotor Account of Vision and Visual Consciousness". *Behavioral and Brain Sciences* 24:5, 883–917.

Noë, A. (2004). *Action in Perception*. Cambridge, MA: MIT Press.

Posner, M. I (1980). "Orienting of Attention". *Quarterly Journal of Experimental Psychology* 32, 3–25.

Styles, E.A. (2006): *The Psychology of Attention*. Second Edition. New York: Routledge.

Suchman, R.G.; Trabasso, T. (1966). "Color and Form Preference in Young Children". *Journal of Experimental Child Psychology* 3, 177–187.

Thompson, E. (2007). *Mind in Life. Biology, Phenomenology, and the Sciences of the Mind*. Harvard: Harvard University Press.

Treisman, A.; Gelade, G. (1980). "A Feature-Integration Theory of Attention". *Cognitive Psychology* 12:1, 97–136.

Varela, F.; Thompson, E.; Rosch, E. (1991). *The Embodied Mind. Cognitive Science and Human Experience*. Cambridge, MA: MIT Press.

Verbeek, P.P. (2005): *What Things Do: Philosophical Reflections On Technology, Agency, and Design*, trans. by R.P. Crease. Pennsylvania: The Pennsylvania State University Press.

Wang, X. J. (2002). "Probabilistic Decision Making by Slow Reverberation in Cortical Circuits". *Neuron* 36, 955–968.

Watzl, S. (2017). *Structuring Mind: The Nature of Attention and How It Shapes Consciousness.* Oxford: Oxford University Press.

Wehrle, M. (2015). "'Feelings as the Motor of Perception'"? The Essential Role of Interest for Intentionality". *Husserl Studies* 31:1, 45–64.

Wehrle, M.; Breyer, T. (2015). "Horizonal Extensions of Attention: A Phenomenological Study of the Contextuality and Habituality of Experience". *Journal of Phenomenological Psychology* 47:1, 41–61.

Wehrle, M. (2013). *Horizonte der Aufmerksamkeit: Entwurf einer dynamischen Konzeption der Aufmerksamkeit aus phänomenologischer und kognitionspsychologischer Sicht.* München: Wilhelm Fink Verlag.

Wehrle, M. (2010). "Die Normativität der Erfahrung – Überlegungen zur Beziehung von Normalität und Aufmerksamkeit bei Edmund Husserl". *Husserl Studies* 26, 167–187.

Part 1: **Attention and Access**

Diego D'Angelo
Chapter 1
Introspection and Access: Some Conceptual Remarks on Attention and the Sense of Agency

Abstract: In my contribution I will try to address the role that attention plays in our sense of agency. In order to achieve this, I will proceed in the following steps. The first section of the paper will review current empirical research showing that some level of attention is required for action control and, conversely, that action can modify attention (for an overview see Bruya 2010). Moreover, I will show that the relative position and movements of bodily organs can boost or impair our attention span, referring to some illustrative results in cognitive sciences (such as Yu, Smith and Pereira 2007, Thura et al. 2008, Yu and Smith 2012).

In the second section, I will claim that such findings need to be taken seriously when reflecting philosophically about the concept of attention. Attention is not merely the intellectual capacity of directing the focus of one's mind toward something, but implies a feedback relation to action. Therefore, I will introduce and define the concept of attention-action-feedback: according to this model, whenever we pay attention to something, bodily action is involved.

In the third part I will tentatively spell out the consequences of this model for the question concerning our sense of agency (for an overview see Pfister 2019). We have a sense of agency when we pay some level of attention to what we are doing. But paying attention does not imply introspection or a purely 'mental state': paying attention means being bodily active, so that a phenomenology of attention does not need concepts such as introspection or mental state.

Some closing remarks will sum up the main arguments of the paper and point out further directions of inquiry.

Introduction

> "What is left over if I subtract the fact that my arm goes up from the fact that I raise my arm?"
> Wittgenstein, *Philosophical Investigations*, §621

In this paper, I will discuss the role that attention plays for our sense of agency. I aim to show that an embodied account of attention allows a description of the sense of agency without having to resort to introspection.

The framework of embodied attention can be spelled out in the claim that attention necessarily involves corporeality – i.e., there is no purely intellectual act of paying attention – and that attention constitutively modifies our experience – i.e., not-attended-to phenomena are experienced as different from phenomena we pay attention to (for more details on this, cf. D'Angelo 2020). Usually, we tend to take attention as a mostly cognitive, rational capacity at our disposal, over which we can exert at least some degree of control. Within this framework, attention is usually understood as the directing of one's own mind to things in the surrounding world, and the effect of attention, often framed within the metaphor of a spotlight (Posner's neurobiological approach, among many other, provides a good example of a spotlight model of attention: Posner et al. 1980), is to bring these things stronger 'into focus', 'to light', or to make our experience of them clearer and more distinct.

But even within the limits of everyday experiences, away from the lab, simple examples show that attention is not purely cognitive, but rather involves action and corporeality. It is sufficient to think about the experience of hearing someone screaming. In an everyday context, hearing someone screaming implies a turning of attention that happens mostly in a passive way; my reaction will usually imply some bodily movements (at least of the head or of the eyes) in order to recalibrate my position in space as to put the source of the scream into the centre of my experiential field; I will also probably experience some kind of emotional involvement and so on. Another everyday example is the experience of listening to or playing music. In this regard, we often feel the need to accompany listening to music with some kind of movement (tapping our fingers, moving the head or straight-forward dancing); furthermore, the experience of the same piece of music from the perspective of the listener differs greatly from the experience of the very same piece of music by the performer who actively plays the instrument. These and other examples show that, at least in everyday contexts, movements of the body play a crucial role in attentional behavior.

The leading question of this paper thematizes the implications such an embodied approach to attention has for the sense of agency. 'Sense of agency' designates a pre-reflective awareness that my actions are my own (for some introductory remarks on the sense of agency, cf. Haggard 2017). Do we need to be attentive in order to feel like we are the authors and the owners of our actions?[1] Or is there a sense of agency even without any kind or level of attentiveness? let us take an example in order to better understand the philosophical and scientific importance of such questions. Imagine I am walking down the road and I am completely immersed in thinking about this paper. For this reason, I fail to notice that a cat is sleeping on the pavement and I accidentally step on its tail. Let us assume for the sake of the argument that I am usually kindly inclined toward our feline companions. In that case I may feel like saying (or communicate in some way) to the cat that I am sorry to have hurt its tail and that I did not mean to.

In such a case there is no intention (I did not mean to step on the tail of the cat), and attention is directed somewhere else (I was thinking about this paper and precisely for that reason I did not see the cat). Yet, the fact that I still feel responsible for what happened seems to indicate that the sense of agency, that is to say, the feeling that I am the owner of my action, is independent from the intentionality of the action as well as from the attention paid to the action itself. A similar problem arises in cases of 'weak agency' (cf. Ryberg Ingerslev 2020), where I repeat the same action over time without realizing it and not knowing why I do it.

According to the received view, attention and sense of agency are indeed two separate problems in psychology and in philosophy of mind. Whether I am paying attention or not, the classical approach claims, my actions are my actions and I am held responsible for them because I ought to pay attention to my actions but failed to do so. This 'ought' arises from the fact that I am assumed to have a sense of agency related to whatever action I carry out (for a review on attention and sense of agency, cf. Hon 2017).

The crucial question in order to understand the relation between sense of agency and attention concerns the way we can access these experiences: how do we assess our sense of agency and our own attentiveness while carrying out a certain action? The trivial answer to this question is introspection: through analysis of and reflection on our mental states we come to say that under some description we were attentive or inattentive, and that we caused the action or not. This amounts to saying that we have access to our agency primarily through introspection. Within such a traditional framework stressing the role of introspection for the sense of agency, different approaches have attempted to account for how exactly this works. For example Chambon, Sidarus and Haggard (2014) contrast the standard view, in which results of an action are retrospectively compared with the intended

---

[1] I understand sense of agency to be a broader phenomenon than sense of control, insofar as we can know that we are the authors of an action without having control of that action, for example when we make errors.

outcome, with a prospective view, in which sense of agency is based on predictions of future states and of action outcomes. Indeed, approaches in which post-hoc computations via introspections are involved seem to be the most widespread type of account of the sense of agency. According to Buhrmann and Di Paolo,

> the most popular accounts of the sense of agency, the experience of being the author and initiator of our own actions, are in-the-head approaches. They involve comparisons between intended and actual states of the body and the world, as if experiencing oneself to be the agent of an action were first and foremost a question of verifying whether these states match or not. (Buhrmann and Di Paolo 2017, 209)

The question I will address in the following pages concerns the conceptual consequences of bypassing introspection in order to focus on the distinctive features of our experience of agency as such. Reviewing some of the literature claiming that a least some of the neural and psychological processes at work while carrying out actions can give rise to a sense of agency without the need for retrospective or prospective scrutiny, I will support the embodied understanding of attention, that can shed some conceptual light on this problem; in particular, I will take a look at forms of embodied attention that can help in spelling out the link between carrying out an action and being attentive. New insights have indeed emerged in the laboratory and in real-life settings that contrast the received view by showing both that attention is required for action and how action can in turn influence attention.[2]

My main claim will be that it is possible to decouple attention and sense of agency, like in the example above, only if attention is understood as a purely intellectual faculty accessible via introspection. However, granted that attention and motricity[3] are co-dependent and that attention should be understood as embodied,

---

[2] A discussion of unreflective actions (cf. for example Far 2015) and of Situated Action Theories (cf. for example the seminal paper by Vera and Simon 1993) claiming that action planning and representations are irrelevant for human activity would exceed the aims of this paper; nonetheless, is is worth noting that an embodied attention theory of the sense of agency does not exclude and does not presuppose representation, making it potentially a strong ally against cognitive and informational theories of action.

[3] In order to have a handy concept for both bodily actions and body-posture, I would like to introduce the concept of motricity. This has to be understood as a very general motor function, including both actual movements and potential movements arising from a specific body position and body posture. The concept of motricity indicates therefore both the position of the body in the surrounding world and its movements and actions. I will therefore use the concept of 'motricity' in order to indicate actions, action planning, and body posture as well as the position of the body in space. The concept of motricity can therefore be linked to the sensory-motor approach of O'Regan and Noë (2001), as well as to the phenomenological descriptions

then introspection is no longer needed, since it is possible to rely purely on the bodily experience of attention viz. of the sense of agency. Just as we do not need introspection to know if we are walking or swimming, but we just do it, we also do not need introspection to know if we are paying attention. To sum the main claim in a nutshell: an embodied and enactive conception of attention allows a description of the sense of agency without resorting to introspection, but rather as a direct experience of one's activity.

In order to spell out my argument, I will proceed in the following two steps. The first section of the paper will review current empirical research showing that some level of attention is required for action control and, conversely, that action can modify attention. Moreover, with reference to some illustrative results in cognitive sciences, I will show that the relative position and movements of bodily organs can boost or impair our attention span. Here, I will claim that such findings need to be taken seriously when philosophically investigating the concept of attention. Attention is not merely the intellectual capacity of directing the focus of one's mind toward something, but instead it involves a feedback loop in relation to action. Therefore, I will introduce and define the concept of 'attention-motricity-feedback' as further specification of a theory of embodied attention. According to this model, whenever we pay attention to something, bodily action is involved.

In the second part I will attempt to elucidate the consequences of this model for the question concerning the sense of agency (for an overview, cf. Pfister 2019). We have a sense of agency when we pay some level of attention to what we are doing. However, paying attention does not imply introspection or a purely 'mental' state. Paying attention means being bodily active, so that having a sense of agency neither requires introspective access to ongoing experiences nor is based on calculations regarding action outcome. Access to agency only needs embodied attention, and it will become clear how both attention and sense of agency modify our access to the body and to the world.

---

of corporeality and movement provided by Merleau-Ponty (2012). I do not, however, mean the concept of motricity to imply a particular philosophical approach to problems in philosophy of mind and in philosophy of perception; I use the concept of motricity as a purely indicative concept. Therefore, this concept is also not identical with the concept of corporeality, since it excludes other aspects of the living experience of the body such as pain and emotions in order to focus on the body and its actions.

1 Attention and Motricity: Results from Empirical Research

In this section I will claim that paying attention involves a feedback loop with motricity. More specifically, I will show that attention implies actions or certain body postures, and that actions and body postures modify attention. In order to do this, I will review some of the current empirical research on the feedback relations between attention and body posture, as well as between attention and action. This will allow me not only to provide a quick summary on the state of the art, but also to review the concepts currently in use. The overall goal is to show that a strong relationship between attention and motricity is suggested by recent literature in empirical research. More precisely, I will claim that the relationship between attention and motricity can be characterized as a feedback relationship. This amounts to saying that attention impinges on and is influenced by the movements or by the position of bodily parts, and, conversely, that the actions we make and the positions we assume in space are influenced by and impinge on our attention. It can be said, as an overall description of the current state of the art, that although empirical research has consistently shown that attention is linked to action, these kinds of findings still need to be accounted for at a philosophical level.

Since the beginning of nineteenth- and twentieth-century psychological and physiological investigations of attention, empirical research has mostly focused on action-based mechanisms of attention. Mostly under the heading of 'attention and performance', a series of contributions have tried to understand how paying attention to what we do can increase performance and how, conversely, being distracted by other elements in our environment can impair the success of our actions. This research approach (for a review, cf. Pashler et al. 2001) focuses not so much on action itself, but on the outcome of action. When dealing with performance, researchers are interested in parameters such as the quantity of actions that can be done per unit of time (speed) or the precision of the outcome (such as in experiments concerning the selection of an appropriate target).

However, although empirical research convincingly shows that paying attention to a task can directly improve success, we are all familiar with what seems to be the exact opposite phenomenon. This has motivated more recent studies to abandon the performance-oriented approach in order to describe attention more holistically. Just to mention two examples, the concept of flow (cf. Csíkszentmihályi 1996) and research on effortless attention (cf. Bruya 2010) shows that if somebody masters a highly complex skill (say, playing the piano), it looks like their actions are effortless and their attention is wandering. A skilled performer does not seem to need the same amount of attention and concentration that a novice needs. More precisely, it has been shown that, after consistent practicing, carrying out the required movements needs less muscular energy or overall physical effort (cf. Lay et al. 2002; Sparrow et al. 1999; for an early review, cf. Sparrow and Newell 1998). In addition, it has also been shown that the mental effort associated with movement production decreases with practice (some classical examples can be found in Abernethy 1988; Leavitt 1979; Smith and Chamberlin 1992). Do skills therefore contradict the idea that attention boosts performance? In order to shed some light on these problems an important conceptual distinction has been introduced by Wulf, Höß, and Prinz (2018). As Gabriele Wulf and Rebecca Lewthwaite aptly explain, they "draw [. . .] a subtle but important distinction between [. . .] internal (body-related) and external (movement-effect-related) content of performer's thought" (Wulf and Lewthwaite 2010, 76). With a small reformulation one could claim that it is necessary to distinguish between attention directed to the body and attention directed toward the effects of bodily movements and actions. Indeed, the expert piano player is thoroughly focused on listening to the music they play, rather than on singular muscular movements. In order to have a 'flowing' musical performance, the performer must not pay any overt attention to single movements of the fingers. The novice, on the contrary, needs to pay attention to the fingers hitting the right key, and they are therefore not able to focus externally on the quality of the performance itself. As every music student knows, music teachers insist that musicians acquire the ability to listen to themselves while playing, that is to say, the ability to focus one's attention on the outcome of the performance instead of, say, on the right position of the fingers on the keyboard. In brief, by distinguishing different kinds of attentional directedness – that is, by distinguishing between an externally directed attention and an internally directed attention – we can claim that the connection between attention and performance remains stable. Attention increases performance, provided it is externally directed attention.

If the idea that paying attention to something increases performance and action outcomes is common-place both in philosophical literature and in psychological research, not as trivial is the converse claim that actions can influence the way we pay attention to something. However, this is precisely what emerges from recent findings. This is probably best exemplified by how saccadic movements are linked to attention. According to Zhao et al. 2012:

> The preparation of saccades can evoke a variety of attentional effects, including attentionally-mediated changes in the strength of perceptual representations, selection of targets for encoding in visual memory, exclusion of external noise, or changes in the levels of internal visual noise. The visual changes evoked by saccadic planning make it possible for the visual system to effectively use saccadic eye movements to explore the visual environment.
>
> (Zhao et al. 2012, 40)

According to Zhao and colleagues, therefore, pre-saccadic shifts are responsible for changes in our capacity for visual selection. This happens before we actually do move our eyes. Hsing-Hung Li, Jasmine Pan and Marisa Carrasco (2019) have recently showed that "visual performance and neural responses for the saccade target are enhanced" by presaccadic shifts of attention, but they also pointed out that "it remains unknown whether presaccadic modulations on feature information are flexible, to improve performance for the task at hand, or automatic, so that they alter the featural representation similarly regardless of the task". Indeed, they go on to show that presaccadic attention not only enhances the target of the saccadic movements, but "actually modifies its featural representation" (Li et al. 2019, 1).

We can conclude then that the preparation of certain movements influences the way in which attention is paid to certain targets. Even more clearly, Harrison, Mattingley and Remington (2012) have stated that experiments show "a systematic bias of attention in the direction of the saccade," thereby further underscoring and confirming the claim of at least some kind of dependence of attention on motricity. These and other similar experimental results have brought to a renewed interest for the so-called "premotor theory of attention", proposed by Giacomo Rizzolatti and colleagues already in the 1980s (cf. Rizzolatti et al. 1987, as well as more recently Rizzolatti and Craighero 2005). This theory claims that attention, both overt and covert, relies on the preactivation of the oculomotor system (cf. Szinte et al. 2019) in order to reach the target. If this is correct, then we can assume that attention depends to some extent on the planning of motor actions.

A similar claim has been put forward in a series of experiments designed and carried out by Davood Gozli and colleagues. They sum up their results in a very straight-forward way: "[During attentive processes] self-caused features can be selected faster" (Gozli et al. 2016, 464). Consistently with sensorimotor accounts of vision, the authors claim that self-caused features are processed faster even if they are less salient than other features in one's surroundings. Gozli and colleagues notably claim that aspects of our surrounding environment which are caused by us (for example, the drawing on a piece of paper I just made) tend to be selected faster, that is to say, they are earlier accessible then other features of the environment. However, whereas accessibility for attention is directly proportional to the role of involved actions, salience (i.e., the vividness with which we perceive or experience such features) is inversely proportional to actions: something we made ourselves is less salient, although more directly accessible, than other properties of the surrounding world. Within an embodied account of attention, this result can be easily explained: the drawing as a result of our own actions is already coupled with attention and therefore readily accessible, but – since we already know it very well and our interest will be, under certain circumstances, diminished – it does not 'stand out' from the environment as something unknown that as such attracts our interest and the attention connected to this very interest. However, attention is linked to actions and to motor dispositions of the body in a variety of other ways. In particular, different research findings from the late 1990s have shown that attention depends in important respects on the position of the body in space and on its posture. For example, Tipper, Howard and Houghton (1998) have shown that attention is based on a hand-centered spatial frame of reference, claiming therefore that the body modifies the availability and accessibility of attention as a resource. They write that "research [. . .] demonstrates that when the behavioural goal is to reach for an object, the frame of reference in which the objects are represented and on which selective inhibition mechanisms act, is hand-centered" (Tipper et al. 1998, 1386).

If we take those findings seriously, we must sum up the current state of the art – at least partially, as far as the claims put forward in this paper are concerned – by saying that attention and motricity are shown to influence each other in a variety of ways. Indeed we can claim that attention has motoric components. When we pay attention to something, the motricity of our body conditions our capacity of paying attention, and paying attention to something conditions the motricity of our body.

It would be possible to expand these analyses in order to encompass other aspects of corporeality not included in the concept of motricity. For instance, the experience of pain is deeply related to attention: watching, say, a thrilling or a passionate movie can make us forget about a slight pain, say, on the knee; conversely, paying attention while in pain can be in some circumstances almost impossible. Just to discuss one single example, Veldhuijzen and colleagues (2006) have shown also at the experimental level that pain and attention interfere in a variety of ways:

> Our findings indicate that highly demanding attentional task performance and pain processing interfere as a result of difficulties in allocating attention. The clinical relevance of this finding is that performing a highly demanding task might distract attention from pain.
> (Veldhuijzen et al. 2006, 11)

What's more, not only attending to a task, but even attending to virtual stimuli can influence our experience of pain (cf. Keefe et al. 2012). At this point, a hypothetical attention-motricity-feedback model can be formulated. According to this model, whenever we pay attention to something, bodily components in the broad sense of motricity previously spelled out – that is, as a concept encompassing both actual actions and body posture – are involved. As for now, this is still an unwarranted claim, since so far it has only been showed that in certain situations, under certain conditions, attention is dependent on and influences motricity, so that the concept of attention-motricity-feedback can only come to be employed in a determinate amount of cases and is not applicable to all the ways in which we pay attention, both in the lab and in everyday contests. What about, for example, following a scientific talk in which the speaker is explaining their theory with such a quiet voice that the smallest movement of my hands would make it impossible for me to catch the next words? Or what about classical lab settings in which the subjects are requested to pay attention in a purely visual manner by sitting still on their chairs?

In order to show that attention and motricity are co-dependent, I will single out a series of two conditions that need to be met in order for attention to work properly. Let us start by analyzing the situation in which I am trying to pay attention to a scientific talk. It seems trivial to point out that even in this case, some minimal motricity as well as corporeality is involved in the process. First of all it can be noticed that – retrieving a problem already mentioned before – my body needs not to be in pain (or at least to suffer only a small, manageable amount of it) in order to be really able to pay attention to the ongoing talk; therefore, we could say that one condition for attention is (1) that *the body be sound*. Moreover – and going on with supposedly trivial points – (2) *my basic needs must have been already satisfied*. My bowels need to be empty, my stomach full, I must not be thirsty, excessively tired, ill and the like. Otherwise, instead of directing the attention toward the scientific talk, the body will reclaim attention to itself. In order for us to have access to attention, the body must fulfill certain criteria that allow it to slip into the background, leaving free sight for surrounding things in my environment. Precisely as in the case of the musicians listening carefully to their own music and not paying attention to the movements of their finger, performance (here, understanding the scientific talk) is dependent on outwards directed attention, which is in turn dependent on the capacity of the body to slip into the background. Both these conditions also apply – as every experimenter knows well – to lab settings; participants with body pain or who for a variety of reasons – such as illness and many more – are distracted by their own body are not appropriate subjects for attention studies. Granted that (1) and (2) need to be fulfilled in order to be able to pay some amount of attention, then it is necessarily true that attention depends on the body. More precisely, attention is dependent on the way in which our body is felt or experienced by us in a certain situation, thereby confirming the assumptions of the attention-motricity-feedback hypothesis.

2 Sense of Agency

It seems reasonable now to point out that, if attention is embodied and enactive, and most notably if between attention and motricity there exists a feedback relation, then this must have some consequences for the current debates on the sense of agency. Is the feedback between attention and action in any way directly felt or experienced, thereby impinging on the way in which the subjects think of themselves as the owners of their actions?

Sense of agency can be defined, following the seminal paper by Patrick Haggard, as "the feeling of making something happen. It is the experience of controlling one's own motor acts and, through them, the course of external events" (Haggard 2017). We can differentiate a sense of agency involved in actually doing a certain action or movement, and a sense of agency in motor-planning. I will focus, in the following, firstly on the sense of agency experienced in actual movements, and only after I will go into details about the sense of agency experienced in motor-planning.

There are different competing theories on the table concerning the sense of agency. In the following I will limit the discussion to the so-called ideomotor theory, as this is one of the newest and most promising approaches to the problem of how the sense of agency has to be understood.

The crucial feature of the ideomotor theory is getting rid of the classical stimulus-response framework that constitutes the main line of most sense of agency theories. The ideomotor theory notably claims that the subject is not just simply the receiver of stimuli that can be processed in order to plan and carry out a certain action. Researchers supporting the ideomotor theory avoid, instead, speaking of a passive reception of stimuli, and insist that the stimuli we perceive are actually the result of our intentions. Hommel, for instance, writes: "Most theoretical accounts of the selection, planning, and control of human action follow the seemingly self-evident custom to begin their theoretical analysis with the stimulus, which is then taken to trigger, and to some degree 'explain', action-related processes". However, Hommel goes on, "most stimuli we are exposed to are actually generated by our own actions [. . .] we do not passively await stimuli to get us going but actively generate and seek the stimulus we intend to perceive. In the beginning is the act" (Hommel 2013, 113). Hommel's straightforward definition has been abundantly quoted in many papers focusing on sense of agency. However – at least from a philosophical point of view – it seems to have a conceptual flaw. I will briefly discuss this point and how we can accommodate the problem.

The definition given by Hommel claims that the stimulus is generated and sought after according to what the subjects intend. Therefore it seems that the actual source (the actual beginning of the process) is not the action carried out in the generation and search for the stimulus, but the intention to perceive. Since the act of seeking and generating stimuli follows a certain intention, the passage should more correctly say: in the beginning is the intention. Intention leads selective attention to seek and create stimuli for action. In the usual way of describing experience, intentions come temporally first, and actions are the second steps. The outcome is temporally in third place.

But if the theory I presented in the first part of this paper concerning attention and motricity is correct, then we can see that Hommel is actually right. In other words, action can be viewed as the origin of the process if we assume an attention-feedback loop. Hommel's theory could be then schematized as follows:

Intention (idea) → Action → Perceived Stimulus

At this point, ideomotor theorists would add one last element to the schema and assume that via introspection – that is, via attention directed toward one's own mental states – we can recognize our own sense of agency. This would mean in a nutshell that if I find among my mental states the intention to carry out this particular action or to perceive this particular stimulus, then I am the author of this action. Ideomotor theorists explicitly state that the core tenets of their theory stem from William James, who was very keen on pointing out that attention has crucial corporeal aspects (cf. D'Angelo, forthcoming). According to James, we can describe the intentions that underlie actions via introspection, so that we get the following schema:

Intention (idea) → Action → Perceived Stimulus → Introspection (attention paid to one's own mental states) → Sense of Agency

Now the question is: how do attention and its doings come into the schema? The ideomotor theorist's approach follows from a very classical conception of attention, one in which attention is an intellectual capacity used only after the action has been carried out in order to analyze mental states and solve the problem of my agency. But this is not the picture of attention that is emerging in later years and that I have been spelling out in this paper. According to this new line of work, attention is already involved at the level of action; carrying out an action implies some level of attention and, in turn, influences the way in which one pays attention – and therefore the way I experience my surrounding world, that is, my access to phenomena. Therefore one could write more precisely:

Intention (Idea) → *Action/Attention* → Perceived Stimulus → Introspection → Sense of Agency

In this (obviously highly oversimplified) schema one can notice that ideation (the creation of an intentional idea by the subject) is the real starting point of the chain of events, and this should suit ideomotor theories well. Usually, one could claim that sense of agency is born out of a particular feeling given when the idea at the start of the chain in some way coincides with what we experience, and that this feeling is produced via introspection (Blakemore et al. 2002; Hughes et al. 2013; Wegner and Wheatley 1999).

But one can also notice a certain tension in the chain, because at a closer look it becomes clear that attention seemingly plays a double role. On the one hand, attention is involved in the actions and events that bring about the perceived stimulus (i.e., at step two of the chain). On the other hand, attention is also necessarily involved in introspection, if we continue to use the established definition given above of introspection as paying attention to one's own mental states. Therefore the question arises if the theory would not be better off without such a redoubling of the attention. Can we apply Occam's razor and simplify the model?

If attention is embodied, then attention is related to action and embodiment in such a way that action and embodiment are both modifiers of and are modified by attention. But the fascinating and crucial property of attention is, as shown before when discussing premotor theories of attention, that it is involved and dependent not only in and on action, but also in and on motor-planning. A series of recent contributions (cf. Chambon et al. 2014 for a review) have clearly shown that, indeed, the same psychological and physiological mechanisms that make it possible to select between alternative courses of action play a crucial role in sense of agency. In particular, Sidarus et al. 2017 claim that sense of agency is based on neural signaling and processing at the moment in which actions are initiated, but also on neural signaling and processing related to the outcome of the action. Moreover, they even claim that the more difficult the selection of action is, the more the sense of agency is reduced:

> This suggests that the signals related to action selection which influence S [sense of agency, DD] could be better described as relating to confidence in selecting or having selected the appropriate response, and not only to selection fluency as has been previously described. Our results therefore link prospective sense of agency to the processes of action monitoring and cognitive control. (Sidarus et al. 2017, 1)

Sense of agency is therefore not (exclusively) based on an introspective comparison between intention (ideation) and outcome, but selecting an action itself gives rise directly to sense of agency. Indeed, the framework of embodied attention can help in conceptually understanding what is at stake. Action selection is not a purely mental activity of focusing on abstract calculations in the effort to assess the best course of action. Rather, attention is a pre-motor activity and is therefore experienced as such. Just as we do not need introspection to know if we are walking or swimming, we *will not need introspection* to know if a certain action was indeed our action. In the most cases,[4] *it is the way in which we experience* something that tells us immediately if we were paying attention, and we immediately know if that particular action has been our action. This is the case because we experience it to be so. Agency is experienced differently than unconscious movements, and this difference rests on the fact that agency is based on attention, whereas unconscious movements are not directly attended to. This is the same difference as the one between walking down the road and breathing.[5]

To sum up the main claim: An embodied conception of attention allows a description of sense of agency without resorting to introspection, inasmuch as it can describe how attention influences our access to our experience of the surrounding world and of the actions we carry out in it. We have a sense of agency when we pay some level of attention to what we are doing. But paying attention does not imply introspection or a purely 'mental' state: paying attention means being bodily active, so that a phenomenology of attention does not need concepts such as introspection in order to explain sense of agency.

The idea that we do not need introspection for a sense of agency has already been spelled out in research – the only novelty of my approach consists in assigning to the concept of attention a more prominent place. As Buhrmann and Di Paolo write, for example, "a sense of the bodily self as an agent, in this view, corresponds to what we experience during the ongoing adventure of establishing, losing, and re-establishing meaningful relations between ourselves and the world. This is what an enactive account of the sense of agency should endeavor to explain" (Buhrmann and Di Paolo 2017, 210).

Indeed, the strongest argument for the experiential (as opposed to reflective-introspective) character of sense of agency derives from a series of experiments showing that sense of agency immediately modifies our experience of

---

[4] I am not claiming that this has to be the case for all kinds of action or for all kinds of attention, but only that this is the case in the majority of experiences we have and that this could therefore become a paradigm around which to orient further research into specific exceptions. Surely examples of covert attention binding to actions performed with different levels of unawareness would need to be excluded from this picture. A satisfying classification of types of actions and types of attention will be presented in a subsequent paper.

[5] The same contrast between experiencing a sense of agency and introspectively knowing about it is spelled out by Gallagher 2007, although in reference to schizophrenia and without discussing explicitly the concept of attention.

our body and of our surrounding world. Just as attention modifies and is influenced by motricity, so is sense of agency as well: both attention and sense of agency modify our access to the body and world.

In recent years, empirical findings have suggested that a sense of agency changes our access to motricity, thereby reinforcing its similarities to attention. Synofzik, Vosgerau, and Newen (2008) propose a crucial distinction in the field of research on sense of agency. In their seminal paper they distinguish between judgment of agency and feeling of agency as the two components of the sense of agency. Whereas the judgment of agency is a reflective, cognitive activity of the subjects that consciously try to categorize a certain action as their own, the feeling of agency "refers to the subjective experience of fluently controlling the action one is currently making, and is nonconceptual" (Haggard and Tsakiris 2009, 243). As Haggard and Tsakiris go on to show, experiments both in psychology and neuroscience demonstrate that the feeling of agency modifies our experience of our body as well as our experience of the surroundings. In other words, just as attention conditions the access to the body and the world, so does the sense of agency. Haggard and Tsakiris spell out three ways in which these modifications of access can come to take shape. Firstly, the feeling of agency modifies the way we perceive time. The results of actions thought to be our own are experienced as closer in time as actions that are not our own. Secondly, sense of agency modifies the intensity of sensory input: if an event in the world is perceived to be caused by us, we will experience it more saliently. And finally, the feeling of agency modifies our access to what we have called motricity and to the way in which the subjects understand and experience their own body. The latter aspect can be easily exemplified if we think about the typical thriller movie scene in which the murderer looks at their hand after the deed screaming 'what have I done?!'. Our own body seems to become alien to us precisely in the moment in which we try to find out if we have really done what we have done.

What is at stake is therefore not just a parallelism between attention and sense of agency, but the theoretical recognition that, if an embodied account of attention is at our disposal, we can dispense with introspection and claim that agency on attentively carried out actions is just as directly experienced as we experience motricity itself.

Conclusion

So, what about me inadvertently squeezing the cat's tail? Usually, we would claim that I squeezed the cat's tail because I was not paying attention. More accurately, one should say that, on the contrary, I stepped on the cat's tail because I was paying attention only to the paper I had to write. The fact that my attention was channelled on something different made it possible that I could step on the cat's tail without noticing it. The question of agency arises precisely only because I was paying attention to something else. In the moment in which the cat screams, I realize what happened, and I immediately know that I stepped on it, without any kind of reflection, introspection or post-hoc reconstruction.

I hope to have provided in this chapter a first hint on how the concept of attention, if understood in an embodied fashion, can shed some light on the way we access our experiences and come to think that we are the owners of our actions. Indeed, the distinctive ways in which the sense of agency – here understood as a feeling of agency, in contrast to any judgment of agency – modifies our access to experience are common to attention as well. Even if the present paper contents itself with having shown similarities between the two fields of inquiry, I hope to have been able to point to a fruitful direction for further research: sense of agency and attention are deeply interwoven with each other, and rely so heavily on motricity, that an embodied conception of attention can shed light on some of the most important and obscure features of the sense of agency. How exactly this is so will need to be the topic of future work.

References

Abernethy, B. (1988): "The Effects of Age and Expertise upon Perceptual Skill Development in Racquet Sport". *Research Quarterly for Exercise and Sport* 59, 210–221.

Blakemore, S. J.; Wolpert, D. M.; Frith, C. D. (2002): "Abnormalities in the Awareness of Action". *Trends in Cognitive Science*, 6(6) 237–242.

Bruya, B. J. (ed.) (2010): *Effortless Attention: A New Perspective in the Cognitive Science of Attention and Action*. Cambridge, MA: MIT Press.

Buhrmann, T.; Di Paolo, E. (2017): "The Sense of Agency. A Phenomenological Consequence of Enacting Sensorimotor Schemes". *Phenomenology and the Cognitive Sciences* 16, 207–236.

Chambon, V.; Sidarus, N.; Haggard, P. (2014): "From Action Intentions to Action Effects: How Does the Sense of Agency Come about?". *Frontiers in Neuroscience*, https://doi.org/10.3389/fnhum.2014.00320

Csíkszentmihályi, M. (1996): *Creativity: Flow and the Psychology of Discovery and Invention*. New York: Harper Perennial.

D'Angelo, D. (2020): "A Phenomenology of Embodied Attention". *Phenomenology and the Cognitive Sciences* 19, 961–978.

D'Angelo, D. (Forthcoming): "Attention and Action. William James on the Genesis of Intentions". *Journal of the British Society for Phenomenology*.

Far, A. (2015): "Conceptuality of Unreflective Actions in Flow: McDowell-Dreyfus Debate". *Journal of General Philosophy* 1/2, DOI: 10.5176/2345-7856_1.2.14

Gallagher, S. (2007): "Sense of Agency and Higher-Order Cognition: Levels of Explanation for Schizophrenia". *Cognitive Semiotics* 2007, 32–48.

Gozli, D.G.; Pratt, J. (2011): "Seeing while Acting: Hand Movements Can Modulate Attentional Capture by Motion Onset". *Attention, Perception, & Psychophysics* 73, 2448–2456.

Gozli, D.G.; West, G.; Pratt, J. (2012): "Hand Position Alters Vision by Biasing Processing through Different Visual Pathways". *Cognition* 124, 244–250.

Gozli, D.G.; Aslam, H.; Pratt, J. (2016): "Visuospatial Cueing by Self-Caused Features: Orienting of Attention and Action-Outcome Associative Learning". *Psychonomic Bulletin & Review* 23, 459–467.

Haggard, P. (2017): "Sense of Agency in the Human Brain". *Nature Reviews Neuroscience* 18, 196–207.

Haggard, P.; Tsakiris, M. (2009): "The Experience of Agency: Feelings, Judgments, and Responsibility". *Current Directions in Psychological Sciences*, https://doi.org/10.1111/j.1467-8721.2009.01644.x

Harrison, W.J.; Mattingley, J.B.; Remington, R.W (2012): "Pre-Saccadic Shifts of Visual Attention". *PLoS ONE* 7, e45670.

Hommel, B. (2013): "Ideomotor Action Control: On the Perceptual Grounding of Voluntary Actions and Agents". In: W. Prinz; M. Beisert; A. Herwig (eds.): *Action Science. Foundations of an Emerging Discipline, Cambridge*, MA: MIT Press, 113–135.

Hon, N. (2017): "Attention and the Sense of Agency. A Review and Some Thoughts on the Matter". *Consciousness and Cognition* 56, 30–36.

Hughes, G.; Desantis, A.; Waszak, F. (2013): "Mechanisms of Intentional Binding and Sensory Attenuation: the Role of Temporal Prediction, Temporal Control, Identity Prediction, and Motor Prediction". *Psychology Bulletin* 139:1, 133–151.

Keefe, F. J.; Huling, D. A.; Coggins, M. J.; Keefe, D. F.; Rosenthal, M. Z.; Herr, N. R.; Hoffman, H. G. (2012): "Virtual Reality for Persistent Pain: A New Direction for Behavioral Pain Management". *Pain* 153:11, 2163.

Lay, B. S.; Sparrow, W. A.; Hughes, K. M.; O'Dwyer, N. J. (2002): "Practice Effects on Coordination and Control, Metabolic Energy Expenditure, and Muscle Activation". *Human Movement Science* 21:5-6, 807–830.

Leavitt, J. L. (1979): "Cognitive Demands of Skating and Stickhandling in Ice Hockey". *Canadian Journal of Applied Sport Sciences*, 4:1, 46–55.

Li, H.; Pan, J.; Carrasco, M. (2019): "Presaccadic Attention Improves or Impairs Performance by Enhancing Sensitivity to Higher Spatial Frequencies". *Scientific Reports*, https://www.nature.com/articles/s41598-018-38262-3.

Merleau-Ponty, M. (2012): *Phenomenology of Perception*, trans. by D. E. Landes. London: Routledge.

O'Regan, J. K.; Noë, A. (2001): "A Sensorimotor Account of Vision and Visual Consciousness". *Behavioral and Brain Science*, 24:5, 939–1031.

Pashler, H.; Johnston, J. C.; Ruthruff, E. (2001): "Attention and Performance". *Annual Review of Psychology* 52, 629–651,

Pfister, R. (2019): "Effect-Based Action Control with Body-Related Effects: Implications for Empirical Approaches to Ideomotor Action Control". *Psychological Review* 126:1, 153–161.
Posner, M.L.; Snyder, C.R.; Davidson, B.J. (1980): "Attention and the Detection of Signals". *Journal of Experimental Psychology* 109:2, 160–174.
Rizzolatti, G.; Craighero, L. (2005): "The Premotor Theory of Attention". In: Itti, L.: Rees, G.: Tsotsos, J. K. (eds.): *Neurobiology of Attention*. New York et al.: Academic Press, 181–186.
Rizzolatti, G.; Riggio, L.; Dascola, I.; Umiltá C. (1987): "Reorienting Attention across the Horizontal and Vertical Meridians: Evidence in favor of a Premotor Theory of Attention". *Neuropsychologia* 25: 31–40.
Ryberg Ingerslev, L. (2020): "On Possible Self-Understanding in Cases of Weak Agency". *Frontiers in Psychology* 25:11, https://doi.org/10.3389/fpsyg.2020.558709
Sidarus, N.; Vuorre, M.; Haggard, P. (2017): "How Action Selection Influences the Sense of Agency: An ERP study". *NeuroImage* 150, 1–13.
Smith, M. D.; Chamberlin, C. J. (1992): "Effect of Adding Cognitively Demanding Tasks on Soccer Skill Performance". *Perceptual and Motor Skills* 75:3, 955–961.
Sparrow, W. A.; Hughes, K. M.; Russell, A. P.; Le Rossignol, P. F. (1999): "Effects of Practice and Preferred Rate on Perceived Exertion, Metabolic Variables and Movement Control". *Human Movement Science* 18:2–3, 137–153.
Sparrow, W. A.; Newell, K. M. (1998): "Metabolic Energy Expenditure and the Regulation of Movement Economy". *Psychonomic Bulletin & Review* 5, 173–196.
Synofzik, M.; Vosgerau, G.; Newen, A. (2008): "Beyond the Comparator Model: A Multi-Factorial Two-Step Account of Agency". *Consciousness and Cognition* 17:1, 219–239.
Szinte, M.; Puntiroli, M.; Deubel, H. (2019): "The Spread of Presaccadic Attention Depends on the Spatial Configuration of the Visual Scene". *Scientific Reports* 9, 14034 https://doi.org/10.1038/s41598-019-50541-1
Thura, D.; Hadj-Bouziane, F.; Meunier, M.; Boussaoud, D. (2008): "Hand Position Modulates Saccadic Activity in the Frontal Eye Field". *Behavioural Brain Research* 186:1, 148–153.
Tipper, S. P.; Howard, L. A.; Houghton, G. (1998): "Action-Based Mechanisms of Attention". *Philosophical Transactions of the Royal Society London B: Biological Sciences* 353:1373, 1385–1393.
Veldhuijzen, D. S.; Kenemans, J. L.; De Bruin, C. M.; Olivier, B.; Volkerts, E. R. (2006): "Pain and Attention: Attentional Disruption or Distraction?". *Journal of Pain Research* 7:1, 11–20.
Vera, A. H.; Simon, H. A. (1993): "Situated Action: A symbolic Interpretation". *Cognitive Science* 17:1, 7–48.
Wegner, D. M.; Wheatley, T. (1999): "Apparent Mental Causation. Sources of the Experience of Will". *American Psychologist* 54:7, 480–492.
Wittgenstein, L. (2009): *Philosophical Investigations*. London: Wiley-Blackwell.
Wulf, G.; Höß, M.; Prinz, W. (1998): "Instructions for Motor Learning: Differential Effects of Internal versus External Focus of Attention". *Journal of Motor Behavior* 30:2, 169–179.
Wulf, G.; Lewthwaite, R. (2010): "Effortless Motor Learning?: An External Focus of Attention Enhances Movement Effectiveness and Efficiency". In: B. Bruya (ed.): *Effortless Attention: A New Perspective in the Cognitive Science of Attention and Action*. Cambridge, MA: MIT Press, 75–101.
Yu, C.; Smith, L. B. (2012): "Embodied Attention and Word Learning in Toddlers". *Cognition*, 125(2), 244–262.

Yu, C.; Smith, L. B.; Pereira, A. (2007): "From the Outside-in: Embodied Attention in Toddlers". In: F. Almeida e Costa, L. M. Rocha, I. Harvey, & A. Coutinho, A. (eds.): *Advances in Artificial Life*. Berlin & Heidelberg: Springer, 445–454

Zhao, M. et al. (2012): "Eye Movement and Attention: The Role of Presaccadic Shifts of Attention in Perception, Memory and the Control of Saccades". *Vision Research* 1:74, 40–60.

Short Biography

Diego D'Angelo, PhD, is currently a postdoctoral researcher and a lecturer in philosophy at the University of Würzburg, Germany. He got his PhD at the University of Freiburg im Breisgau with a thesis on Edmund Husserl's phenomenology of perception, now published by Springer (*Zeichenhorizonte. Semiotische Strukturen in Husserls Phänomenologie der Wahrnehmung*). His main research topics include phenomenology, perception, and attention.

Susanne Schmetkamp
Chapter 2
Aesthetic Attention and Change of Perspectives

Abstract: Although attention is obviously an essential property of both consciousness and perception, it has not been taken seriously in philosophical aesthetics. My paper strives to fill the gap by focussing on *aesthetic attention* as a specific form or kind of attention itself. My aim is to define aesthetic attention more clearly than it has been done so far; sometimes it is conceptualised as 'disinterested attention', which just seems to be a more elaborate term for the Kantian notion of disinterestedness or distance. And sometimes it is compared with absorption or immersion (what is definitely not the Kantian view point, though). My thesis, however, is that aesthetic attention is something in between aesthetic distance or disinterestedness and immersion; both *presuppose* a form of attention: namely attention as a specific commitment and stance towards the object which we not only *perceive* (like I perceive a chair) but which we also *intend* to perceive in another way than everyday life objects and which we uncouple from any practical interests. But we are still committed. To perceive, experience, and appreciate something aesthetically means that we attend to it differently: concentrated, within a temporal frame, and without a practical objective. Moreover, instead of selecting and evaluating the respective features of the object with regard to practical interests, in aesthetic perception we distribute our attention. However, we are at the same time focused or even immersed, experiencing the 'here and now', as John Dewey puts it in his theory of *Art as Experience*. That said, aesthetic attention does not only deploy different forms of attention in a different manner, such as focused or distributed attention, but it also seems to demand a specific *kind* of attention. Possibly the aesthetic domain offers something akin to attention training similar to the way in which meditation evidently does: a form of deep attention which differs from our everyday attention but is at the same time valuable for it.

Note: The research and the writing for this article has been conducted thanks to funding by the Swiss National Science Foundation (SNSF). The topic is part of my broader research project on the "Aesthetics and Ethics of Attention" at the University of Fribourg (Switzerland). Parts of it have been presented at different conferences and workshops. I am thankful for helpful comments, especially by members of my team, colleagues at the Philosophy Department and the editors of this volume.

Introduction

A paper on *aesthetic attention* can or should approach its topic from two vantage points: namely philosophical aesthetics and philosophy of attention. From these two angles, my aim is to elucidate the concept of *aesthetic experience* and its interrelation with attention. Attention is traditionally defined as the capacity of our mind by which we focus on some things and thereby withdraw from others (James 1890). The cognitive sciences and philosophy of mind in particular have concentrated on this *focused* attention, metaphorically expressed by the 'spotlight' metaphor: Attention has been compared with a spotlight that projects a beam of light onto only one specific object or limited region of space at a time; whereby other things are shielded, out of focus, so to say (Posner et al. 1980). However, as phenomenologists since Husserl, Gurwitsch, and Merleau-Ponty have stressed, attention has not only such a focus like a beam of light; it also has a context and a margin that finds its way into consciousness. They describe it rather like a continuum or a 'thematic whole' of features of a scene (be it perceptual or intellectual), where some of the features are present and relevant, others are around and semi-relevant and some are almost absent and irrelevant (Arvidson 2010). When we focus on something, we also recognize the context in which we are embedded with a specific perspective and take the perspectivity and affordances of our environment into account for our actions (Wehrle and Breyer 2016). Another contemporary philosopher puts it like this: Attention prioritizes the manifold stimuli of the environment and structures the mind into foreground and background (Watzl 2017).

Additionally, attention does not just come in degrees of focus; it is sometimes divided or distributed. This is because we often have to react to a plurality of stimuli quasi-simultaneously and we also live in a society in which the cognitive mode of hyper attention – an attention that switches quickly between different inputs – has replaced a mode of deep attention (Hayles 2007). But as trends and studies in Yoga and Meditation practices have also shown, there seem to be both the possibility and the need for other forms of attention and awareness, which differ from the standard focused attention as well as from forms of divided or hyper attention. Especially with regard to the information flood and the increasing demands of flexibility, mobility and economic achievements in Western meritocracies people seem to look for a relief from the typical cultures of attention.

That said, attention can also come in the mode of attentiveness or mindfulness, and this again seems to differ from the default mode of focused attention toward a particular object. And last but not least, we can speak of attention in the sense of a claim on us by other people. In the following, I want to consider another type of attention, which differs in some respects from these other forms and thereby offers a specific realm of experience, namely *aesthetic attention*. The hypothesis of the paper is that aesthetic attention is a form which is an integral part of aesthetic experience, be it with art, nature, sport or other potential aesthetic objects. This mode of attention is characterized not just by its object: that is, it is directed toward something that we experience aesthetically. It is also characterized by an aesthetic quality in itself. Aesthetic attention, so to speak, feels different from other forms of attention: it is an intensified and unified experience of attentiveness, distanced from direct self-concern and mere interest. This has to do with the fact – or so I would argue – that it is not (merely) task-related but intrinsic, and it comes with a change of perspective: we see something differently when we attend to it aesthetically. This is a specific way of a world's and self's disclosure and it has both an ethical impact and a relation to mindfulness.

In what follows, I will give, first, an example of aesthetic attention towards nature (1). I will then distinguish versions, modes, and degrees of attention and try to situate aesthetic attention within this picture (2). Afterwards, I will provide a sketch of the relation of attention and perspective-shifting (3). I will then discuss a debate on aesthetic attention (4) and one current approach to it put forward by Bence Nanay (5). My own account will be outlined in the sixth section (6), before concluding (7).

1 Aesthetic Attention – An Example

Imagine you are walking through a forest, captured entirely by some stressful thoughts. You are thinking about your job obligations, your family, partnership and friendship expectations, some future plans, some unfulfilled needs of your own, etc., and how you can manage to integrate them in your life. Although you are (more or less) actively deliberating, you also feel passively absorbed: everything seems to come at once, you are overwhelmed by this stream of consciousness, you are distracted from the outer world. The bodily effects of this situation are that you walk faster than normal, as if racing with your thoughts, trying to shake them off; you become short of breath, your body feels boxed in. You do not recognize much of your concrete surroundings, though on a subpersonal level you might perceive the world as an enemy rather than as a friend. Now imagine you are all out of breath and need a break. There is a tree with a big trunk. You lean against it, close your eyes, breathe out, and calm down. Suddenly you recognize that your focus will change: it shifts away from your tense inner thoughts toward the sights, scents, and sounds of the environment. You hear the wind in the leaves, you feel the knobby trunk in your back, the soil under your feet. Was it your mind that was wandering before, so are your senses now – distributed to what appears to you. You are not focused on something specific, nor were you before. But you are not detached either. While previously you might have seemed absorbed in your thoughts and in this regard isolated from your outer world, you did voluntarily consider some pros and cons of your world and actively shaped your perspective on it. You were both concentrated inwardly and your attention was divided between this and that. Now you are open-minded, grasping different inputs, jumping from one to the other, but all in all somehow present. You feel freed, released, playful, your chest is open, your outlook is far-reaching without fixating on anything concretely.

What forms of attention are exercised in these different cases? How do the two situations differ from one another and from other forms of attention? In both cases, your attention might be described as being both vigilant and sensitive in one way or another, namely internally or externally. You are highly attentive, catching and seeking what is happening either in or around you. However, the difference is that in the first case you are totally locked in your thoughts and your urgent need to find solutions to some (seemingly) pressing problems in your life. All your thinking is aimed at finding some exit from your inner conflict. But instead of finding a way out, you are becoming more and more bogged down, dealing with a complex network of thoughts, ideas, plans, problems, questions and so on. This makes your chest feel constricted, your whole body tenses. The second case, by contrast, exemplifies not only a change of focus or perspective; it also represents an attention that differs qualitatively from the other occurrence. With this attention, we are open and present to what appears to us, without longing for a direct solution or salvation. We are fully attentive but not merely focused on a particular task or object. Instead it is a form of deep intrinsic attention for the here and now and its plural appearances, properties, and affordances. However, it is not the same as the classic understanding of aesthetic contemplation, disinterestedness, or absorption. By contrast, in aesthetic attention we are in fact interested and engaged. It does not suggest escapism from the world, but rather a specific form of a mediating, interconnecting commitment to it, based on a passive and active change of view: by aesthetic attention we experience a particular intertwinement with the world, free from practical and functional interests and demands. By doing so we cultivate an important mode of attention that can also be deployed in non-aesthetic contexts: a form of flow, presentness, affective openness, and mindfulness towards what appears to us or what forces us to change our perspective. But before we go into detail here, let us sketch the phenomenon of attention from a more general perspective and see where the specific aesthetic attention might fit into this picture.

2 Versions, Modes, and Degrees of Attention

Very broadly defined, attention in general is an integral part of our consciousness and experience in our interaction with the world. It is by way of attention that we navigate our environment, being able to focus on those things which are relevant for our actions, our needs, or those which are so salient that they catch our attention. Attention oscillates between passivity and activity, immediate and derived interest, involuntary and voluntary reaction (Waldenfels 2004; James 1890, 416). We can be passively caught by something which is salient or which changes something in our routine: a loud noise or a sudden thought, for instance. Waldenfels describes attention as an "intermediary" event that occurs because of a perturbation: something in our environment is different or demands a response. Usually, we are able either to actively and deliberately attend to it or to turn away. In this sense it is a *response* to the affordances of our environment in relation to our embodiment within the world (Magrì 2019; Wehrle and Breyer 2016). In any case, as soon as we are addressed by something, a response is demanded. Another key feature is that attention comes along with withdrawal: When we attend to something, we get distracted from something else as soon as our attention is caught by an object or when we turn to something attentively. As William James famously put it:

> It is the taking possession by the mind, in clear and vivid form, of one out of what seem several simultaneously possible objects or trains of thought. Focalization, concentration, of consciousness are of its essence. It implies withdrawal from some things in order to deal effectively with others. (1890, 404)

But as James also stressed, completely focused attention – that is, concentrating solely on one object – is a rather rare phenomenon. The same goes for distracted attention in its extreme form. Between these extremes lies a 'span', a continuum. Now, thinking about something like a specific 'aesthetic attention' entails the suggestion that there are different versions, degrees, and modes of attention. By 'versions' I mean that we can distinguish between perceptual, intellectual, and imaginative attention (Watzl 2017). Perceptual attention implies visual, auditory, haptic, proprioceptive and other sensual forms of attention such as smelling or tasting: what we are attending to with our senses. In intellectual attention, we focus on, e.g., thoughts, memories, beliefs. Imaginative attention is concerned with mental imagery processes. By 'modes' I mean that our attention can be not only focused, but also distributed or dispersed, deep or hyper. We can be deeply focused on one thought without recognizing anything (or hardly anything) around us – interestingly enough, those people who are completely absorbed by an intellectual task (let us say by solving a philosophical problem) seem detached and absentminded to observers. The *internal* attention then is so strong that no external attention is executed.[1] Or our attention can be distributed to several stimuli at a time or in quick succession. By 'degrees' I mean that we can be more or less fully attentive to something or someone: while looking at a painting with half concentration we might think about something the painting reminds us of. The question that comes up here is whether aesthetic attention is another version, mode or degree of attention, or whether aesthetic attention takes up all these various forms but is yet different from mere practical or instrumental attention. My hypothesis is that aesthetic attention is mostly executed with full and deep attention, or with at least a form of immersion. In aesthetic attention, we experience the world differently; it changes something in our relation to the world.[2] The qualitative aspect of our embeddedness in the world is changed here (cf. Husserl 1973).

Aesthetic attention fits into this picture in the sense of a unified or holistic *attentiveness*, a manner of being attentive to a specific interaction between human beings and their environment. In my view, aesthetic attention is a specific case of a unified attentiveness which, however, can also be found in practical situations. Christopher Mole – a proponent of an adverbialist approach to attention (or attentiveness) – claims that an agent who is performing a task is attentive if and only if the cognitive recourses in the background set are not occupied by activity that does not serve the task (Mole 2011, 51). In other words, one should not be distracted by other tasks at the same time. In Mole's account, he deploys the analogy of an orchestra whose members play in unison. As long as everyone who is playing is playing nothing but the same tune, it is an instance of unison (Mole 2011). According to this view, attention is more a manner than a capacity. Being attentive to an artwork, for instance, would imply that all our cognitive processes and resources (such as thinking, imagining, remembering) are focused on the task of 'looking at an artwork' or 'experiencing something

---

[1] See the contribution to this volume by Luis Sandoval and Betzamel López González.
[2] I ought to set out here what I mean by 'aesthetic': I have a rich, non-minimalist, and pluralistic concept of aesthetic experiences. There is not *one* aesthetic experience, but, depending on the object, there is a plurality of different experiences. Some are more perceptual, while others are more intellectually; some are felt as beautiful, others as overwhelming. That said, I do not reduce 'aesthetics' to 'beauty', but instead refer to the broader term, traditionally conceptualized in Baumgarten's Aesthetics, that refers to our sensual perception and knowledge.

aesthetically'. Although Mole's approach is meant as a general framework for attention, I think that aesthetic attention might be the paradigmatic case of such a unified attentiveness, or, to put it the other way around: when we practice full attention in this sense it gets an aesthetic quality. It is experienced as something complete, valuable, pleasant. The image of our resources playing in unison, then, fits very well with the aesthetic quality of attentiveness. What Mole has in mind is a holistic form of attention or attentiveness that is, in my view, compatible with the holistic and pluralistic approach of aesthetic experience put forward by John Dewey. According to such a pragmatist account, aesthetic experiences can occur with almost everything or everywhere, but they are distinct from ordinary experiences (Dewey 2008; Seel 2005). Later I will show that one difference lies in the condition of aesthetic attention not being merely instrumental but intrinsic and thus valuable for its own sake. But unlike the classic conception of aesthetic disinterestedness, this notion of aesthetic attention entails engagement and thus is not a mere passive absorption, detached from the rest of our experiences. On the contrary, it is an active commitment.

3 Attention and Perspectivity

Attention is constitutively interrelated with perspectivity, and so is, in my view, an aesthetic experience. A perspective is a subject's point of view or first-person viewpoint, not only in a perceptual or local sense but also in a metaphorical sense as a 'worldview' or attitude. It can be meant literally in the sense that every being (or even every object) has a certain local position within the world. We see from here to there from a particular (perceptual, bodily) standpoint, which changes as soon as we change our location. However, this fact also has an epistemological implication and then goes beyond the mere positional feature: our knowledge of the world is always bound to our epistemic position: how we (get to) know the world; our knowledge is situated (Haraway 1988). This leads to the more metaphorical sense of perspective as a specific worldview or attitude toward the world: How we 'see' the world also implies how we interpret it, evaluate it, feel in it, react to it, etc., subjectively in fact. Yet, what we're talking about here is not just a passive encounter, nor are we completely locked into this perspective. On the contrary, we are able to actively take on a perspective on an object, be it real or in thought, that is to say, we are also able to attend to it voluntarily – at least to a certain degree. And we can also change perspectives actively, and by doing so, our emotions and mood might change radically (Schmetkamp 2017). On the other hand, there are some constraints. Due to ontological reasons, we cannot get rid of our first-person perspective, nor can we take over another's perspective as a first-person viewpoint (Nagel 1974). However, we are able to take some distance from our self and try to focus our attention on something from another second- or third-person perspective: 'see' the world through someone else's eyes. This is what happens in aesthetic experiences: we change our view.

4 Aesthetic Experience

The insight that aesthetic experience is in some way connected to a change of perspective, especially in a metaphorical sense, is not a new one, though it is often not articulated in terms of perspectivity and attention. It is one of the main ideas within theories of aesthetic experiences that they are free from practical interests; already in this minimal respect, it seems to imply a change of perspective: If we consider a tree in its pure form, its concrete beauty, its literal and metaphorical power, etc., and not as a mere practical thing (that shelters one from rain, for instance, or offers an orientation on our walk), then we consider it an object without any determined functions. Instead of dealing with it instrumentally, we value it intrinsically.[3] The change of perspective that comes about via an aesthetic experience implies that we experience or discern something in a radical new or different way and that we thereby value this process for its own sake. That said, I take up a non-minimalistic approach rather than a minimalistic approach (Levinson 2016; Carroll 2005). That is, it is not only the perception of some formal aesthetic properties that makes an experience with something an aesthetic experience. It is *the way in which* something is experienced: our attitude or – in my wording – our perspective towards it: we are open to see something in another way than usual.

Of course, the question of what aesthetic experience is opens up such a vast field of research and as such cannot be sufficiently explained here. What I will try to do, though, is to elaborate a concept of aesthetic experience that involves what I call aesthetic attention. Yet, we should provide at least a brief sketch of how aesthetic experience can be characterized (very broadly). Some typical, though debated, characteristics of aesthetic experience are (some form) of pleasure or a specific emotion, an attitude of disinterestedness, a particular form of understanding, and some exclusivity and experiential value that are

---

[3] This is not sufficient for it to be an aesthetic object – a similar feature is attributed to moral objects.

not to be found in other, non-aesthetic experiences. Aesthetic attention is sometimes considered to be just another word for the attitude of disinterestedness or for contemplation. This view has been widely criticized on account of its exclusive and Romantic impact that risks separating aesthetics from other realms (for example, ethics and epistemology) (Dickie 1964). Others expand the notion to almost every object and practice (Dewey 2008; Seel 2005). According to the pragmatist John Dewey we can have aesthetic experiences not only with art, but also with nature, sports, everyday practices (like homework), or a mathematical proof. Dewey's account defines aesthetic experience as a holistic experience – or what he calls 'an' experience. This involves full or deep attention. By contrast, non-aesthetic experiences are inchoate, distracted, and dispersed (Dewey 2008, 36). "In contrast with such experience, we have *an* experience when the material experienced runs its course to fulfillment [. . .]. Such an experience is a whole and carries with it its own individualizing quality and self-sufficiency" (Dewey 2008, 42).

A famous debate on *aesthetic attention* was initiated in the 1960s when (among others) Jerome Stolnitz (1961) argued for an aesthetic attitude which has disinterestedness at its center and George Dickie (1964) argued against such an approach. Discussing the theories from the seventeenth and eighteenth century in British philosophy, Stolnitz strengthened the concept of aesthetic "disinterestedness" as a distanced and self-less attitude, involving an observer's "sympathetic attention" toward an object "for its own sake alone" (Stolnitz 1960, 34–35). "Disinterestedness" hereby means that the percipient is not motivated to act or has no desire to possess the object. "Sympathetic" means to "accept the object on its own terms to appreciate it" (Stolnitz 1960, 36). Thus, when we are disinterestedly and sympathetically attending to an object, we do not treat it instrumentally but accept it for its own sake. Another famous term in this context is that of contemplation, which refers to a specific mode of perception without analyzing the respective object and/or experience. Traditionally, this was experienced in the perception of beauty (harmony, proportionality, etc.) and/or the sublime. Aside from the typical proponents like Shaftesbury, Hutcheson, and Kant, Stolnitz presents the theory of Archibald Alison who combined disinterestedness with attention. For Alison, aesthetic attention is a different state of mind than practical attention. In aesthetic perception, the spectator must "withdraw [. . .] attention from the particular objects of [his] thought" and "abandon" himself to the scenery (Stolnitz 1961, 137). Dickie argues against such theories of disinterestedness, distance, and aesthetic attitude, claiming that it is just a form of (more or less) full attention that is at stake when we perceive an art object adequately. What others describe as not being distanced enough when simultaneously thinking of some self-oriented goals, for instance, instead of being fully dedicated to what is presented (e.g., in a theater), Dickie considers just another conception of focused attention (1964, 57). Dickie offers an example wherein 'Jones' listens to music for an examination, whereas "Smith" listens to the same music with no further purpose. Jones's attention is divided. Now, Dickie sees no difference between these two experiences. According to Dickie, there is only one way of attending to the music. As I will show later, there is one current approach which argues, contra Dickie, that there are indeed different ways of listening attentively to the same music (Nanay 2016). But what is more important to me is the following point: Although I think that Dickie offers an important argument against theories of disinterestedness, he underestimates the phenomenology of full attention versus divided attention. It makes a difference whether while listening we think about what we need to remember for our examination about the music, or we are completely present and open to what appears to us. In both cases, we are focused on the music, but the task-related attention feels different, since it is bound to practical purposes. However, depending on the situation and the object of aesthetic experience, I would agree with Dickie that our experience can be mingled. Dickie himself gives the example of a film critic who watches a movie not only with an attitude of enjoyment or aesthetic disinterestedness, but also from an analytical perspective. Hence, the critic has a further ulterior purpose, namely to judge the quality of the movie (Dickie 1964, 61). Yet, she is attentive to it. She attends to the properties and contents of the film, which are relevant for the aesthetic experience and appreciation. The same is true when it comes to moral reflection on an artwork. In the following, I will present a contemporary approach that tries to overcome Dickie's critique of attention by arguing that aesthetic attention is a combination of focused and distributed attention.

5 Focused and Distributed Attention as Aesthetic Attention

In his book *Aesthetics as Philosophy of Perception* (2016) Bence Nanay ascribes a key role to attention in some paradigmatic cases of aesthetic experience. He combines two modes in which we can attend to artworks and other possible aesthetic objects: a distributed mode and a focused mode. In aesthetic experiences, or so his argument goes, we exercise our attention differently than usual. Nanay claims that in experiencing a painting, for instance, it is constitutively implied that we focus on it as an object and at the same time distribute our attention across various properties of the painting and the represented scene. By contrast, in standard modes of attention (that is, in performing everyday tasks), we focus on a limited set of properties of one or more perceptual objects, when, for instance, we sort all black socks in the laundry from the white ones. The change of modes in aesthetic experience, though, is linked to a change of world perception: we see the world differently, just as described in the examples of the tree and the bird. Nanay now claims that the world appears differently *because* we exercise focused and distributed attention in another manner. The reason would lie both in the fact that our attention is exercised differently, and in the fact that we are not interested in the object for practical reasons (for instance, we do not use the painting as an instrument to perform a task, say, hitting a burglar with it). However, I think the difference only lies in this lack of practical interest and, moreover, in the specific *kind* of attention or attending that is displayed here in a more holistic sense: a non-practical or task-unrelated attention that has aesthetic content and aesthetic quality. Aesthetic attention, thus, is special not because two modes are exercised differently, but rather because it is a distinctive deployment of attention that is not merely instrumental but intrinsic. It comes along with a broader view and different perspective or attitude. It is a form of deep attention that is not the same as focused, task-oriented attention. Before I give some more arguments, I want to discuss Nanay's other examples of focused and distributed attention (Nanay 2016, 24). Usually (i), we would distribute our attention on objects and focus only on a few properties (or a single property), namely those which are connected to a specific interest or task. Imagine looking at two paintings by Vermeer. You are interested in Vermeer's famous technique of using ultramarine pigments, for instance to understand it or to use it, and compare the two paintings in this regard. In this moment, you distribute your attention to different objects (two paintings) though at the same time you focus on a particular property (ultramarine pigments or the technique). Following Nanay, this would be a common way of exercising our attention but not aesthetic attention. Other more or less everyday exercises of attention are when (ii) our attention is distributed with regards to objects *and* properties. Nanay gives the example of waiting in a doctor's office and having nothing to do (reading, for instance): our attention "wanders aimlessly" (Nanay 2016, 24). A third case (iii) is that our attention is focused with regard to one object and focused with regard to a specific property of this object. An example could be when we cross the street and focus on an approaching car and its property of moving quickly. In Nanay's view, only a fourth case represents the phenomenon of aesthetic attention: (iv) focused attention to an object and at the same time distributed attention across its properties. The object of this focused attention could be a landscape; the properties are the different parts of the landscape, its forms, its scents, etc. Although I think

Nanay's approach marks a pivotal starting point for a concept of aesthetic attention or an aesthetics of attention, there are some objections to be made:

Firstly, what Nanay describes as distributed attention could also be compared to the cognitive mode of hyper attention. Hyper attention is typically characterized as the ability to switch quickly between different stimuli and tasks (Hayles 2007, 187). The opposite is deep attention: the cognitive mode of concentration on a single object for long periods of time, "ignoring outside stimuli while so engaged" (Hayles 2007, 187). Very often hyper attention is related to some degree of vigilance and alertness: the urgency to grasp everything that seems relevant for a given situation or task. We could imagine focusing on a landscape and being hyper attentive to what appears to us, distributing our attention from one stimulus to another – yet without having an aesthetic experience with it. Hyper vigilance is related to the need to not miss anything; it is a form of tension. The difference between focused/distributed attention with aesthetic impact and focused/hyper attention without aesthetic impact does not lie in the fact that we focus on one object and distribute our attention between different properties of the whole scene. The difference is one of attitude and how this is experienced in mind and body. Attention with aesthetic quality is not just a different token or interrelation of modes of attention. It is another type that feels different, touching us in a different way as embodied beings. Being hyper attentive can be overwhelming and can merge into exhaustion, where we reach the limits of our receptivity. Having an aesthetic experience can also be overwhelming, for example in relation to the sensual inputs or epistemic impact. But all in all, it should feel valuable. Obviously, this does not depend on the interaction between focused and distributed (or hyper) attention; rather, it depends on how we experience something beyond mere practical utility.

A second objection is that the differentiation of non-aesthetic cases (i–iii) and aesthetic attention cases (iv) is not convincing. Case (i) obviously has a clear goal. In the Vermeer case, it is a scientific or craft-related goal to discover something about the ultramarine pigments. Case (iii) also seems task-related. However, I think it is very easy to give counterexamples. To do so, I will introduce another concept, which is constitutively related to attention and displays the passive-active character of our attentional interactions with our world: the concept of affordance. 'Affordances' (Gibson 1986) traditionally are the more or less objective features of objects in our (human and non-human animals') world and their opportunities for action. One example (in the human world) is the chair which affords human beings (and some animals) the ability to sit down, but maybe not the ability to play air guitar with it. The concept of affordances usually captures the pragmatic interaction between an organism (human being or animal) and its environment with regard to function and need, respectively. However, whether we acknowledge the respective affordance of an object also depends on other criteria than the object's features: It also depends on our (cultural, social, personal) interests, practices and habits, our context, our aims, and on our respective skills (Rietveld and Kiverstein 2014; Wehrle 2010).[4] The chair also invites us to stand on it when we want to reach a book at the top of a high bookshelf and are able to stand up. However, we could also easily imagine playing with our children, pretending to be musicians without any real instruments. With its seat and back, the chair might well serve as an air guitar after all (albeit a fanciful one). Or maybe you feel invited to consider the specific design and feel the surface of the seat without seeing it as a practical object at all – since you are a person with a sensibility for a perfectly shaped design. What happens in these respective cases is – or so I would argue – that you change your perspective and in doing so, you perceive different affordances. But in all these cases, I would argue, you focus on the object and focus on one property (to sit on it, to play with it, or to look at its smooth surface); only in the third case is your perception of the chair released from mere practical utility.

Case (ii) can also represent either a form (token) of aesthetic attention or a form of inattention. This depends on the form of mind wandering that is at stake here. When we are sitting in a doctor's office and just let our view (or mind) wander, we are not interested in something specific, nor do we aim at performing a particular task – indeed, what else could we mean by 'aimlessly'? So, what is this? An experience with aesthetic content? Or an unconscious, non-autonomous form of mind wandering where we have no control over what's happening? I think it can be both, and when it is an active engagement of daydreaming, it might as well have aesthetic quality.

Finally, what bothers me is not that Nanay's conceptualization and scheme are not persuasive with regards to the difference between aesthetic experiences and everyday exercises of attention. As I have said, Nanay's approach signals an important contribution that helps to shape the concept of attention. But the whole picture is too schematic and, moreover, falls short of the specific point of an aesthetic experience. It is too schematic insofar as our attention occurs in even more variety and plurality than depicted. And the account misses the point, since aesthetic attention is not just another mode or token of exercising this faculty, but is also a distinctive type with another function than task-

---

[4] Rietveld and Kiverstein develop a broader concept of affordances than usual and claim that the affordances the environment offers "are a good deal more extensive than has standardly been recognized" (Rietveld and Kiverstein 2014).

related attention. The central question is not whether it is focused or distributed or both. Nor is it a question of whether it is perceptual or intellectual or emotional. The crucial point is that aesthetic attention can imply all of this. (And I think, too, that we should not reduce aesthetic attention to direct perceptual attention. Aesthetic attention can also occur on a mere cognitive level: how we think and evaluate with regard to the aesthetic object. And of course our imagination plays a role in this context, too.) But even more crucially, aesthetic attention is of intrinsic value and is therefore distinct from other exercises of attention: it feels different. Nanay himself stresses the uniqueness of aesthetic attention and experience (Nanay 2016, 117). Here, he highlights that in aesthetic experience, we treat something (an object) differently: akin to encountering something or someone for the first time. Aesthetic attention, thus, has an existential and ethical impact. But why should this depend on the distribution or focusing of attention? The normative idea with this encountering for the first time rather refers to some open-mindedness, and thus to an attitude of recognition, sensibility, and, as Waldenfels has conceptualized it, responsivity (Waldenfels 2004). Insofar as this attentiveness is not merely task-related or goal-oriented, it is much more based in the present than in the future (Seel 2005). Its focus lies on appearances, in a passive-active manner. In this regard, it recalls both the aesthetic theory proposed by Dewey and some normative accounts of attention (Murdoch 1970; Debus 2015). The difference between aesthetics and ethics in this regard, though, is that in aesthetics it is an invitation to *play* with the appearances and affordances, whereas in ethics it is not (just) a game; when we are really and personally confronted with a serious moral problem in reality; where we have to make a decision and cannot just play it through imaginatively without thinking about any acts and consequences. In moral contexts we are asked to act, namely by other (real) people. In aesthetic contexts we are kind of relieved from such demands – at least primarily. In the following section, I will sketch out how I think we might better describe the interrelation between attention and aesthetic experience than in terms such as focused or distributed attention.

6 Aesthetic Attentiveness

As explained, attention can be focused or distributed, deep or hyper. Focused attention has a theme, but it is also aware of the context and the margin (Arvidson 2010; Watzl 2017). That focused attention is aware of the context is not the same as saying that we can distribute our attention over a scene: it belongs to the general phenomenology of attention; it is contextualized in a field of perception

(Wehrle and Breyer 2016). Some aspects of attention are, for instance, voluntary, others are involuntary. Thus, we can deliberately focus on an object, a sound, a smell, etc., but our attention can also surprisingly and partly be attracted by a noise, a light, a phenomenal change in our environment and the like. These dimensions point to, on the one hand, controlling factors of attention – we can direct and regulate attention – and, on the other, the immersive and sometimes manipulative power of things which literally arrest our attention. The relation between these ambiguous aspects of attention and aesthetic objects is obvious. We can deliberately focus on an aesthetic object (for instance, a painting); but, in turn, aesthetic objects can also catch our attention involuntarily (for instance, a film and its cues). Additionally, our attention can also occur in degrees: we can be more or less *attentive*. Consider watching a film only incidentally, for instance while doing homework, in contrast to viewing it in a cinema and thereby paying full attention to it. Which of these more or less attentive activities are at the same time aesthetic experiences?

Furthermore, how and to what we attend depends on personal situations, contexts, and emotions. If you are in a good mood because your life is going well, you might attend to different features of an artwork than if you are in a bad mood. Maybe in one case you attend to the light sides of a Caravaggio painting, while in the other case you attend to the dark sides. But in many cases you do not *notice* any of these qualities since you are locked in a mere functionalistic view on the world, insensitive for any sensual or aesthetic qualities. But attention or, let us say: specific attention grabbing features can change our view in a way that we see these qualities.

The philosopher and author Iris Murdoch – one of few who have sketched a normative concept of attention in her ethics – uses a compelling example. She describes how she is looking out of the window "in an anxious and resentful state of mind" (Murdoch 1970, 82). Suddenly she recognizes a kestrel and stops thinking and instead considers the hovering flight of the bird. At that moment her intellectual attentional focusing is disrupted by the visual stimulus (the hovering of the kestrel) and is modified not only in the sense that it is replaced by perceptual attention, but also insofar as such modification is connected to some change in her perceiving the world, in a more encompassing way: "In a moment everything is altered" (Murdoch 1970, 82). With regard to my concept of perspective, we could also say: the perspective or view on the world has altered. Or even more: the horizon has been widened. For Murdoch, the occurrence has also an ethical impact: the aesthetic experience (with the kestrel) is intertwined with a degree of distancing from a selfish perspective and toward a broader point of view. The example demonstrates how a literal change of focus can accompany a metaphorical change of perspective in the sense of another broader worldview which overcomes the former narrow view. Furthermore, it shows that this change is not simply a passive process; of course Murdoch's attention is 'suddenly' caught by something around her; but she also decides to consider the flight of the bird.

Aesthetic attention might even constitute the paradigmatic case through which we can see how attention not only shapes our field of perception and consciousness, but also discloses the world to us in another way. When we attend to something aesthetically, we do not perceive another world; rather, we perceive the same world differently. We encounter a new perspective that is not reducible to a visual versus intellectual field, but involves a transformative power of seeing the world in a new, more encompassing manner. Hence, this experience is linked to a change and opening up of new perspectives.

As suggested above, I think it is rather something like a specific form of *attentiveness* that we actually have in mind (or should have in mind) when talking about aesthetic attention. Focused and distributed attention are then executions based on that general kind of attention. My point is that attention is necessary for aesthetic experiences, but it is not a decision between focused or distributed attention. Attention arises in so many modes, and one could also imagine that some artworks or aesthetic experiences come with some form of inattention or of dispersal, while being based overall on a passive-active engagement.

My suggestion is to distinguish four different aspects of aesthetic attention (or attentiveness) in aesthetic experience:
1. Attention is a general condition of the perception of an object of aesthetic experience;
2. Aesthetic attention implies some form of a unified experience;
3. Aesthetic attention implies a change of perspective in both a literal and metaphorical sense: as a modified view and attitude.
4. Aesthetic attention is compatible with a moral attitude.

A<small>D</small> 1: The first aspect is evident and trivial. In order to acknowledge an artwork as an artwork – a painting, a play, a movie, a novel – we have to direct our attention to it. That means we have to look at it (read it, watch it), in order to perceive it not only as a vague input among others but as something we actively single out or prioritize within our perceptual field. This is due to the fact that, as mentioned above, attention means to focus on some things and withdraw from others: to look at an artwork requires not to look at your smartphone at the same time, pragmatically speaking. Thereby, attention is not only directed by the stimuli but also influenced by our subjective being-in-the-world and by our respective experience. Hence, following Husserl attention is more than mere perception; attention goes one step further, so to say (Wehrle 2010). Beyond this descriptive requirement, there is another normative one. We have to pay attention to something more specific within the artwork – show some sense and sensibility for the formal properties, the content, perhaps its moral vision. Attention, in this general mode, seems to be something like a gateway to aesthetic experience, judgment, and understanding of the artwork. But it already implies a discern for the features which are relevant for the aesthetic experience. In other words, if we only look at the painting's frame, for instance (where the frame is *not* part of the work), then we do not pay attention to the aspects which *are* relevant for the work, and which the artwork affords us to receive in order to get the point of the object as an artwork. This is an attention to the aesthetically relevant aspects.

AD 2: As John Dewey writes in his influential book *Art as Experience*, one condition of aesthetic experience is (besides *experiencing* something at all) concentration and duration. Aesthetic experience has a temporal and spatial course, a scope with a beginning, a middle, and an end. Thereby, it is the intensified, structured, and framed version of ordinary experiences (Dewey 2008, 11).[5] What is even more striking is that, according to Dewey, aesthetic experience is not a *detachment from* the world; instead, it is the experience of the unity of experience. In experiencing something aesthetically, we experience it as a whole with a duration and a conclusion. In order to experience it that way, we have to be fully attentive and not detached. Only when attending to an object without being distracted by other things that do not contribute to the experience are we able to experience it in a way that Dewey calls *an* experience or *aesthetic* experience. As Dewey writes very presciently, in our accelerated life we experience most things only on the surface: "No one experience has a chance to complete itself because something else is entered upon so speedily" (Dewey 2008, 46). In aesthetic experience there is time and space for concentrated attention without distraction. However, this is not reducible to some focused or distributed attention, nor is it reducible to perceptual attention. (We can also experience a mathematical proof in an aesthetic manner (Dewey 2008, 44)).

In terms of attention, the crucial point is that we stay tuned and that we want to fulfill and complete the experience. For Dewey, there lies an interest and value in this specific experience, but what is striking here is that we are committed. It is not an experience of a mere passive contemplation; it is a dynamic and active 'flow' toward the presence of an object (artwork, nature) or practices. "Attentive observation is certainly one essential factor in all genuine perception including the esthetic [sic]. But how does it happen that this factor

---

5 Dewey has quite an elaborate concept of experience as such which cannot be explained here (cf. Dewey 2008, 36–59).

is reduced to the bare act of contemplation?" (Dewey 2008, 257). And although the aesthetic experience and attentiveness are different from practical task-related actions, they are not separated completely from ordinary experiences. There is a continuity of experiences.

AD 3: This leads to the change of perspective. Experiencing something in an aesthetic way demands something different from us than attending to it in a non-aesthetic, functional or practical way. Hence, just like many aesthetic theories argue (including Dewey's), aesthetic experience implies *perceiving* something (e.g., an object, a viewpoint, an emotion, an idea, other persons, the world) differently, from another perspective or in another way than usual. For instance, we see the 'Fountain' by Marcel Duchamp not as a urinal which men could use as a toilet, but as a joke or a comment on the art world or as a challenge to our understanding of art in general. What is needed here is not a mere perceptual attention, but rather an intellectual attention toward the concepts of this early conceptual artwork. But attention to its modified affordances is needed. We attend to the artwork in a way that is detached from our usual discursive logics. The aspect of disturbance or perturbation in the attentional sphere, which was mentioned at the beginning of this paper with reference to Waldenfels, is vital here. When we focus our attention only on the usual affordances, we get confused; our view is perturbed by the refunctioning of the urinal. But the perturbation at the same time depends on our usual understanding of this object.

To understand it as an artwork, that is, to change our focus, we still have to reconnect it to its usual function. What is at stake here is not a perceptual attention in the first instance, but an intellectual attention. Perceiving only its formal qualities (and only to 'play' with it disinterestedly) would miss the point. We still focus on something, but not from a practical or usual point of view. In the case of the urinal, it is not (or not only) its practical use as a toilet, but the fact that it is signed, the contextual fact that it is presented upside-down and isolated from its usual environment, and the art-historical fact that it was supposed to participate in an art exhibition. When we look at Duchamp's urinal, we somehow have to open our mind in order to recognize those features and the whole context. We must take them into our (intellectual) attentional account, as it were (this is not achieved by the fact that we distribute our perceptual attention on the features, such as the form of the urinal or the shiny white color). To attend to something aesthetically entails a specific way of attending, one that is freed from mere practical demands and yet addressed by the aesthetic affordances of the respective object.

AD 4: The fourth aspect is related to the aforementioned open-mindedness and change of attitude. In aesthetic attention, we obviously *treat* the object differently. This dimension of "attentional commitment" or *attentiveness* has a lot in common with moral respect, which implies, at least in the Kantian sense, that we should never treat others as a mere means but always as ends in themselves. Though having a lot in common with the other three dimensions, it is still distinct. In being attentive toward its aesthetic features and affordances, we respect its autonomy – which is not to say that it is totally unrelated to human practices, but rather that it is not (merely) instrumentalized as a means to an end (such as a practical interest). Aesthetic attention differs from our usual deployment of attention, since it offers us the chance to relinquish our self-oriented goals. It is the mere exercise of a passive-active encounter with the world within a more or less framed and specific experience.

7 Conclusion

In this paper, I have sketched out a concept of aesthetic attention as an integral component of aesthetic experiences in all their plural forms, be it with nature, art, sports, or ordinary activities. In contrast to classical theories of disinterestedness, I have stressed that aesthetic attention is a form of attentiveness in which we are open-minded toward appearances and affordances, going beyond mere practical interest. Aesthetic attention is thereby accompanied by a change of perspective in a literal and metaphorical sense that is a change of seeing something newly or differently and as a change of attitude or worldview. In contrast to traditional theories, however, I have also underlined that the aesthetically attentive subject is not one of passive contemplation, but is actively committed. She interacts with her environment, but differently than usual. Again, the double structure of passivity and activity, involuntary and voluntary action, is vital here. To take up the tree example from the beginning: there is this passive moment when we are caught by something (the scent, the light) and our focus changes. But we must also take up the 'invitation' actively and decide whether to respond and turn our attention to something other than our selfish thoughts. Aesthetic attention differs insofar as it has itself an aesthetic, unified quality. I have argued that it can be mingled with interests, but that it also attends to its objects intrinsically. The experience is something in-between involuntary affection, voluntary control, and 'fluidity' to what occurs. In sum, aesthetic attention is a combination of rapt attention and free-floating immersion.

References

Arvidson, S. P. (2010): *The Sphere of Attention. Context and Margin*. Dordrecht: Springer Netherlands.
Carroll, N. (2005): "Aesthetic Experience: A Question of Content". In: K. Matthew (ed.): *Contemporary Debates in Aesthetics and the Philosophy of Art*. Hoboken: Wiley-Blackwell, 69–97.
Debus, D. (2015): "Losing Oneself (in a good way): On the Value of Full Attention". *European Journal of Philosophy* 23:4, 1174–1191.
Dewey, J. (2008): *The Later Works 1925–1953: 1934: Art as Experience*. Volume 10. Carbondale: Southern Illinois University Press.
Dickie, G. (1964): "The Myth of the Aesthetic Attitude". *American Philosophical Quarterly* 1:1, 56–65.
Gibson, J. (1986): *The Ecological Approach to Visual Perception*. Abingdon-on-Thames: Taylor and Francis.
Haraway, D. (1988): "Situated Knowledges: The Science Question in Feminism and the Privilege of Partial Perspective". *Feminist Studies* 14:3, 575–599.
Hayles, K. (2007): "Hyper and Deep Attention: The Generational Divide in Cognitive Modes". *Profession* 2007:1, 187–199.
Husserl, E. (1973): *Experience and Judgement*. Evanston: Northwestern University Press.
James, W. (1890): *The Principles of Psychology*. Vol. 1. New York: Henry Holt and Co.
Levinson, J. (2016): "Toward an Adequate Conception of Aesthetic Experience". In: J. Levinson: *Aesthetic Pursuits. Essays in Philosophy of Art*. Oxford: Oxford University Press, 28–46.
Magrì, E. (2019): "Situating Attention and Habit in the Landscape of Affordances". *Rivista Internazionale Di Filosofia e Psicologia* 10:2, 120–136.
Mole, C. (2011): *Attention is Cognitive Unison. An Essay in Philosophical Psychology*. Oxford: Oxford University Press.
Murdoch, I. (1970): *The Sovereignty of Good*. Abingdon-on-Thames: Routledge and Kegan Paul.
Nagel, T. (1974): "What is it like to be a bat?". *The Philosophical Review* 83:4, 435–450.
Nanay, B. (2016): *Aesthetics as Philosophy of Perception*. Oxford: Oxford University Press.
Posner, M. I.; Snyder, C. R.; Davidson, B. J. (1980): "Attention and the Detection of Signals". *Journal of Experimental Psychology* 109:2, 160–174.
Rietveld, E.; Kiverstein, J. (2014): "A Rich Landscape of Affordances". *Ecological Psychology* 26:4, 325–352.
Schmetkamp, S. (2017): "Gaining Perspectives on our Life: Moods and Aesthetic Experience". *Philosophia* 45: 4, 1681–1695.
Seel, M. (2005): *Aesthetics of Appearing*. Palo Alto: Stanford University Press.
Stolnitz, J. (1960): *Aesthetics and Philosophy of Art Criticism. A Critical Introduction*. Boston: Houghton Mifflin.
Stolnitz, J. (1961): "On the Origins of 'Aesthetic Disinterestedness'". *The Journal of Aesthetics and Art Criticism* 20:2, 131–143.
Waldenfels, B. (2004): *Phänomenologie der Aufmerksamkeit*. Frankfurt: Suhrkamp.
Watzl, S. (2017): *Structuring Mind. The Nature of Attention and how it Shapes Consciousness*. Oxford: Oxford University Press.

Wehrle, M. (2010): "Die Normativität der Erfahrung – Zur Beziehung von Normalität und Aufmerksamkeit bei E. Husserl". *Husserl-Studies* 26:3, 167–187.

Wehrle, M.; Breyer, T. (2016): "Horizontal Extensions of Attention: A Phenomenological Study of the Contextuality and Habituality of Experience". *Journal of Phenomenological Psychology* 47:1, 41–61.

Short Biography

Susanne Schmetkamp, PhD, is Assistant Professor at the Philosophy Department of the University of Fribourg (Switzerland). She leads the SNSF-Research Group "Aesthetics and Ethics of Attention". Her main interests and fields of expertise are: attention, empathy, perspectivity, aesthetic experience, love, and recognition. Recent publications are "Understanding A.I. – Can and should we empathize with humanoid robots?", in: *Review of Philosophy and Psychology*, 11, 2020, 881–897, DOI: https://doi.org/10.1007/s13164-020-00473-x. And "Theorien der Empathie – Zur Einführung", Hamburg: Junius, 2019. Susanne Schmetkamp also works as an author and presenter. She lives in Zurich with her family.

Felipe León
Chapter 3
Attention in Joint Attention: From Selection to Prioritization

Abstract: This chapter investigates the relationship between attention and joint attention by focusing on one seemingly plausible way of characterizing the central role of attention. On this view, basic cases of perceptual attention are fundamentally a matter of selecting one item and filtering out unattended items. Drawing on contributions from philosophy of mind and classical phenomenology, I propose that joint attention research can benefit from adopting a richer and more nuanced view of attention. On this second view, conscious perceptual attention is not a matter of selecting and filtering out, but rather of re-organizing and re-articulating the experiential field into a foreground-background structure. I consider three benefits of the second view: (i) a more context-sensitive approach to the question of how to account for the transition from solitary to joint attention; (ii) a potentially fruitful examination of that transition in terms of thematic modifications; and (iii) a better appreciation of the co-constructed aspect of joint attentional interactions.

Introduction

Given the terminological proximity between the notions of attention and joint attention, one might think that there are conspicuous and widely investigated connections between them. However, apart from a few exceptions, the interrelations between attention and joint attention have remained largely unexplored.[1] The aim of the present chapter is to contribute to filling in this gap, by

---

[1] For brief explorations of the notion of attention in the context of joint attention research, see (Tomasello 1995, 104; Call and Tomasello 2005, 60; Eilan 2005, 19; Hobson 2005, 185–187; Hobson and Hobson 2011, 115–116; Reddy 2005, 104). For the reception of the notion of joint attention in the context of attention research, cf. Section 2.

Acknowledgements: Thanks to Diego D'Angelo, Maren Wehrle and Oddy Stone for helpful comments on earlier versions of this chapter. This project has received funding from the European Research Council (ERC) under the European Union's Horizon 2020 research and innovation programme (grant agreement No. 832940).

https://doi.org/10.1515/9783110647242-004

focusing on the question of what might research on joint attention learn from attention research. My main point will be that there is one intuitive and seemingly plausible way of characterizing the central role of attention that turns out to be of limited help in joint attention research. On this view, perceptual attention is fundamentally a matter of selecting one item – a target of attention – and filtering out unattended items. Drawing on contributions from classical phenomenology and philosophy of mind, I will suggest that joint attention research can benefit from adopting a richer and more nuanced view of attention. According to this second view, conscious perceptual attention is not a matter of selecting and filtering out, but rather of re-organizing and re-articulating the experiential field into a foreground-background structure.

The chapter proceeds as follows. In Section 2, I elaborate on the lack of communication between research on attention and on joint attention. I suggest that the disconnection between the literatures has been motivated partly by the history of the involved notions, as well as by some operative theoretical assumptions that have influenced the research carried out about them. Section 3 elaborates on the notion of joint attention, and it introduces one central question that arises in joint attention research: how to account for the transition from solitary to joint attention. I go on to consider different ways of cashing out the idea of attentional selection, and I find them unconvincing to explain that transition.

In Section 4, I turn to some contributions about attention from Maurice Merleau-Ponty and Aron Gurwitsch. Influenced by Gestalt psychology, both authors suggest that attention has a structuring and organizational power. They refrain from conceptualizing the relation between a target of attention and the immediate context in which the target is experienced in merely aggregative terms. They propose instead that such relation should be characterized in terms of the interdependence between a prioritized foreground and a de-prioritized background. In Section 5, I come back to the question of what research about joint attention might learn from attention research, in light of the preceding discussion.

1 On the Notions of Attention and Joint Attention

Both attention and joint attention are relatively undertheorized topics in philosophy, at least compared with more canonical topics, such as consciousness and perception. While there is no shortage of empirical work on attention in psychology and the cognitive sciences, there has been comparatively less philosophical work on it (Watzl 2017, 4–5). Given this under-theorization, it is not very surprising that the conceptual connections between attention and other topics, including joint attention, have also remained quite unexplored. Something similar could be said about joint attention, although for different reasons. The notion of joint attention was coined as a distinct research topic in the mid 1970's, in the context of investigations in developmental psychology about how infant and caregiver manage joint reference in pre-speech communication (Bruner 1974; Scaife and Bruner 1975; Bruner 1977; see Siposova and Carpenter 2019, 260). Importantly, research on joint attention didn't arise as an expansion of research on attention, or when research about attention had reached some widely agreed-upon results. Rather, it emerged independently, motivated by the need to conceptualize a phenomenon that had fallen out of the radar of psychological research.

Scaife and Bruner began their 1975 *Nature* paper – one of the seminal papers on joint attention – with the observation that "[l]ittle is known about how visual attention of the mother-infant pair is directed jointly to objects and events in the visual surround during the first year of the child's life" (Scaife and Bruner 1975, 265). On a widely held characterization, which builds on the work of Bruner and colleagues (and which has been influential also beyond developmental psychology), joint attention is a triadic relation between two subjects and an object, in which the subjects' attending to the object is 'out in the open' or 'mutually manifest' for them. Simply put, this means that co-attenders are both aware of attending jointly or together to the relevant object (Eilan 2005, 1; Campbell 2011, 417; Campbell 2018; Peacocke 2005, 303; Carpenter and Liebal 2011; Moll and Meltzoff 2011b, 290; Schilbach 2015, 132; Eilan ms.).[2,3]

---

[2] Psychological research has provided increasing evidence of the crucial role that joint attention plays in a variety of domains, including socio-cognitive development (Trevarthen and Hubley 1978), early language acquisition (Dunham et al. 1993; Tomasello and Farrar 1986), and the development of perspective-taking (Moll and Tomasello 2007; Moll and Meltzoff 2011a, 2011b). Joint attention has also been important in research on autism (Loveland and Landry 1986; Hobson 2002; Mundy 2016), in comparative and evolutionary psychology (Carpenter and Call 2013; Tomasello 2014), joint action research (Böckler and Sebanz 2013), and, more recently, in the literature on collective intentionality (Rakoczy 2018; León et al. 2019).

[3] The phenomenon identified by Bruner and colleagues had not been overlooked in classical phenomenology – although it hadn't been called joint attention. Consider the following passage from Merleau-Ponty's *Phenomenology of Perception*: "My friend Paul and I point to certain details of the landscape [. . .] Paul and I see the landscape 'together,' we are co-present before it, and it is the same for the two of us not merely as an intelligible signification, but also as a certain accent of the world's style, reaching all the way to its *haecceity*" (2012, 428. Cf. also 1964, 17). In the lecture *Einleitung in die Philosophie*, from 1929, Heidegger discusses one

Joint attention typically emerges around the first year of life (Carpenter et al. 1998). It differs not only from dyadic mutual attention – also called 'attention contact' (Gómez 1994) – in which two individuals attend to one another in a face-to-face interaction – but also from other triadic social phenomena (Carpenter and Liebal 2011). These include *parallel attention*, in which two individuals – unbeknownst to each other – attend to the same object at the same time, *gaze following*, in which one individual follows another's gaze to discover what the second is attending to, and *social referencing*, in which one individual tracks the affective reactions of another to a target of attention. What singles out joint attention with respect to these phenomena is that co-attenders experience a characteristic 'openness' or 'mutual manifestation' of the common object of their attention.[4]

Moving beyond the history and basic contours of the notions of attention and joint attention, research on these topics has been influenced by some operative theoretical assumptions that haven't been conducive to promote the dialogue between the two bodies of literature. Consider, first, that while joint attention is an inherently social phenomenon – one cannot jointly attend to something on one's own – investigations about attention tend to presuppose that attention is a domain-general capacity that, in principle, could be isolated from any social context. The operating assumption here appears to be that, at a suitable level of abstraction, whether one attends with (or to) other people or in solitude to environmental objects like chairs and apples doesn't make a real difference for clarifying the very nature of attention as a psychological phenomenon. Consequently, it may be natural to assume that, at a basic level, attention is neutral with respect to the contexts in which it is exercised, whether those contexts are social or not.

Concededly, assuming that attention is neutral in this sense is compatible with recognizing the existence and potential relevance of joint attention (as well as of dyadic mutual attention). This is something that several researchers working on attention are keen to do (Mole 2011, 170; Mole 2017, 63–64; Watzl 2017, 15–16; Wu 2014, 9). Nonetheless, the question of whether research on

---

example of linguistically mediated joint attention (involving presumably more than two co-attenders): "Looking at this piece of chalk, we all now make – with-one-another – the statement: 'This piece of chalk is white'. [. . .] The present-at-hand [*Das Vorhandene*], with which we are, is something common, it is the same for many, so that these many, based on this 'sameness for them', become a 'we'" (Heidegger 2001, 89, 97; my translation).

4 Not everyone would agree with this characterization of joint attention. There are different points of disagreement. First, the characterization I adopt is a so-called 'rich' characterization of joint attention. For 'lean' theorists, gaze following and focusing on the same object may be sufficient conditions for joint attention (Butterworth 1995). Second, some authors suggest that dyadic mutual attention is a form of joint attention (Reddy 2008).

joint attention can in any way illuminate research on attention is typically left unconsidered.⁵ That this is a question worth asking is indicated by research suggesting that early infant attention is mostly *exogenous* – not under volitional control, and mostly captured by external objects and events – and that the *endogenous*, voluntary attention exhibited by adults is a capacity ontogenetically shaped and stabilized in interactions with caregivers (Posner and Rothbart 1998, 1921; Wexler 2006, 101). If the social scaffolding provided by a caregiver is important for the ontogenetic development of attention, the phenomenon – familiar from adulthood – of voluntarily directing one's perceptual attention to something has a developmental trajectory in which social interactions and social forms of attention (including joint attention) seem to play an important role.

Turning now to joint attention research, the question of how to clarify the 'jointness' or 'openness' of joint attention has been of paramount importance ever since Bruner's work (see Eilan et al. 2005; Seemann 2011). The issue of "what puts the jointness into joint attention?" (Hobson 2005) and, relatedly, of how to clarify the "meeting of minds" (Bruner 1995) that happens in joint attention, has been intensively discussed.⁶ Some developmental psychologists and philosophers have proposed that communication is critical for establishing joint attention (Carpenter and Liebal 2011; Eilan ms.). Others have underlined the role played in joint attention by affective engagement and interpersonal identification (Hobson and Hobson 2011).

But the relevant point for my present purposes is that the pronounced interest in the 'jointness' of joint attention appears to have overshadowed other aspects of the target phenomenon. Most notably, the very notion of attention at stake in debates on joint attention has been rarely scrutinized.⁷ Consider a recent paper by Siposova and Carpenter, which offers a detailed and helpful taxonomy of 'social attention levels' (Siposova and Carpenter 2019). As part of

---

5 In attention research, joint attention has been related to a "folk-psychological matrix" of attention, where such matrix refers to "the complex web of ways ordinary people think, feel, and talk about mental phenomena" (Watzl 2017, 14). Joint attention has also been mentioned in connection with the explanatory roles that a theory of attention would have in the development of our understanding of other minds (Mole 2017, 62).
6 The openness of joint attention has proven to be notoriously difficult to account for. As Rakoczy writes, "[i]t is not sufficient that each of them [i.e. the co-attenders] looks at the same target, nor that, asymmetrically, one sees the other looking somewhere and follows her gaze to the same target. It is not even sufficient, more symmetrically, that each looks at the same target while knowing that the other does so as well [. . .]. Rather, in some intuitive sense that conceptually proves notoriously difficult to spell out, both have to attend to the same target in joint and coordinated ways." (Rakoczy 2018, 409). For discussion, see León (2021).
7 For some exceptions, cf. footnote 1.

their proposal, they claim that a necessary pre-condition for engaging in each of the levels of social attention that they identify is "individual attention".[8] Yet, when characterizing this notion, they say very little about the very notion of attention that is at issue.[9] I suggest that this is an important omission. We won't fully understand what joint attention is by focusing only on the 'jointness' component of joint attention. We also need to factor in the attention component.

But why, one might ask, have researchers of joint attention been generally uninterested in the question of how to best conceptualize attention, and the implications that this might have for joint attention research? I suggest that at least part of the reason is the reliance on a widespread and intuitive characterization of attention as *selection*. Let me illustrate this by considering some representative characterizations of joint attention:

- "there is present from a surprisingly early age a mutual system by which *joint selective attention* between the infant and his caretaker is assured" (Bruner 1974, 269; my emphasis).
- "what underlies infants' early skills of joint attention is their emerging understanding of other persons as intentional agents; [. . .]. When infants begin to view others as intentional they begin to comprehend that: Other persons may *attend selectively (intentionally)* to some things in the environment and ignore others; Other persons may intend for them to *selectively attend* to some things in the environment and ignore others" (Tomasello 1995, 103–104; my emphasis).
- "[t]here are two quite separate parts to the communicative process in basic joint attention. *There's the joint selection of a target.* [. . .] And then there are our open reactions to it, shared as we look at one another" (Campbell 2018, 122; my emphasis).

A common feature of these characterizations is the appeal to the notion of selection. Although such an appeal might seem harmless and perhaps unavoidable, in the following I would like to scrutinize it and problematize it. I will do so by exploring different ways of understanding the notion of attentional

---

[8] Joint attention, as I have characterized it, corresponds to "shared attention" in Siposova & Carpenter's taxonomy (Siposova & Carpenter 2019, 263).
[9] "*Individual attention* simply means attending to something while engaging with the environment from a first-person perspective only. The attender is completely independent of others (no others need even be around), she has individual knowledge about the object of attention, and she does not take the perspective of anyone else present or change or connect with their attention state." (Siposova and Carpenter 2019, 261).

selection.[10] To narrow down the discussion, I will focus on one specific theme that arises in the context of joint attention research: how to account for the transition from solitary to joint attention. Or, to put it differently: what exactly happens when a subject shifts from attending to an object in solitude to attending to it jointly with someone else? I will propose that a promising route to answer this question would start by endorsing a view of attention that, instead of building on the contrast between selection and ignorance – as suggested by Tomasello in the quote above –,[11] conceptualizes attention as an ongoing activity of re-organization and re-articulation of experiences.

To anticipate a possible objection, I do not take this issue to be a mere terminological dispute. That is to say, the issue is not whether the term 'selection' may be used or not for characterizing attention. Even if one were to do so, the important question would then be what one means by attentional selection. Although this is not a question that has been explored in joint attention research, research on attention provides interesting resources to elaborate on it.

2 Shifting to Joint Attention and Attention as Selection

Joint attention is neither in development, nor in adulthood our default way of engaging with the world and others. Rather, engaging in joint attention requires that somehow one makes a transition to it. This can happen in different ways. Sometimes a target of solitary attention becomes a target of joint attention. For example, one subject may actively draw the attention of another subject towards a target that the first subject is already attending to. Suppose, for example, that a person standing next to you is attending to the moon. She establishes eye contact with you, and points to the moon. You follow the pointing gesture, make eye contact again, and as a result of this interaction you and the other subject become aware of attending to the moon together.

---

10 This exploration is not intended to be exhaustive.
11 Consider also the following characterization of attention, which foregrounds the contrast between selection and exclusion: "Attention is an intentional phenomenon in the sense that *it involves the individual intentionally focusing on one aspect of their current experience to the exclusion of others*. A person may see an apple and then choose to focus their attention either on its shape, its colour, its edibility, or any of an infinite number of aspects. *To understand another individual's visual attention is to understand that they have made an intentional choice about what to include and what to exclude in their visual experience.*" (Call and Tomasello 2005, 60; my emphasis)

In another kind of scenario, a salient environmental stimulus attracts the attention of the two subjects without any intervening goal-directed behavior of either subject to attend together to the target object. The following example illustrates this case. You are sitting at a meeting, and a loud fire alarm unexpectedly goes off.[12] You and the person sitting across from you make eye contact, and you both become aware of attending together to the fire alarm. The first example illustrates what has been called in the literature 'top-down' joint attention, in reference to the goal-directed and voluntary character of the action of drawing someone's attention to something. The second case exemplifies 'bottom-up' joint attention, which is non-deliberate and stimulus-driven (Tomasello 2008; Carpenter and Liebal 2011, 170–171; Kaplan and Hafner 2006).[13] Although the two scenarios differ in how joint attention is brought about, the outcome of both situations is that two subjects become aware of attending together to a relevant target object.

Now, consider the following example of top-down joint attention.[14] You are sitting on a park bench watching a swan, and someone comes to sit next to you. After establishing eye contact with the other person, you point to the swan, to direct her attention to it. She looks at the swan, and then looks back at you, in perfect understanding of what you are both looking at. No verbal communication needs to have taken place. According to Campbell (2005, 287), in a situation like this there is a 'shift' from solitary attention to joint attention to the swan. In agreement with Campbell, I will assume that the shift in question is an experiential matter, in the sense that the way in which the swan appears to you as a target of joint attention is qualitatively different from the situation in which you were attending to the swan in solitude. Typically, a prolonged episode of joint attention to the swan will involve an exchange of significant glances, alternated with moments in which the co-attenders watch the swan again.

But what exactly goes on in a situation like this? How does the transition from solitary to joint attention happen? Campbell suggests the following:

> [w]hat is distinctive of the case in which you and another are jointly attending is that (a) you are monitoring the direction of the other person's attention, and (b) one of the factors controlling the direction of your own attention is the direction of the other person's attention.
>
> (Campbell 2005, 287)

---

12 I borrow the example from Eilan (ms., 14).
13 This distinction maps onto the distinction between "endogenously driven" and "exogenously driven" joint attention (Campbell 2018, 122). I focus here on cases of joint attention involving two co-attenders, and I leave aside the interesting topic of the distribution of joint attention across sensory modalities (Botero 2016; Núñez 2014).
14 I borrow (and slightly modify) an example from Campbell (2005).

Going back to the idea that attention is fundamentally a matter of selection, it may be natural to articulate this suggestion along the following lines. In oscillating attention back and forth between the other co-attender and the object, each co-attender is alternating between two acts of selection: the selection of the swan and the selection of the other co-attender. Whatever else might be involved in the shift from solitary to joint attention, the alternation between the attentional selection of the target object and of the co-attender would have to be factored into the analysis of joint attention.[15] How should this proposal be assessed? I do not think it takes us very far. The main reason is that the notion of attentional selection is too unspecific and potentially misleading.

To begin with, consider that, for it to be minimally plausible, the view that attention is a matter of selection would have to specify what type of selection is at stake. If attention is about selection, it cannot be just *any* type of selection. To rule out the least plausible option, it couldn't be the same type of selection executed by an object sorter. An object sorter – for example, a sorting machine that discriminates objects based on the material they are made of – can be highly selective, but that doesn't mean that it attends in the relevant sense (Wu 2014, 13). More importantly, an object sorter doesn't have a conscious perspective on the world, yet this seems to be clearly involved in the joint attention scenarios described above. The 'openness' of joint attention, and the way in which each co-attender relates to the common target and to the other co-attender are not a matter of subliminal or unconscious perception, even though joint attention is surely underpinned by a number of unconscious processes. Rather, the co-attenders' attending to one another and to the common target object are phenomena with an experiential character. For it to play a sufficiently robust role in the elucidation of joint attention, the view that attention is a matter of selection would have to hold, minimally, that attentional selection (or one central aspect of it) is conscious selection.

One way to approach the notion of conscious attentional selection is by considering the famous 'spotlight' metaphor of attention. On one reading of this metaphor, attention operates like a flash of light focused on one object, while leaving the rest in darkness. How should this view be assessed? On the

---

[15] The question about the transition from solitary to joint attention is particularly relevant for an account like Campbell's, according to which joint attention is a primitive (non-reductively analyzable) relation, to be understood in terms of the notion of "co-consciousness" of a target (Campbell 2018, 122, 124). How to account for the transition between the individualistic consciousness of a target, and the "co-consciousness" of it? This is not a question that Campbell elaborates on. For discussion of this point, in connection with developmental issues, see León (2021).

face of it, it has little support in its favor. A critical weakness of the spotlight metaphor is its phenomenological inadequacy. When one turns one's perceptual attention from one thing to another, or when one alternates attention between different things, it is not as if everything else apart from one's current focus of attention is left in darkness. Rather, whatever it is that one attends to, there are a number of other things that one is also conscious of. While they are not in one's focus of attention, they are not suppressed from one's experience either. Were those things really suppressed from one's experience, it would be difficult to make sense of the idea of *turning* attention to something.

At this point, one might say that attentional selection doesn't have to be modeled on the spotlight metaphor, or that the latter may be understood differently. For example, one might say that, irrespectively of whether the spotlight metaphor is adequate, the critical feature of attentional selection is the filtering of stimuli and information. This idea has been historically important in research on attention, particularly in cognitive psychology and neuroscience, in connection with the view that attentional selection is closely tied to limits in information-processing capacities (Mole 2017). According to one traditional view of cognition, a cognitive system is an information-processing device that has limited resources for processing information coming from outside of the system. If one accepts this picture, the idea that some information has to be filtered out via (low-level) attentional selection appears to gain in plausibility.[16]

But the view that attention is selection to avoid information overload doesn't come without shortcomings (for discussion see Watzl 2017, 106–107). Consider that, even if an individual had unlimited processing capacities, attention would still be needed for action guidance (van der Heijden 1992, 243). That is to say, even supposing that there were no limits in the amount of information that an individual could process, it seems that attention would still play an important role in enabling the individual to navigate its environment. Moreover, the idea that the function of attention is selection of information to avoid information overload cashes out the notion of attentional selection at a fairly low level, which is not illuminating to clarify the experiential character of joint attention.

That being said, the idea that attention is fundamentally a matter of selection doesn't have to be tied to the view that the function of attention is to manage

---

16 William James writes that attention "implies withdrawal from some things in order to deal effectively with others" (1890, 404), and that in attending to an object, "one principal object comes then into the focus of consciousness, others are temporarily suppressed" (1890, 405). This last remark contains *in nuce* the idea that attentional selection happens at the expense of what is not selected. On this view, attention operates by extracting what is attended to, and filtering out – or "temporarily suppressing", as suggested by James – the rest.

capacity limitations. To employ Watzl's (2017, 107) helpful terminology, an alternative to the "information-pruning conception" of the role of attention is an "organizational conception", according to which the central function of attention is to organize the mind. There are different ways of cashing out the organizational conception of attention. One way to do so is also in terms of the notion of selection. On this view, attention is about selection because of constraints that subjects have to meet in order to navigate the *abundant* information that is available to them. To put it differently, it is because subjects can process multiple inputs, that there must be a way for them to link those inputs with relevant outputs, and thereby successfully navigate their surroundings.

Considered from this perspective, attention doesn't protect a subject from cognitive overload, but rather allows flexible behavior by assisting in managing "capacity excess" (Mole 2017, 33). A clear representative of this view is the "selection-for-action" theory of attention (Wu 2014). According to this theory, given the many perceptual inputs that a subject is exposed to, and the many possible behavioral outputs that it can carry out, it has to face the problem of how to couple relevant inputs with relevant outputs. The selection-for-action theory of attention holds that the way in which subjects solve this "Many-Many Problem" is by means of selecting a target for action. On this view, attention is individuated by its central functional role: "S's attention to X is S's selection of X for action." (Wu 2014, 96)

Without going into the details of this proposal, the view that attention is subject-level selection-for action provides a more substantive development of the idea that attention is a matter of selection. It goes beyond the 'spotlight' metaphor of attention, with its misleading connotations, and also beyond the information-pruning conception of attention. But is the selection-for-action theory warranted in identifying attention with selection for action? Watzl has argued that the selection-for-action view faces a number of problems. I will here focus on three considerations advanced by Watzl, which put pressure on the selection-for-action theory of attention.[17]

First, attention often *explains* that one can select an object for action: "often you are only able to select an object to reason about it or act on it because you have focused your attention on it" (Watzl 2017, 111). For example, if a subject focuses her attention on the perceived sound of the saxophone in a jazz band and reasons about it, "attention to the sound plays an explanatory role: it in part explains why she is able to select the sound as the target of her reasoning"

---

[17] For a fuller discussion, see Watzl (2017, 110–113).

(Watzl 2017, 111). But if attention partly explains why a subject can select a target for action (such as reasoning and thinking about it), attention cannot be identified with selection for action.

Secondly, a subject can select an object as the target of an action even if the subject does not attend to the relevant object. For example, if one is engaged in a face-to-face conversation with someone, and one's perceptual attention is focused on one's interlocutor, one will typically still be able to take a glass of wine that is within one's reach. If one succeeds in doing so, one will have selected the glass for the action of reaching, even though one hasn't focused attention on the glass (Watzl 2017, 111). This indicates that selection-for-action is not sufficient for attention.

A third consideration that, according to Watzl, puts pressure on the selection-for-action view is that selecting and attending have different temporal profiles. Whereas attending is an ongoing and temporally extended process, selecting something as a target of action is a particular event. Consider that, in principle, it should be possible to count how many times one has selected something in a certain period of time. But the idea of counting how many times one has attended to an object in a period of time is much less compelling (Watzl 2017, 112). The complex temporality that attention can exhibit appears to be downplayed in the selection-for-action view.

Taken together, these considerations make the selection-for-action theory less attractive as a candidate proposal for elaborating on the idea that attention is a matter of selection. But where to go from here? Watzl's own proposal is that the organizational conception can be cashed out in a quite different way. He calls it the 'priority structure account' of attention (Watzl 2017). Roughly put, the central idea of this view is that attention is the activity of regulating priority structures. Attention, in its different varieties, is an activity that organizes the mind into parts that are central and prioritized and parts that are peripheral (Watzl 2017, 3, 70).[18]

Consider again the previous examples, in light of this proposal. The idea that attention is an ongoing process of prioritization can explain that attention eventually makes possible to select objects for acting upon them, depending on how one's experiences – of, say, perceiving the sound of the saxophone vis-à-vis the sound of the piano – are prioritized. Moreover, when one attends to a conversation partner, one is shaping and organizing one's experiences in a certain way. The action of reaching for a glass on the table, while keeping the attentional focus on one's partner, can succeed because one's perceptual experience of the glass plays a role in a certain priority structure (the perception of the glass

---

18 See Watzl (2017) for a book-long treatment and defense of the prioritization view.

is de-prioritized with respect to the perception of one's interlocutor). Finally, prioritization, in contrast to selection, is a fluid, ongoing and temporally extended process, that can prima facie make better sense of the complex temporality of attention (Watzl 2017, 60).

Leaving aside many aspects of Watzl's wide-ranging theory, I believe that the idea that attention is fundamentally a matter of ongoing prioritization is potentially promising for joint attention research. By giving center stage to the context in which attention takes place, the prioritization view can potentially bring to the investigation of joint attention a rich and articulated framework for conceptualizing the transition from solitary to joint attention. It has potential to make questions about the 'meeting of minds' (Bruner 1995) in joint attention more tractable than they have been so far, and thereby also to make the 'openness' of joint attention a less elusive phenomenon.[19]

However, there is one aspect of attentional engagements that plays an important role in joint attention, but which is considerably downplayed in Watzl's proposal. While Watzl emphasizes that attention is an activity (Watzl 2017, 38), he takes it to be primarily a *mental* activity, which may only in some cases involve the body (Watzl 2017, 45–46). More specifically, he proposes that bodily movements are not necessary for perceptual attention. Consider, for example, the example of listening to someone in a face-to-face situation (Watzl 2017, 42). On Watzl's analysis, listening to someone is an activity that requires effort, that one can try to do, and that one can also fail at. Listening to someone is a way of perceptually attending to her or him, which differs from merely hearing what that person is saying. At the same time, Watzl notes (2017, 42), one might listen to someone even if one doesn't move at all while performing the activity of listening.

One might agree with Watzl that attention need not *always* involve bodily movements, but the question is whether consideration of such cases supports what might be taken to be a dis-embodied picture of attention as a purely mental activity. Part of the difficulty is some ambiguity in the talk of how the body might (or not) be involved in attention. Watzl writes:

> [i]n some cases attending involves the body: there is overt visual attention (where you *do* move your eyes), tactile attention ('feeling'), gustatory attention ('tasting'), and olfactory attention ('sniffing'). But such involvement of the body need not be present in *all* forms of perceptual activity. (Watzl 2017, 45–46)

---

19 For a perceptive description of the elusiveness of joint attention, cf. Rakoczy's quote in footnote 6.

This suggests that, for Watzl, the relevant kind of involvement of the body is constituted by proper bodily movements, or "overt, bodily behavior" (Watzl 2017, 42), such as when we move our eyes, hands, tongue and nose in perceptually attending to something. However, even in the absence of such bodily movements, the body is involved quite centrally in prototypical cases of perceptual attention. After all, going back to the example above, the listener will have to occupy a suitable spatial position with respect to the speaker (and vice-versa), and that position will be anchored in a body with respect to which the perceptual surroundings will be organized in a certain way. My point is that, even in the absence of overt bodily behavior, the body is structuring the whole situation. Moreover, since the priority structure account appears to operate with a stark distinction between mental activity and bodily activity,[20] one might wonder how, if at all, the two types of activity might be integrated in attentional engagements. It doesn't seem controversial that we attend to the world, at least primarily, by exploring it and moving around in it. But even when we do not perform overt bodily movements, the body remains the reference point of perceptual attentional engagements. To downplay the role of the body in the analysis of attention is to downplay the primary medium through which we attend to the world of perception.

My interim conclusion is twofold. Watzl's priority structure account improves on shortcomings of views of attention that give center stage to the notion of selection. At the same time, it neglects the link between attention and active bodily engagement with the world. This matters for joint attention, because overt and expressive bodily behavior – for example, in the form of pointing gestures and sharing looks – arguably play a central role in its establishment (Carpenter and Liebal 2011). But I suggest that the take-home message from the foregoing discussion is not that the proposal that attention is fundamentally a matter of ongoing prioritization is not applicable to joint attention. Rather, I suggest that such a view can be pursued in a different direction, which foregrounds the role of the body in attention. Work on attention in classical and contemporary phenomenology has highlighted the active and bodily aspects of attention, while at the same time advancing ideas that resonate well with some tenets of the priority structure account.[21] In the next section, I turn to some resources from classical phenomenology, in order to outline a

---

20 "We exercise agency in arranging the parts of our mind in just the way we exercise agency in moving the parts of our body." (Watzl 2017, 140)

21 Although Watzl (2017, 208–209) cites Gurwitsch, Merleau-Ponty, and Sartre as precursors of his priority structure account, it is worth underlying how different are the views on embodiment held by Watzl and by the phenomenologists.

phenomenological view of attention that pursues the prioritization view, although in a direction that differs from Watzl's approach.[22]

3 Phenomenology of Attention: Contributions from Merleau-Ponty and Gurwitsch

In the Introduction to the *Phenomenology of Perception*, Merleau-Ponty provides a concise, yet remarkably rich discussion of attention. He suggests a distinction between two views of attention, which are dependent on different ways of understanding the role of the body in perceptual experience. On the first view, the notion of attention is employed in order to accommodate the idea that the external world provides information to the sensory organs, information that is "decoded in such a way as to reproduce in us the original text" (Merleau-Ponty 2012, 8). This view faces the challenge of how to explain cases of perceptual ambiguity and illusion, such as the Müller-Lyer illusion.[23] If there is a constant correspondence between external stimuli and sensory registrations, why can there be discrepancies between the two? To answer this question, this view invokes the notion of attention as an "auxiliary hypothesis" (Merleau-Ponty 2012, 7). Briefly put, the suggested answer is that there is indeed a constant correspondence between stimuli and registrations, but in the case of perceptual illusions we simply fail to notice and attend to this. Were we to scrutinize our experience enough, we find the correspondence that the illusion masks (Merleau-Ponty 2012, 7). On this view of attention, the body is thought of as a "transmitter of messages" (Merleau-Ponty 2012, 10) coming from the outside world and registered by our sensory organs, and the role of attention would be to clarify the presupposed "constant connection" between stimuli and registrations.

Merleau-Ponty also sketches a quite different view of attention. This second view builds on the idea that "the external world is not copied, but constituted" (Merleau-Ponty 2012, 9). On this second view, as he writes, "[a]ttention first presupposes a transformation of the mental field [. . .]. To pay attention is not

---

[22] For discussions of some of the phenomenological resources about attention, in connection with contemporary debates, see Wehrle and Breyer (2016), Depraz and Perreau (2010), and D'Angelo (2018).
[23] In this well-known visual illusion, two lines of the same length appear to be of different lengths, depending on whether arrowheads at the ends of each line are pointing towards each other or away from each other.

merely to further clarify some preexisting givens; rather, it is to realize in them a new articulation by taking them as *figures* [. . .]" (Merleau-Ponty 2012, 31–32). Taking something as a figure means to articulate it as a foreground against a particular background. But attention could not accomplish this if one endorses the first view, taking attention to simply clarify the assumed correspondence between external inputs and sensory registrations. Merleau-Ponty's (2012, 32) suggestion is that the first view of attention is mistaken, and that we should recognize instead that attention has a creative power.

How should this proposal be understood? Surely, not in the sense that attention would create its objects *ex nihilo*. A more plausible way of understanding Merleau-Ponty's proposal is in terms of the phenomenological concept of horizon. An object attended to is, in one way or another, within a horizon of perceivability. In becoming attended to, it appears articulated against a particular background. Its presence in the horizon of experience, as a more or less determinate object, is "overthrown", to make place for a newer determination (Merleau-Ponty 2012, 33). In this sense, attention accomplishes an active reorganization and re-articulation of the field of experience. The whole process is described by Merleau-Ponty as "the miracle of consciousness": "to make phenomena appear through attention that reestablish the object's unity in a new dimension at the very moment they destroy that unity" (Merleau-Ponty 2012, 33).

Now, for it to be compelling, the second view of attention sketched by Merleau-Ponty cannot rely on a conceptualization of the body as a "transmitter of messages" from the outside world. Rather, it has to acknowledge – as Merleau-Ponty elaborates *in extenso* in his work – that the body structures our experience of the world. When we attend to perceptual objects, these are experienced as oriented with respect to our perceiving body, and the active exploration of a perceptual environment is itself a way of shaping the latter through our bodily engagement.

While Merleau-Ponty's discussion of attention highlights the organizational and embodied character of the latter, perhaps one of the most interesting points that can be gathered from his discussion is that it calls into question the general input-output framework that is operative in some of the conceptualizations of attentional selection mentioned in the previous section. Whether one thinks of attention as a filter, or as selection-for-action, the underlying picture of the mind within which attention would be slotted is one in which an individual receives and processes information coming from the external world, and on the basis of suitable processing of that information, generates outputs and thereby acts upon the world. The question then becomes where exactly to locate attention in this circuit, which has been aptly called the "sandwich model of the mind" (Hurley 2001). Is attention at play in filtering and controlling which

Chapter 3 Attention in Joint Attention: From Selection to Prioritization — 81

inputs are allowed to pass for further processing? Or is attention playing a role in allowing subjects to behave flexibly in view of many possible outputs they can generate, i.e., different courses of action?

In an alternative picture, inspired by Merleau-Ponty's emphasis on the subject's embodied situatedness in the world, attending doesn't consist in filtering out and suppressing information, but rather in organizing, articulating, and literally shaping our experience of the world in a certain way. The creative power of attention concerns an ongoing activity that may be aptly qualified as 'sense-making' (Thompson and Stapleton 2009), which is not premised on the division and separation of mind and world, but rather on the dynamic coupling and interaction between the two.

Merleau-Ponty's discussion of attention can be supplemented with some contributions from Aron Gurwitsch.[24,25] In his dissertation, Gurwitsch discusses critically the idea that attention operates like a 'beam of light', and he highlights the phenomenological inadequacy of this proposal. As he writes,

> [c]onsciousness is, in general, not the presence of a content surrounded by a chaotic manifold of any other contents whatever; and thematic consciousness does not consist, as one usually asserts of attention, in a beam of light being cast upon a certain content while a chaotic confusion of other contents fills the regions of shadow and darkness. We must beware of taking literally the metaphor of the 'illuminating light' of attention.
>
> (Gurwitsch 2009, 223)

Even more than Merleau-Ponty, Gurwitsch insists on the organizational character of attention: "turning to, and being turned to, a theme cannot be accounted for in terms of distribution of illumination, but rather in terms of organization of the field of consciousness [. . .]. It is not a matter of obscuring or brightening but is one of organization" (Gurwitsch 2009, 226). The distinction that he introduces in later work between the *theme*, the *thematic field*, and the *margin* of consciousness aims at capturing, at a formal level, the basic elements of this organization:

> Every field of consciousness comprises three domains or, so to speak, extends in three dimensions. First, the *theme*: that with which the subject is dealing, which at the given moment occupies the 'focus' of his attention, engrosses his mind, and upon which his

---

[24] In the following, I touch upon a few aspects of Gurwitsch's discussion of attention. I leave aside other important aspects, such as his challenge to the idea that the notion of attention picks out a unitary phenomenon, and his non-egological conception of consciousness (Gurwitsch 2009).

[25] Gurwitsch had a direct influence on Merleau-Ponty, who attended Gurwitsch's lectures in Paris in the early 1930's (Moran 2019, 15).

mental activity concentrates. Secondly, the *thematic field* which we define as the totality of facts, co-present with the theme, which are experienced as having material relevancy or pertinence to the theme. In the third place, the *margin* comprises facts which are merely copresent with the theme, but have no material relevancy to it. (Gurwitsch 2010, 53)

According to Gurwitsch, attention does not impose order on a chaotic stream of sensations or experiences. Experiences are already organized, insofar as the triad of theme-thematic field-margin is an "autochthonous" (Gurwitsch 2010, 28) feature of the field of consciousness. In this sense, the organization of the field of consciousness is not a sporadic or incidental event, but rather an ongoing process of *re*-organization and *re*-configuration that is constantly at play. Through that process, objects of attention are brought to the foreground of consciousness, while other objects recede into the background. Importantly, whereas Gurwitsch's notion of margin comprises elements that are not of direct pertinence to the theme, the relation between theme and thematic field is supposed to be much closer. The elements of the thematic field are not merely put next to one another, co-given in an aggregative fashion. They are structured in a certain way in virtue of the theme, to which priority is given: "The thematic field is not a conglomeration of any contents whatever, not like a box in which sundry things can be put and from which they can be taken out [. . .]." (Gurwitsch 2009, 224) In contrast to the "and-connection" of a mere conglomeration or aggregation, Gurwitsch (2009, 224) operates with the concept of a "Gestalt connection". It is the theme which provides a specific orientation to the thematic field:

> Whether or not a cogiven item belongs to the thematic field, how it is inserted in this field, which place it has there, etc. – all this depends upon its relation to the theme. The ground (thematic field) is organized around the figure (theme). There is always given a thematic field organized and oriented with respect to this theme. Whatever is experienced as pertaining to the thematic field has 'directedness to the center'. (Gurwitsch 2009, 225)

One might wonder whether Gurwitsch's conceptual triad of theme-thematic field-margin is an invariant and universal feature of all conscious experiences (Arvidson 2000, 5). But more relevant for my purposes is to consider the way in which he conceptualizes the relationship between theme and thematic field. Apart from suggesting that the theme is point of reference for the thematic field, he also holds that the latter allows the former to emerge: "The relation of theme to thematic field is reciprocal; it is a correlation." (Gurwitsch 2009, 228)[26]

---

26 As Gurwitsch underlines, "[i]t is erroneous to speak of the 'theme *simpliciter*' without mentioning its relation to, and its insertion into, the thematic field, just as it is also inappropriate to speak of the thematic field without taking into consideration its organization and orientation with reference to the theme" (Gurwitsch 2009, 227).

Gurwitsch's suggestion is that the theme, as the focal point of consciousness, emerges as articulated within a context of objects relevant to it. To be sure, what counts as relevant or not will depend on the specifics of a situation. But the general point is that, far from being an obstacle or mere distracting 'noise' for a proper apprehension of the theme, it is the thematic field which makes possible the emergence of the theme as the focal point of consciousness. (This doesn't mean, though, that a theme is necessarily tied to one particular thematic field.) At the same time, what counts as theme will have an impact on delineating the objects that are to be taken as constituents of the thematic field. The theme is a point of reference for the thematic field.

Gurwitsch takes the proposed correlation between theme and thematic field as a platform for examining different ways in which theme and thematic field can be modified. In his dissertation, he distinguishes between three series of 'thematic modifications' (Gurwitsch 2009, 241). The first series is characterized by the invariance of the theme and the modification of the thematic field, such as when the thematic field is enlarged, narrowed down, or replaced (Gurwitsch 2009, 247). In a second series of modifications, a theme loses its status as theme, and is replaced by a new theme (Gurwitsch 2009, 257). Finally, in the third series of modifications, the theme is affected in a "deep-reaching way", something that may happen as a result of various processes – Gurwitsch (2009, 262) considers 'restructuration', 'singling out', and 'synthesizing'. All in all, Gurwitsch's distinction between theme-thematic field-margin seeks to capture the dynamicity and rich texture of attention, and it supplements Merleau-Ponty's observations concerning the embodied and structuring power of the latter.

4 Back to Joint Attention: From Attentional Selection to Attentional Prioritization

It is time to come back to my initial question: what might research on joint attention learn from attention research? Let me summarize my answer in the following three points.

In the first place, I contend that part of the reason why the transition from solitary to joint attention might appear puzzling is the reliance on an unproblematized understanding of attention as selection. Shifting to joint attention, it seems, must somehow involve oscillating one's selective attention between a target object and another co-attender. How to get from these two dyadic relations to triadic joint attention is far from clear. I have argued that we shouldn't take for granted the selectivity of attention. The notion of attentional selection is very broad and potentially misleading. The organizational conception of attention suggested by Merleau-Ponty and Gurwitsch, and which finds important parallels in Watzl's proposal, points towards a more holistic and context-sensitive understanding of attention. If the central function of attention is to reconfigure and re-organize the field of experience into a foreground-background structure, the transition to joint attention would have to be understood in terms of that capacity. The question of how to understand the transition from solitary to joint attention would lead to the question of how an object can become a center of orientation around which co-attenders can organize and prioritize their perceptual experiences in a convergent and integrative fashion. That being said, while the organizational conception of attention may help to address this question, it is clearly not sufficient. This brings us back to the 'openness' of joint attention. One element that is missing from the picture, and that is being increasingly recognized in the literature, is the role of communication in establishing 'jointness'.[27]

One might rightly point out that joint attention theorists who appeal to the notion of selection would straightforwardly reject any view suggesting that perceptual attention operates in a vacuum, or that unselected features of the environment are chaotic or hidden in darkness, as one reading of the 'spotlight' metaphor of attention might suggest. Nothing of the sort is suggested by Bruner, Tomasello, or Campbell, and it would be implausible to attribute to them such a view. But even if one accepts that attentional selection necessarily takes place in a certain context, the follow-up (and more interesting) question is how exactly the context is brought into the picture. At this point, the question of how to understand the relationship between the attended object and the context of attention becomes pressing. Is it an aggregative relation, or a correlation, as suggested by Gurwitsch? The important point is that by operating with an unproblematized notion of individual attention in terms of selection, research on joint attention risks taking on board a somewhat atomistic picture of how conscious attention operates. A potential problem that arises from this is to miss on a relevant distinction between thinking of attention in terms of the atomistic selection of particular target objects, and in terms of a holistic re-organization of experience.

---

27 The recognition of the role of communication in establishing the 'jointness' of joint attention is an interesting, although perhaps not very surprising development in the joint attention literature (Carpenter and Liebal 2011; Eilan ms.; Campbell 2018). Consider that the idea that the target of joint attention is "mutually manifest" (Eilan 2005, 1) for co-attenders, an idea endorsed by several theorists to characterize the 'jointness' or openness of joint attention, is explicitly taken by Elain from Sperber and Wilson's (1986) relevance theory of intentional communication.

Secondly, although Gurwitsch doesn't discuss joint attention, he provides some resources to analyze the transition from solitary to joint attention via his notion of thematic modifications (cf. Arvidson 2006, 61, 68). Consider that, in some cases, the thematic field is expanded or enlarged while the theme remains constant, such as when one realizes that someone else is also watching the swan that one has been watching for a while. Part of what goes on in such a situation is that the other person's attention is integrated into the thematic field of the attended object. Insofar as the other person is apprehended as directed to the object that one is also prioritizing, the other acquires relevance for the theme, which is itself enriched by the other's presence. This, in turn, prepares the ground for establishing communicative connection with that person, and engaging in joint attention. The transition to joint attention can also be less smooth. While the theme remains constant, the thematic context can be replaced (Arvidson 2006, 68). It can change from being non-social to being social, say, because of the sudden appearance of a co-attender who hasn't been registered before as part of the thematic field.[28]

Moreover, even in scenarios (such as the example of joint attention to the fire alarm, mentioned in Section 2) in which it might be less clear that the target of joint attention was attended to in solitude before joint attention sets in, a specific thematic modification seems to take place. When you and the person sitting across from you make eye contact upon hearing the fire alarm, part of what is plausibly going on is that you and your co-attender are prioritizing your experiences, and possibilities of engagement with the world in a certain fashion. Joint attention opens up the possibility for a "shared action space" (Pezzulo et al. 2013), in which agents reframe and re-calibrate a situation in light of the presence of a co-attender, in order to achieve joint goals. For example, attending together to the fire alarm might motivate you and your co-attender to immediately try to make space in the room where you are sitting, so that everyone can safely evacuate. The third series of thematic modifications mentioned by Gurwitsch is also of interest for research on joint attention, insofar as part of

---

[28] It seems less promising to analyze joint attention in terms of a shifting back and forth between two different *themes* (the target object and the other co-attender): "For example, the other becomes thematic and the object is part of the thematic context, and then the object becomes thematic and the other is part of the thematic context, and so on" (Arvidson 2003, 113). The reason is that this analysis in terms of alternation of attention would decompose joint attention into a set of two dyadic attentional relations. Relatedly, this analysis doesn't fit well with the idea that the target of joint attention is prioritized throughout the joint attentional interaction: even when co-attenders communicatively look at each other, the target of attention remains that *about which* they communicate.

what can attract and sustain the attention of co-attenders is that a certain theme undergoes modifications of various sorts. The more general point is that analyzing the transition to joint attention in terms of various thematic modifications presupposes the rich picture of attention developed by Gurwitsch.

The last point I would like to mention concerns Merleau-Ponty's idea that attention has a creative power, which I proposed can be related to the notion of sense-making. I suggest that this idea can also be fruitfully carried into joint attention research. Consider that joint attention is a type of social interaction that is essentially anchored in the common world shared by co-attenders. A rich conceptualization of joint attention, one that gives center stage to its 'openness', suggests that this convergence of perspectives on a common and public world is not merely a matter of a probabilistic expectation that the other is attending to the same particular item of the world as oneself. Rather, joint attention discloses the publicity and overtness of the world of perception.

But a picture of attention based on the contrast between selection and exclusion of information faces the challenge of how to get from such selection to the idea that the world of perception is shared with others in a sufficiently robust sense, at least in the sense that it affords basic communication about it. From the perspective of the organizational conception put forward by Merleau-Ponty and Gurwitsch, the target of joint attention is not out there, 'waiting' as it were to be registered and selectively highlighted by the co-attenders. Rather, the triangulation in which co-attenders participate has an active and co-constructed aspect, because it actively shapes their experiences of the world in a certain way. Through joint attention, co-attenders confer intersubjective validity to a common world, and lay the ground for enriching it with new, co-created meanings, as they may arise from their reactions and – verbal or non-verbal – comments about the relevant target of their joint attention.

5 Concluding Remarks

The interrelations between attention and joint attention merit more consideration than has been given to them so far in the literature. As one step in this direction, I have argued that a broad understanding of attention as conscious selection is of limited help for joint attention research, and that the latter would do better in turning to the prioritization view suggested by Merleau-Ponty, Gurwitsch, and Watzl. A key idea of this conception is that attention is not a matter of picking out an item and filtering out the rest, but rather of shaping and re-organizing experiences in a certain way. I have considered three benefits of the second view: (i) a more context-sensitive approach to the question of how to account for the transition from solitary to joint attention; (ii) an examination of that transition in terms of thematic modifications, and (iii) a better appreciation of the co-constructed character of joint attentional interactions.

References

Arvidson, P. S. (2000): "Transformations in Consciousness: Continuity, the Self and Marginal consciousness". *Journal of Consciousness Studies* 7:3, 3–26.
Arvidson, P. S. (2003): "A Lexicon of Attention: From Cognitive Science to Phenomenology". *Phenomenology and the Cognitive Sciences* 2, 99–132.
Arvidson, P. S. (2006): *The Sphere of Attention: Context and Margin*. Bern: Springer.
Böckler, A.; Sebanz, N. (2013): "Linking Joint Attention and Joint Action". In: J. Metcalfe; H. S. Terrace (eds.): *Agency and Joint Attention*. Oxford: Oxford University Press, 206–215. https://doi.org/10.1093/acprof:oso/9780199988341.003.0013
Botero, M. (2016): "Tactless Scientists: Ignoring Touch in the Study of Joint Attention". *Philosophical Psychology* 29:8, 1200–1214. https://doi.org/10.1080/09515089.2016.1225293
Bruner, J. (1974): "From Communication to Language – A Psychological Perspective". *Cognition* 3:3, 255–287. https://doi.org/10.1016/0010-0277(74)90012-2
Bruner, J. (1977): "Early Social Interaction and Language Acquisition". In: H. R. Schaffer (ed.): *Studies in Mother-Infant Interaction*. Oxford: Academic Press, 271–289.
Bruner, J. (1995): "From Joint Attention to the Meeting of Minds: An Introduction". In: C. Moore; P. J. Dunham (eds.): *Joint Attention: Its Origins and Role in Development*. Hillsdale: Lawrence Erlbaum Associates, 1–14.
Butterworth, G. (1995): "Origins of Mind in Perception and Action". In: C. Moore; P. J. Dunham (eds.): *Joint Attention: Its Origins and Role in Development*. Hillsdale: Lawrence Erlbaum Associates, 29–40.
Call, J.; Tomasello, M. (2005): "What Chimpanzees Know about Seeing, Revisited: An Explanation of the Third Kind". In: N. Eilan; C. Hoerl; T. McCormack; J. Roessler (eds.): *Joint Attention: Communication and Other Minds: Issues in Philosophy and Psychology*. New York: Oxford University Press, 45–64.
Campbell, J. (2005): "Joint Attention and Common Knowledge". In: N. Eilan; C. Hoerl; T. McCormack; J. Roessler (eds.): *Joint Attention: Communication and Other Minds: Issues in Philosophy and Psychology*. New York: Oxford University Press, 287–297.
Campbell, J. (2011): "An Object-Dependent Perspective on Joint Attention". In: A. Seemann (ed.): *Joint Attention: New Developments in Psychology, Philosophy of Mind, and Social Neuroscience*. Cambridge, MA: MIT Press, 415.
Campbell, J. (2018): "Joint Attention". In M. Janković; K. Ludwig (eds.): *The Routledge Handbook of Collective Intentionality*. London: Routledge/Taylor & Francis Group, 115–129.
Carpenter, M.; Call, J. (2013): "How Joint Is the Joint Attention of Apes and Human Infants?". In: J. Metcalfe; H. S. Terrace (eds.): *Agency and Joint Attention*. Oxford: Oxford University Press, 49–61.

Carpenter, M.; Liebal, K. (2011): "Joint Attention, Communication, and Knowing Together in Infancy". In: A. Seemann (ed.): *Joint Attention: New Developments in Psychology, Philosophy of Mind, and Social Neuroscience*. Cambridge, MA: The MIT Press, 159–181.

Carpenter, M.; Nagell, K.; Tomasello, M.; Butterworth, G.; Moore, C. (1998): "Social Cognition, Joint Attention, and Communicative Competence from 9 to 15 Months of Age". *Monographs of the Society for Research in Child Development* 63:4, i.

D'Angelo, D. (2018): "A Phenomenology of Creative Attention. Merleau-Ponty and Philosophy of Mind". *Phänomenologische Forschungen* 2018/1, 99–116.

Depraz, N.; Perreau, L. (2010): *L'attention*. Vol. 18. Paris: Éd. Alter.

Dunham, P. J.; Dunham, F.; Curwin, A. (1993): "Joint-Attentional States and Lexical Acquisition at 18 Months". *Developmental Psychology* 29:5, 827–831.

Eilan, N. (2005): "Joint Attention, Communication, and Mind". In: N. Eilan; C. Hoerl; T. McCormack; J. Roessler (eds.): *Joint Attention: Communication and Other Minds: Issues in Philosophy and Psychology*. New York: Oxford University Press, 1–33.

Eilan, N. (ms.): *Join Attention and the Second Person*. Accessed on November 3, 2021. https://warwick.ac.uk/fac/soc/philosophy/people/eilan/jaspup.pdf

Eilan, N.; Hoerl, C.; McCormack, T.; Roessler, J. (eds.) (2005): *Joint Attention: Communication and Other Minds: Issues in Philosophy and Psychology*. New York: Oxford University Press.

Gómez, J. C. (1994): "Mutual Awareness in Primate Communication: A Gricean Approach". In: S. T. Parker; R. W. Mitchell; M. L. Boccia (eds.): *Self-Awareness in Animals and Humans*. Cambridge, MA: Cambridge University Press, 61–80.

Gurwitsch, A. (2009): *Studies in Phenomenology and Psychology*, ed. by F. Kersten; L. E. Embree. The Hague: Springer.

Gurwitsch, A. (2010): *The Field of Consciousness: Phenomenology of Theme, Thematic Field, and Marginal Consciousness*, ed. by R. M. Zaner; L. Embree. The Hague: Springer.

Heidegger, M. (2001): *Einleitung in die Philosophie*. Frankfurt am Main: Vittorio Klostermann.

Hobson, R. P. (2002): *The Cradle of Thought*. London: Macmillan.

Hobson, R. P. (2005): "What Puts the Jointness into Joint Attention?". In: N. Eilan; C. Hoerl; T. McCormack; J. Roessler (eds.): *Joint Attention: Communication and Other Minds: Issues in Philosophy and Psychology*. Oxford: Oxford University Press, 185–204.

Hobson, R. P.; Hobson, J. (2011): "Joint Attention or Joint Engagement? Insights from Autism". In: A. Seemann (ed.): *Joint Attention: New Developments in Psychology, Philosophy of Mind, and Social Neuroscience*. Cambridge, MA: MIT Press, 115–136.

Hurley, S. (2001): "Perception And Action: Alternative Views". *Synthese* 129:1, 3–40.

James, W. (1890): *The Principles of Psychology*: Vol. I. New York: Henry Holt and Company.

Kaplan, F.; Hafner, V. (2006): "The Challenges of Joint Attention". *Interaction Studies* 7:2, 135–169.

León, F. (2021): "Joint Attention Without Recursive Mindreading: On the Role of Second-Person Engagement". *Philosophical Psychology* 34:4, 550–580.

León, F.; Szanto, T.; Zahavi, D. (2019): "Emotional Sharing and the Extended Mind". *Synthese* 196, 4847–4867.

Loveland, K. A.; Landry, S. H. (1986): "Joint Attention and Language in Autism and Developmental Language Delay". *Journal of Autism and Developmental Disorders* 16:3, 335–349.

Merleau-Ponty, M. (2012): *Phenomenology of Perception*, trans. by D. A. Landes. London: Routledge.

Merleau-Ponty, M. (1964): *The Primacy of Perception*. Evanston, Ill: Northwestern University Press.
Mole, C. (2011): *Attention is Cognitive Unison: An Essay in Philosophical Psychology*. Oxford: Oxford University Press.
Mole, C. (2017): "Attention". In: *Stanford Encyclopedia of Philosophy*. http://plato.stanford.edu/entries/attention/
Moll, H.; Meltzoff, A. N. (2011a): "Joint Attention as the Fundamental Basis of Understanding Perspectives". In A. Seemann (ed.): *Joint Attention: New Developments in Psychology, Philosophy of Mind, and Social Neuroscience*. Cambridge, MA: The MIT Press, 393–413.
Moll, H.; Meltzoff, A. N. (2011b): "Perspective-Taking and its Foundation in Joint attention". In: J. Roessler; H. Lerman; N. Eilan (eds.): *Perception, Causation, and Objectivity*. Oxford: Oxford University Press, 286–304.
Moll, H.; Tomasello, M. (2007): "How 14- and 18-Month-Olds Know what Others Have Experienced". *Developmental Psychology* 43:2, 309–317.
Moran, D. (2019): "Husserl and Gurwitsch on Horizontal Intentionality: The Gurwitch Memorial Lecture 2018: In Memory of Lester E. Embree". *Journal of Phenomenological Psychology* 50:1, 1–41.
Mundy, P. (2016): *Autism and Joint Attention: Development, Neuroscience, and Clinical Fundamentals*. New York: The Guilford Press.
Núñez, M. (2014): *Joint Attention in Deafblind Children: A Multisensory Path Towards a Shared Sense of the World*. Glasgow: Glasgow Caledonian University.
Peacocke, C. (2005): "Joint Attention: Its Nature, Reflexivity, and Relation to Common Knowledge". In: N. Eilan; C. Hoerl; T. McCormack; J. Roessler (eds.): *Joint attention: Communication and Other Minds: Issues in Philosophy and Psychology*. New York: Oxford University Press, 298–324.
Pezzulo, G.; Iodice, P.; Ferraina, S.; Kessler, K. (2013): "Shared Action Spaces: A Basis Function Framework for Social Re-Calibration of Sensorimotor Representations Supporting Joint Action". *Frontiers in Human Neuroscience* 7, 800.
Posner, M. I.; Rothbart, M. K. (1998): „Attention, Self–Regulation and Consciousness". *Philosophical Transactions of the Royal Society of London. Series B: Biological Sciences* 353:1377, 1915–1927.
Rakoczy, H. (2018): "Development of Collective Intentionality". In: M. Janković; K. Ludwig (eds.): *The Routledge Handbook of Collective Intentionality*. London: Routledge/Taylor & Francis Group, 407–419.
Reddy, V. (2005): "Before the 'Third Element': Understanding Attention to Self". In: N. Eilan; C. Hoerl; T. McCormack; J. Roessler (eds.): *Joint Attention: Communication and Other Minds: Issues in Philosophy and Psychology*. New York: Oxford University Press, 85-109.
Reddy, V. (2008): *How Infants Know Minds*. Cambridge, MA: Harvard University Press.
Scaife, M.; Bruner, J. (1975): "The Capacity for Joint Visual Attention in the Infant". *Nature* 253, 265–266.
Schilbach, L. (2015): "Eye to Eye, Face to Face and Brain to Brain: Novel Approaches to Study the Behavioral Dynamics and Neural Mechanisms of Social Interactions". *Current Opinion in Behavioral Sciences* 3, 130–135. https://doi.org/10.1016/j.cobeha.2015.03.006
Seemann, A. (ed.) (2011): *Joint Attention: New Developments in Psychology, Philosophy of Mind, and Social Neuroscience*. Cambridge, MA: MIT Press.
Siposova, B.; Carpenter, M. (2019): "A New Look at Joint Attention and Common Knowledge". *Cognition* 189, 260–274.

Sperber, D.; Wilson, D. (1986): *Relevance: Communication and Cognition*. London: Blackwell Publishers.
Thompson, E.; Stapleton, M. (2009): "Making Sense of Sense-Making: Reflections on Enactive and Extended Mind Theories". *Topoi* 28:1, 23–30.
Tomasello, M. (1995): "Joint Attention as Social Cognition". In C. Moore; P. J. Dunham (eds.): *Joint Attention: Its Origins and Role in Development*. Hillsdale: Lawrence Erlbaum Associates, 103–130
Tomasello, M. (2008): *Origins of Human Communication*. Cambridge, MA: MIT Press.
Tomasello, M. (2014): *A Natural History of Human Thinking*. Cambridge, MA: Harvard University Press.
Tomasello, M.; Farrar, M. J. (1986): "Joint Attention and Early Language". *Child Development* 57:6, 1454.
Trevarthen, C.; Hubley, P. (1978): "Secondary Intersubjectivity: Confidence, Confiding and Acts of Meaning in the First Year". In: A. Lock (ed.): *Action, Gesture and Symbol: The Emergence of Language*. Oxford:Academic Press, 183–229.
van der Heijden, A. H. C. (1992): *Selective Attention in Vision*. London: Routledge.
Watzl, S. (2017): *Structuring Mind: The Nature of Attention and How it Shapes Consciousness*. Oxford: Oxford University Press.
Wehrle, M.; Breyer, T. (2016): "Horizonal Extensions of Attention: A Phenomenological Study of the Contextuality and Habituality of Experience". *Journal of Phenomenological Psychology* 47:1, 41–61.
Wexler, B. E. (2006): *Brain and Culture: Neurobiology, Ideology, and Social Change*. Cambridge, MA: MIT Press.
Wu, W. (2014): *Attention*. London: Routledge.

Short Biography

Felipe León, PhD, is an Assistant Professor at the Center for Subjectivity Research, University of Copenhagen. He holds MA degrees in Philosophy from the National University of Colombia and the University of Copenhagen, and a PhD degree in Philosophy from the University of Copenhagen. León's primary research areas are classical phenomenology, social cognition, and collective intentionality. Recent publications include "Emotional Sharing and the Extended Mind", *Synthese* 196:12, 2019, 4847–4867 (together with Dan Zahavi and Thomas Szanto), and "Joint Attention Without Recursive Mindreading: On the Role of Second-Person Engagement", *Philosophical Psychology* 34:4, 2021, 550–580.

Miguel Segundo-Ortin and Glenda Satne
Chapter 4
Sharing Attention, Sharing Affordances: From Dyadic Interaction to Collective Information

Abstract: Cognitivist approaches to joint attention conceptualize it as a form of triangular interaction, between two agents and one object. When describing the interpersonal dimension of this triangle they frame it as a form of simulation, theorizing or both, involving representations of the other agent's mental states – representation of representations – and inferences.

In this paper, we advocate a different framework for understanding shared attention, the ecological psychology framework that understands attention through the notion of 'affordance'. Affordances are relational and not representational. They are direct relationships between agents and their environments. While some authors have pointed to the notion of 'social affordance' (Heft 2007, 2017; Rietveld and Kiverstein 2014; Moreira de Carvalho 2020) for understanding phenomena related to shared attention, the notion remains general and imprecise. The problem is that the notion is used indistinctively to refer to a number of different phenomena that involve social attention in very different ways. To address this issue, we offer an initial classification of different kinds of social affordances, from dyadic relations between agents, and different forms of triangular interactions, reciprocal and non-reciprocal, that provide direct and indirect information about common environments to one or both agents, all the way to collective affordances that lie at the basis of socio-cultural forms of life. We argue that this account is better placed than the standard cognitivist alternative to account for both shared attention and joint action in a non-cognitively demanding way. In addition, we show how these forms of shared activity are, in turn, fundamental for the acquisition of the socio-cultural norms that come to permeate human perception.

Acknowledgements: MSO's contribution to this chapter was funded by the Nederlandse Organisatie voor Wetenschappelijk Onderzoek VIDI Research Project "Shaping our action space: A situated perspective on self-control" (VI. VIDI.195.116).

https://doi.org/10.1515/9783110647242-005

Introduction

The ecological theory of perception (Gibson 1966, 1979 [2015]) is famous for introducing two radical hypotheses. The first one is that perception is direct, that is, to perceive the environment we do not need to build and manipulate mental representations, but rather to detect (or 'pick up') the information that is available in the ambient.[1] The second is that perception is fundamentally perception of 'affordances'. Affordances are the opportunities for action that a specific environment offers an organism. According to this hypothesis, when we detect information about an object, for example, a glass, the first thing we perceive is the actions that we can carry out with it, such as picking it up, drinking from it, throwing it, etc. The thesis that perception is of affordances also implies that perception and action are complementary processes. This explains why ecological psychologists often speak of 'perception-action' in tandem.

In this context, 'shared attention', that is, the situation in which two agents are mutually aware of themselves attending to the same object, should be conceptualized as a form of 'social' or 'shared affordance.' Yet, ecological psychological literature on this topic has not been very precise when it comes to conceptualize the specific phenomenon that takes place when attention to affordances is shared. The key issue is to offer the nuanced concepts we need to sufficiently specify these phenomena, as to be able to distinguish them from other concepts that are similar and to which they are related in several ways.

Indeed, although ecological psychologists have accumulated a great amount of empirical evidence concerning the perception of affordances in sophisticated motor control tasks (Wagman 2019), some defenders of the theory consider that not enough attention has been paid to the study of the peculiarities of human perception-action. In particular, little attention has been given to the question of whether, and in that case how, social aspects of interaction and cultural norms that permeate human groups affect our perception of affordances. According to Heft (2018), ecological psychology rarely attends to the disposition that human beings have to guide their behavior according to social norms, and, therefore, "[i]t remains to be seen [. . .] whether this approach as articulated thus far can

---

[1] The rationale for claiming that perception is not representational is the nature of perceptual information. According to ecological theory, perceptual information is specific with respect to the affordances of the environment, and it is relational, i.e., it occurs *between* the organism and the environment, not *within* the organism with respect to the environment (for a detailed argument about why the notion of representation is incompatible with ecological theory, cf. Segundo-Ortin et al. 2019).

adequately capture the socio-cultural dimensions of human action and experience" (124; cf. also Brancazio and Segundo-Ortin 2020; Segundo-Ortin 2020).

This paper[2] aims to explore perception-action when this is a matter of an agent relating to another agent. One central case of this is when affordances are the product of socio-cultural norms. We understand socio-cultural norms as patterns of thinking and action, both broadly understood, that have their origin in a practice that prevails in the context of a community of agents (Bicchieri 2006; Satne 2015). These norms emerge from the intersubjective evaluation by the community members of each other's actions, and "express the group's expectations for how anyone who would be one of 'us' should act, on pain of admonishment, punishment, or ostracism" (Tomasello 2019, 254). Our analysis starts by looking at how socio-cultural norms permeate human perception. It then moves forward the ecological psychology account of these phenomena by expanding on the concept of 'social affordance', i.e., what it means to say that an affordance is shared by two agents and in what different ways an affordance can be said to be shared. This analysis is then put at the service to illuminate the question of how socio-cultural norms come to permeate human perception.

In the first section, we consider the way different theorists of ecological psychology have approached theorizing on social affordances, that is, focusing on the relationship between socio-cultural norms and perception-action. In the second section, we address Tomasello's (2014, 2019) account of social learning and the acquisition of these norms by children. The analysis of Tomasello's proposal is relevant, since it is often cited by authors who have tried to address the relationship between the socio-cultural context and the perception of affordances (cf. Heft 2007, 2017). Nonetheless, we claim that Tomasello's cognitivist proposal is incompatible with adopting an ecological perspective of perception and intersubjective action, and that it is problematic in its own terms when one tries to give an account of basic forms of shared attention. Finally, the last two sections outline an alternative to Tomasello's account and present the outline of an account of social interaction, social perception and the learning of socio-cultural norms, and their mutual relations. While this proposal is presented in broad strokes, we claim that it sits well with the main tenets of ecological psychology and thus, it deserves further exploration.

---

2 A previous version of these ideas was written in Spanish for a chapter on social affordances, "Affordances y normas socio-culturales", to be included in the forthcoming collected volume *Affordances y ciencia cognitiva. Introducción, teoría y aplicaciones*, edited by Heras-Escribano, M.; Lobo, L.; Vega, J.

1 Affordances, Social Interaction and Socio-Cultural Norms

Most of our actions are mediated by socio-cultural norms. For instance, when we go to a work meeting, we look for a free chair to sit on, instead of sitting on the floor or on a table. In addition, when we sit down, we adopt a specific position, unlike the one we would adopt at a dinner with friends. Likewise, when we enter an elevator, we try to maintain an 'adequate' distance from others, and we can appreciate when other people do not do it.

As Haugeland explains:

> [W]hen community members behave normally, how they behave is in general directly accountable in terms of what's normal in their community; their dispositions have been inculcated and shaped according to those norms, and their behavior continues to be monitored for compliance (Haugeland 1990, 440)

This fact has not gone unnoticed by some theorists of ecological psychology. For example, Heft (2001, 2007, 2017, 2018) claims that a fundamental aspect of the human condition is that our perception-action is shaped by the socio-cultural context in which we live. Thus, our relationship with the affordances of the environment does not depend only on the detection of information that is present in our natural environment but is affected by the social norms of the community of which we are part (Rietveld and Kiverstein 2014). Likewise, Reed explains that while learning to perceive the affordances of the environment requires discovering the relationship between our action capabilities and the properties of the environment, "[w]hen one learns about norms, one is learning about properties of one's own action (and their objects) with respect to the awareness and activities of others" (Reed 1993, 52).

Following Reed, Heras-Escribano writes:

> [T]he taking of affordances can be affected by certain pressures exerted by social norms. This happens constantly in our everyday situations: We do not eat with our hands for a normative reason, even when we sometimes can grasp the food more firmly with our hands than with a fork and a knife; also, someone in a hurry gives preference to an elder instead of blindly taking the affordance of passthrough-ability at the gates of a subway train. Our social norms and conventions share their space with our individual perception of affordances, and sometimes our norms exert some pressure for not taking certain affordances given some social conventions. (Heras-Escribano 2019, 175)

Furthermore, Gibson (1950) distinguished between 'expedient' and 'proper' action. While the former refers to those actions that are useful to achieve a specific goal, the latter refer to those that are appropriate in the context of a community (Heft 2018, 126). Crucially, in most cases these two criteria do not go together. On the contrary, according to Gibson we frequently act according to what is considered appropriate or adequate in our community or social group instead of doing what is most efficient considering our current goal (Gibson 1950, 153). Furthermore, it is often the case that social norms do not affect all members of the group in a homogeneous way. As Reed (1993, 52) suggests, human communities often have specialized roles related to gender, age, or socioeconomic status, and these roles carry restrictions on what affordances can be used, by whom, and under what circumstances.

But the existence of social norms not only influences how we relate to the affordances existing in the environment. Some affordances are also a product of these norms. To illustrate this, Gibson appeals to mailboxes (Gibson 1975 [2015], 130). According to Gibson, for a metal box located in the middle of the street or on the facade of a building to offer the possibility of sending and receiving letters, there must be a community with a postal system. Costall (1995, 2012) refers to the affordances that depend on the existence of social norms as "canonical affordances". According to Costall, these affordances only exist in relation to a shared socio-normative context, and can only be perceived by those individuals who are aware of the norms that support them.

Social norms thus play a fundamental role both in creating and in shaping our interaction with some affordances. These social norms are seldom verbalized, but they are manifested as embodied habits of perception-action that predispose us to perceive and take advantages of certain affordances instead of others and in particular situations (Heras-Escribano 2019; Segundo-Ortin 2020; Menary 2020; Segundo-Ortin & Heras-Escribano 2021). It is undeniable that human beings are in contact with socio-cultural norms from the moment they are born, but how do we learn to behave according to them? Most authors agree that it is through social interaction that we learn to coordinate our perception-action with respect to these norms (cf. Satne 2015; Krueger 2011, 2013; Reddy 2015; Tomasello 2019).

Theorists of ecological psychology have long noted the importance of social interaction to understand perception. As Gibson explains: "it is a mistake to construct a behavior theory without reference to social interaction, and then to attach it only at the end" (Gibson 1950, 155). As Heft (2007) expounds, adults often guide children's attention to objects and show them, either implicitly, through demonstrations, or in the context of cooperative actions, how to interact with those objects appropriately. According to Heft, "[s]uch intersubjective acts of 'joint attention' [. . .] contribute to the developing patterns of selection in perception-action, a process of *guided attunement*, which forms a crucial part of the child's history as an agent" (Heft 93, emphasis original).

However, these observations by Gibson and Heft have not been generally recognized by ecological psychologists. Instead, as Adolph and Hoch (2019) argue, most ecological psychologists tend to adopt an individualistic approach when studying perceptual learning, which leads them to ignore the role that social interaction plays in this process. In the next section, we focus on analyzing the proposal of Tomasello (2014, 2019) who emphasizes the role of social interaction and joint attention in learning and the acquisition of socio-cultural norms. The figure of Tomasello is important in this regard, for it is often used as a reference by those theorists who challenge the individualistic approach to perceptual learning and aim to understand how social norms and social interaction influence human perception-action (cf. Heft 2007, 2017; Costall 2012; Rietveld and Kiverstein 2014).

2 Learning Socio-Cultural Norms

For Tomasello (2014, 2019), what distinguishes human cognition from that of other primates is the ability to participate in collaborative actions that involve common goals. According to Tomasello et al., "[t]he result of participating in these activities is species-unique forms of cultural cognition and evolution, enabling everything from the creation and use of linguistic symbols to the construction of social norms and individual beliefs to the establishment of social institutions" (2005, 675).

Tomasello explains the learning of social norms by postulating two different cognitive capacities: joint intentionality and collective intentionality. 'Joint intentionality' involves joint attention but goes beyond it. While joint attention implies the mutual awareness that both agents are attending to the same object or affordance, joint intentionality involves on the part of the agents the ability to pursue a common goal, and thus the ability to manipulate the object that is jointly attended to, with respect to the common end that both agents share. Hence, joint intentionality refers to the ability to collaborate with others in short-term face-to-face relationships. These are interactions in which the individuals involved share a common goal and jointly attend to situations or objects in the immediate environment in order to pursue said goal. Importantly, although joint attention is a common phenomenon in many primates, Tomasello thinks that joint intentionality is an exclusively human cognitive capacity (Tomasello 2019, 82).

Joint intentionality, suggests Tomasello, appears between 14 and 18 months of age, when children acquire the ability to form a "joint agent" with others (Tomasello 2014, 39; 2019, 87). It is noteworthy that in this type of collaboration, children are already exposed to social norms (Reddy 2015; Krueger 2013). However, says Tomasello (2019, 250; cf. also Hardecker and Tomasello 2017), children understand these norms as impositions or requirements from the other person (usually an adult), and not as expectations that go beyond the concrete interaction and that apply to all members of the community.

The other great cognitive revolution occurs around the age of 3. At this point, children develop what Tomasello calls 'collective intentionality'. This type of intentionality goes beyond the immediate nature of joint intentional engagements and allows children to pursue long-term collaborations with others. It is important to point out that, while in the type of collaborations typical of joint intentionality children are capable of adopting the perspective of another agent, from the age of 3 they start "'collectivizing' [. . .] perspectives and positing a kind of invariant objectivity that grounds them all" (Tomasello 2019, 77). This ability to adopt an objective perspective is crucial for learning (and expecting others' conformity to) social norms:

> [Collective intentionality] begins around three years of age and transforms joint commitments into collective commitments, second-personal protest into the enforcing of social norms, and a sense of fairness toward individuals into a sense of justice to all in the group. (Tomasello 2019, 251)

That children over 3 years old are capable of correcting the behavior of others for reasons that go beyond their own interest suggests that they are capable of adopting this objective perspective. For example, according to Vaish et al. (2011), children at this age show a tendency to protest when someone shows signs of wanting to break someone else's toy. As Tomasello explains, since the child is not affected by the action of the other directly, her reaction does not constitute a second-person protest, that is, a reaction in which the child seeks retribution for damage or injustice that another has inflicted upon her. On the contrary, "[w]hat she is protesting is a lack of conformity to the group-minded social norm for how one should treat others" (Tomasello 2019, 256).

As we mentioned before, Tomasello is often cited by those authors who, from ecological psychology, aspire to understand how our perception-action is affected by socio-cultural norms. Nevertheless, our thesis is that Tomasello's proposal is in direct conflict with an ecological approach to perception and social action and unsuited in its own terms to give an account of the basic forms of shared attention that lay at the basis of the acquisition of socio-cultural norms.

To see this, we must take a closer look at the way Tomasello understands joint intentionality, the previous and necessary step for the development of collective intentionality. Tomasello (2019, 7; 2014, 38) makes it clear that he understands joint intentionality in terms of the theoretical framework proposed by Bratman (1992, 2014). Accordingly, for there to be joint intentionality, the following three conditions must be satisfied:

If you and I are agents, and *J* is a goal, then:
(1) I must have the goal of doing *J* together with you,
(2) You must have the goal of doing *J* together with me, and
(3) We must have "mutual knowledge, or common ground, that we both know each other's goals" (Tomasello 2014, 38)

According to Tomasello, to account for this sort of social interaction, we must explain how it is possible for an agent to know that another agent knows that they have the goal of carrying out *J* with them. This, he argues, is only possible if the agents have the ability to recursively read each other's minds (Tomasello 2019, 85). For this, I must be able to represent the mental states of the other (in this case her intentions and beliefs), simulate the abductive inferences that she is making about my intention to carry out *J* with her, and infer that she knows the same as I do about the situation in which we are to *J* (Tomasello 2014, 94). All of this involves according to Tomasello, the ability to manipulate recursive inferences indefinitely iterated.[3]

In summary, to explain joint intentionality we have to assume: (i) that the agents involved possess concepts such as 'intention,' 'belief,' 'desire', etc.; (ii) that they are capable of forming representations about the mental states of other agents; (iii) that they are capable of carrying out complex cognitive operations such as simulating the inferences that the other is making about their intention to carry out J with her; and (iv) that they have the ability to know that the other agent knows what they know, by indefinitely iterating recursive inferences to that effect.[4]

Nonetheless, if we take into account that, according to Tomasello, the ability to form joint intentions appears between 14 and 18 months, it seems excessive to assume that children of that age are already capable of carrying out cognitive operations of such complexity. In fact, several authors have argued that attributing these capabilities to young children is implausible (Tollefsen 2005; Michael et al. 2014; Pacherie 2013). These authors propose, alternatively, that the ability to engage in joint actions with others could appear in infancy *before* the development of such complex recursive cognitive abilities (Satne

---

3 Tomasello explains these capacities by combining Theory-Theory and Simulation Theory.
4 Tomasello suggests that some situations offer enough information for the two agents to understand that they have a common goal without the need for recursive inferences, standing in 'common ground'. Yet, according to him, situations in which agents do in fact make those inferences, for example when there is a potential misunderstanding, demonstrate that agents do possess those underlying recursive inferential capacities, and that those can be legitimately presupposed as explanations of jointness (Tomasello 2014, 38).

2016; Satne and Salice 2020). Continuing with this line of thought, some theorists argue that it is precisely the interactions with other agents, especially those interactions that involve language and narratives, that provide children with the cognitive tools that allow them to grasp concepts such as 'intention' and 'belief'. This also allows them to represent other agents the mental states, of increasing complexity, through recursive inferences (Hutto 2008; Hutto and Satne 2015). This means, against Tomasello, that we should be able to explain joint intentionality in a simpler way, without resorting to handling representations of concepts for mental states and recursive inferences from the set off.

Moreover, Tomasello's proposal is in direct conflict with ecological psychology. Although Tomasello (2000; Tomasello et al. 1999) uses the term "affordance" on numerous occasions, and even refers to Eleanor J. Gibson's work on perceptual learning, he assumes that we cannot understand other agents' mental states through direct perception. In contrast, James Gibson (1979 [2015], 127) argues that we can perceive the affordances of other agents in the same way we perceive the affordances of objects: detecting directly, without mediation of inferences or 'theories', sensory information in the environment.[5] Therefore, ecological psychology argues that we do not need to postulate the existence of meta-representations and abductive inferences to explain our ability to understand, at least in basic cases, what others are doing and to cooperate with them to carry out common goals.

From this we can conclude two things. First, that Tomasello is not the best ally for authors, like Heft and Costall, who, from an ecological psychology perspective, try to explain how children learn social norms through social interaction. Tomasello's account of social cognition builds on the notions of meta-representation and recursive inference, something that does not sit well with ecological psychologists' claims about the direct nature of perception (including social perception). Second, ecological psychology's account of perception in terms of affordances, might be in the position to offer novel resources for elucidating how children interact with others and learn collective norms without resorting to an explanation that is excessively cognitively demanding, like the one provided by Tomasello. With this in mind, in the next section we distinguish different types of 'social' affordances. These will be shown to be a crucial set of concepts to understand the phenomena of shared attention in the context of ecological psychology. In the last section, the distinction of 'social affordances' in different kinds will prove to be useful to address the question of how children learn of socio-cultural norms. In contrast to what Tomasello's account assumes, such learning turns out not to be such a cognitively demanding endeavour.

---

5 As we will discuss below this is one case of what we can call 'social affordances'.

3 From Social Interaction to Social Affordances, and Back Again

In a trivial sense, we could say all human affordances have a social character (Costall 1995). This is because, as we have said before, our perception-action of the environment is mediated by the social norms of the community we partake in (Rietveld and Kiverstein 2014). Furthermore, it is almost impossible to find an environment that has not already been transformed by the action of human beings, and that does not bear the "stamp of the social" (Heft 2007, 95). However, in order to advance a more nuanced characterization of the role that social affordances play in perception-action, in this section we will propose a more restricted characterization of the concept of 'social affordance,' as well as distinguish between different types of social affordances.

We can characterize social affordances as those opportunities for action that depend on the presence in my environment of one or more agents with whom I can interact. Following Marsh et al. (2006, 2009a, 2009b), we defend that a study of social affordances requires taking as the minimum unit of reference the O-O-E [Organism-Organism-Environment] system, instead of the classic O-E [Organism-Environment] system of ecological theory. Our aim is to show how an ecological approach to social interaction based on the concept of social affordance can account for joint intentionality without the need to postulate meta-representations and recursive inferences. According to this, our hypothesis is that joint intentionality can be understood as the coordinated exploitation (by two or more agents) of the affordances of the environment, and that this is possible through the direct perception of social affordances.

To begin with, we must note that the environment we inhabit is often populated by other agents, and these agents offer affordances we can perceive:

> The richest and most elaborate affordances of the environment are provided by other animals and, for us, other people. [. . .] Behavior affords behavior, and the whole subject matter of psychology and of the social sciences can be thought of as an elaboration of this basic fact. Sexual behavior, nurturing behavior, fighting behavior, cooperative behavior, economic behavior, political behavior – all depend on the perceiving of what another person or other persons afford, or sometimes on the misperceiving of it.
> (Gibson 1979 [2015], 126–127)

But in what sense does the existence of other agents imply the existence of 'richer' affordances? This could be due to several factors (Reed 1993; Gibson 1979 [2015], 127). First, other agents are not passive. Agents *act*, transforming the environment and generating new affordances for others. For example, it may happen that while we are walking along a busy street, a distracted pedestrian changes her trajectory, approaching us in a straight line, forcing us to avoid her. It can also happen that this same person starts running towards us, aggressively waving her arms, forcing us to flee from her. Second, when we interact with other agents, they can either resist or collaborate with us. Two people can transport an object together, coordinating their movements and their attention, or they can compete to see who carries the object. Both cases imply totally different interactions, and with them, different affordances that can be perceived and used by each of the agents. Furthermore, as Baron (2007) suggests, it often happens that when we collaborate with others we tend to adopt different roles, and these roles determine what affordances are immediately relevant for us.

In what follows, we distinguish four types of social affordances, according to the type of interaction involved and the factors that need to be taken into account to describe each one. These distinctions allow to enrich the concept of social affordance and make it useful to describe various types of perception-action within the O-O-E system.

As Gibson (1979 [2015], 127) suggests, just as there is information in the environment about the affordances of objects, there is also information about the affordances of other organisms. Following Reed (1993), we propose that the most basic case of social affordance is that in which an agent (O_1) is able to perceive the presence of another organism (O_2) by detecting some traces that it leaves in the environment (Figure 4.1). These traces can constitute perceptual information about other organisms, and their detection is essential for the control of action. This type of perception-action is common in the animal kingdom. Some predators are able to follow the trail of their prey by perceiving the smell they leave, for example, and dogs and wolves can determine the proximity of the prey based on the intensity of the smell.

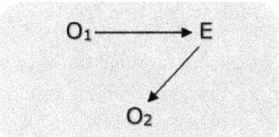

Figure 4.1: The first agent (O_1) perceives the presence of another agent (O_2) detecting some type of trace that it left in the environment (E).

Another kind of case is one in which an agent perceives an affordance of the environment indirectly, that is, through the *action* of another agent (Figure 4.2). For example, it is well known that the primates of the species *Chlorocebus pygerythrus* emit a characteristic sound to alert their conspecifics about the presence of predators. These 'alert calls' are also different depending on the type of predator they identify. Perception of these sounds generate differentiated responses in the other members of the group (Seyfarth et al. 1980). It is important to note that, although this type of situation depends on cooperative forms of interaction between several organisms, we cannot speak of 'joint' action yet, since organisms performing the calls do it even when they are not aware of the presence of others receiving the calls.

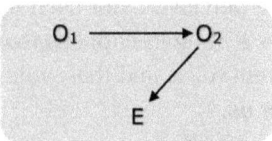

Figure 4.2: O_1 perceives an affordance of the environment (E) by means of O_2.

In the case of humans, direct perception of other organisms' actions is pervasive. Johansson (1973) filmed a series of people performing different physical activities (running, walking, dancing, lifting objects, etc.) in the dark while having different light points attached to their joints. Afterwards, he showed the videos to a series of participants and found that a high percentage of them were able to identify not only the type of activities they were doing, but also how much effort it was taking them to do so (cf. also Runeson and Frykholm 1983). Subsequent experiments show that people can perceive whether the other person intends to carry out this or that action – that is, to exploit this or that affordance – observing their movements as well as whether the movements are performed with the awareness that others are observing them (Runeson 1985; Hodges and Baron 2007; Mark 2007).

In the same vein, Kiverstein (2015) points out that the ability to detect which affordances are relevant and significant for an agent (Figure 4.3) allows humans to become aware of the other agent's mental states (e.g., if she is angry, if she has an aggressive attitude, or if she intends to cooperate with us, etc.) directly (not inferentially). In the words of Gallagher and Hutto,

> in most intersubjective situations, that is, in situations of social interaction, we have a direct perceptual understanding of another person's intentions because their intentions are explicitly expressed in their embodied actions and their expressive behaviors. This understanding does not require us to postulate or infer a belief or a desire hidden away in the other person's mind. (2008, 20; cf. also Krueger 2011; for an overview cf. Satne 2020)

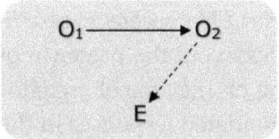

Figure 4.3: O_1 perceives O_2 acting in the environment (E). In this case, the action of O_2 provides information about the mental states of O_2. The dashed arrow represents the action of O_2 on E.

According to Marsh et al. (2006, 2009a, 2009b), "direct social perception" – the name by which they refer to our ability to perceive the intention of the other through their action – is essential for interpersonal coordination. Once I am aware of the intention of the other, I can decide whether or not to collaborate with her. When this collaboration occurs, Marsh et al. suggest, a new unit of perception-action is created – a "plural subject of action" (Richardson et al. 2007) or a "joint agent" in Tomasello's words –, meaning that my perception-action is coordinated with that of the other to achieve a common goal:

> Just as perception and action are mutually and causally coupled to behavioural aims at the individual level – by the detection of information to constrain action and by the control of action to order perception – the perception and action capabilities of the social unit are mutually constrained, ordered, and dynamically coupled [. . .] Each individual's perception is coupled to his or her partner's action as it is to his or her own, and each individual's action alters their partner's perception just as it alters his or her own [. . .] the perceiving and acting of those individuals within the social unit are causally entailed to form a distinct but irreducible system motivated by a mutually perceived goal.
> (Marsh et al. 2006, 20)

To illustrate this idea, Marsh et al. (2006) asks us to imagine two people carrying an object, for example a table. To carry out this action, both individuals have to coordinate their perception-action with respect to both the affordances of the environment and the action of the other (Figure 4.4). The key for this to be possible, they explain, lies in the detection of information that is generated at the level of the O-O-E system (information about social affordances), and not in the knowledge and simulation of the inferences of the other agent (Marsh et al 2006, 22; Hodges and Baron 2007). Through this interpersonal coordination, agents can carry out a common task, exploiting the affordances of the environment in a coordinated way (Marsh et al. 2009a, 2009b; Baron 2007). Something similar, says Reed (1993, 58), occurs when, through gestures, we call the attention of other agents to a specific aspect of the environment. In both cases, the perception of an affordance of the environment is mediated by the perception of the action of another agent and the agents must coordinate their responses for the interaction to be successful.

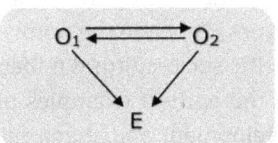

Figure 4.4: To carry out a joint action, O_1 and O_2 have to coordinate their perception-action with the environment (E) and the action of the other simultaneously.

In sum, we hold that the ecological perspective offers resources to understand social interaction without the need to postulate complex cognitive processes as Tomasello does (2014, 2019). According to the analysis offered in this section, the notion of social affordance (in the multiple ways in which these are presented – cf. Figures 4.1–4.4) allows us to account for how it is possible for an agent to perceive the possibility of interacting with others without the need to simulate and infer their mental states. It follows that the ecological theory of perception-action can help to explain the type of cognitive activities characteristic of joint intentionality without assuming that it depends on abilities such as those of representing mental states and making inferences on the basis of these. Once we have this firm foundation to start from, we can see how socio-cultural norms are learned. This is addressed in the next and final section.

4 An Ecological Approach to the Learning of Socio-Cultural Norms

In the previous section, we have shown that the ecological theory of perception-action offers resources for explaining joint intentionality without resorting to meta-representations and inferences. It now remains to be seen how to articulate a vision of the learning of socio-cultural norms compatible with this account. In this section, making use of studies from developmental psychology, we offer some suggestions about the different cognitive abilities that could contribute to the learning of socio-cultural norms. Although the capacities we call upon in what follows are not standardly studied in the ecological theory, we believe that they are perfectly compatible with its fundamental tenets, and, therefore, that they can contribute to extend the ecological approach of perceptual learning to account for the learning of socio-cultural norms. It should be noted that these suggestions do not constitute a complete account of such learning, but rather a platform along which such a theory could be developed.

As we mentioned before, children are in contact with normative social practices from birth. For example, caregivers often carry out actions (gestures, facial expressions, sounds, etc.) with the aim of regulating children's perception-action patterns of response. These actions, Krueger argues, "encode the norms, values, and patterned practices distinctive of their specific socio-cultural milieu [. . .]. These physical interventions are thus arguably the earliest examples of social practices that scaffold the infant's cognitive development and shape the development of their cultural education" (Krueger 2013, 40).

Following the reasoning from the previous section, our hypothesis is that the ability to learn from others in the context of these normative practices arises in development before the ability to form mental representations and make inferences about the mental states of other agents (Satne and Salice 2020; Hutto and Satne 2015; Satne 2016; Gallagher and Hutto 2008). We hold that the perception of social affordances is key to account for these early forms of cultural learning.

Some authors have pointed out that the ability of human beings to dynamically coordinate our movements with those of others constitutes a fundamental aspect for learning social norms (Pacherie 2013; Knoblich et al. 2011). Empirical evidence suggests that this capacity could be innate, present in the so-called "primary intersubjectivity" (Trevarthen 1979), that is, the interactive capacities that a child has from birth on, and almost exclusively, until 6 months of age. For example, babies between 2 and 4 months can identify when an adult intends to pick them up, and adapt their body posture to the way a specific caregiver picks them up even before contact (Reddy et al. 2013; Reddy 2019). If we take into account that the way in which caregivers take children is already governed by socio-cultural norms (Krueger 2013; Reddy 2015), the fact that children are able to coordinate their perception-action to the behavioral habits of their caregivers can be considered a first form of conformity, albeit very basic, to social norms.

As we mentioned in the previous section, the studies by Johansson (1973) and Runeson and Frykholm (1983) suggest that we can directly perceive the intention of another agent to carry out an action by detecting specific patterns in their body movements. This suggests that social interaction could be based on the perception of affordances of other agents (social affordances in our terminology), without the need to carry out abductive inferences or use meta-representations of the mental states of other agents.

Another capacity also present in primary intersubjectivity is to perceive the emotional responses that others have to our actions and adapt to them. This capacity is observed as early as 2 months of age, when children co-ordinately react to the facial expressions of their caregivers (Trevarthen 1979; Reddy 2019; Gallagher 2013). As Adolph and Hoch explain, "infants and caregivers are acutely sensitive to each other's facial gestures and vocalizations and use this social information to update their own actions in real time" (Adolph and Hoch 2019, 26.11). By being able to detect the emotional responses that their actions produce in adults, children begin to acquire a basic knowledge of what types of actions are acceptable in the specific context of that interaction (Gallagher 2013; Kiverstein 2015; Satne 2014). Thus, both the action of others, as well as their gestures, offer sensory information about affordances that the child can perceive and use to coordinate their action (Figures 4.2a and 4.3a).

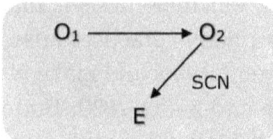

Figure 4.2a: O_1 perceives an affordance of the environment (E) by means of O_2. The action of O_2 is already mediated by socio-cultural norms (SCN), meaning that the perception of E by O_1 is influenced by these SCN.

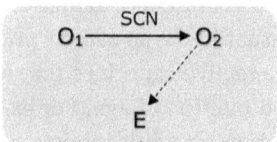

Figure 4.3a: O_1 perceives O_2 acting in the environment (E). The action of O_2 provides information about the mental states of O_2. By perceiving the mental states of O_2, O_1 begins to learn what kinds of actions are acceptable and which are not, which implies a learning of social norms (SCN).

Joint attention also plays a fundamental role in learning social norms. This involves the mutual awareness by the two agents of the presence and features of an object in the environment they are both attending to. According to developmental psychologists, this ability is consolidated between 6 and 9 months of age, when children are able to follow the gaze of their caregivers towards specific aspects of the environment and inquire back into the adults' gaze while exploring their meaning. This ability becomes more acute as the children interact with the contextual use of gestures and vocalizations by adults. Through these interactions, adults educate the child's attention, allowing them to identify those affordances that are relevant for a specific purpose (Krueger 2013; Tomasello 1999, 2000). As the child grows, she no longer needs an adult to guide his attention in order to perceive and respond to these affordances and begins to take the initiative by indicating her objects of interest to others, for example by using his fingers to point at something, or her gaze to guide the gaze of others. Later, children begin to use words to draw their caregivers' attention to something they want. For example, it is common for Spanish-speaking young children to use words like 'water', to indicate food and drink interchangeably. These types of interactions are already part of the so-called "secondary intersubjectivity" (Trevarthen 1979).

Habitualization to these patterns of perception and action is key to children's development in this stage. The child is already adapting and learning social norms by means of interacting with her caregivers and the cultural material environment that surrounds her. These norms, patterns of thought and action, are manifested as embodied habits of perception-action that predispose the child to perceive and take advantage of certain affordances instead of others and in particular situations (Heras-Escribano 2019; Segundo-Ortin 2020).

In secondary intersubjectivity,[6] children learn from adults how to place themselves in pragmatically defined contexts, that is, specific social practices, that embody aims, goals, action-styles, etc. At this point, children begin to incorporate more explicit social norms to the ways they behave – for example, they learn how to use the spoon and the plate, or chopsticks, to feed themselves (cf. Figure 4.4a). The ability to synchronize attention with other agents continues to be key in this stage of their development, and imitation in this period becomes pervasive (Rochat 2012).

Several studies conducted by Tomasello and his team (Tomasello 1999, 2019) show that while chimpanzees are capable of emulating the behavior of others, for example, they are capable of replicating the behavior of others for a specific purpose when they have seen others do it, children tend to imitate the behavior of others, that is, they tend to replicate even arbitrary features of observed behavior. Through imitation, children incorporate specific perception-action patterns that are already fully subject to social norms (Rochat 2015), and begin to expect others to conform to them at around 3 years of age (Rackoczy and Tomasello 2012). In addition, it is important to note that, although imitation does not necessarily imply joint action, it does involve joint attention: the child needs to pay attention both to the affordances of the object and to the way in which the adult interacts with them and, in order to regulate her action for acceptance and adequacy, the child uses adult approval and disapproval as a guide (Satne 2014), which implies mutual awareness of responding to the same affordances, and the aim to do it in the same way (cf. Figure 4.4a below).

Furthermore, studies by Gergely and Csibra (2009) show that only when learning is accompanied by express instructions, this is when observation is accompanied by interactions in which the adult guides the child using language, children generalize the learned perception-action patterns to other contexts. Gergely and Csibra call this form of learning "natural pedagogy", and suggest that it is both innate and fundamental to learning social norms.[7]

---

6 There is no consensus regarding the age at which secondary intersubjectivity starts, but it is generally accepted that its appearance coincides with the emergence of joint attention. While Reddy (2019) places this at 6 months of age, for Tomasello (2019) joint attention does not appear until approximately 9 months.
7 Gergely and Csibra (2009) explain this ability in representational terms. A direction that we resist, as explained above. We suggest that the ability to follow linguistic instructions that

The following scheme (Figure 4.4a) illustrates how socio-cultural norms transform the relation between the observers and with their (common) environment:

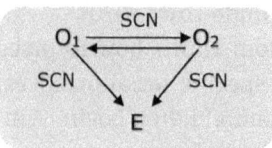

Figure 4.4a: O_1 and O_2 coordinate their perception-action with the environment (E) and the action of the other simultaneously. In this case, the perception-action of both agents is subject to socio-cultural norms (SCN) that determine what affordances are relevant in this specific context.

In sum, although we agree with Tomasello that social interaction, joint attention and joint action are key to learning social norms, our hypothesis is that these can be explained through the direct perception of social affordances and the mutual awareness of common affordances, without the mediation of meta-representations and inferences about the mental states of other agents.[8] We agree with Tomasello (2019, 88) also that linguistic interaction is basic to develop the collective perspective characteristic of social norms. However, we argue that the ability to form representations, be them of the world or of other agents, is not part of the basic cognitive repertoire that allow children to acquire such cultural perspectives. Contrary to the position defended by Tomasello, we hold that it is only when children master the socio-normative practices associated with the use of language that they have the resources necessary for forming objective representations about the world and other individuals (Hutto and Myin 2013, 2017; Hutto and Satne 2015).

---

these authors discuss can be described as a capacity to attune to perception-action patterns in interactive situations, and not as a representational inferential capacity that targets other people's mental states.

8 Note that this idea is in line with recent claims by Heft that joint and collective actions require common awareness of mutual goals and affordances, but that such common awareness "does not require mind-reading, of any sort, but instead what is needed to bootstrap joint and collective processes is access to a common ground for perception-action. [. . .] knowers have a common ground of information" (Heft 2019, 202). The aim of our proposal is precisely to give a sufficiently rich description of what the building of such common ground might consist in in the absence of mind-reading capacities.

5 Concluding Remarks

In this chapter, we have defended not only that joint action is possible without mental representations, but that learning the socio-cultural norms that permeate our perception-action as a whole, does not depend on the existence of such representations. Rather, this socio-cultural learning depends on our habitualization to embodied socio-cultural norms that we learn from other community members from early childhood.

The ecological perspective, we have argued, offers through the notion of social affordance, in its different types, fundamental tools to make sense of this possibility. By repeatedly engaging in social interactions that involve the mutual coordination and perception of social affordances, children learn social norms, consolidating perception-action habits that enable them to behave according to what is considered normal or acceptable within their communities.

References

Adolph, K. E.; Hoch, J. E. (2019): "Motor development: Embodied, Embedded, Enculturated, and Enabling". *Annual Review of Psychology* 70:1, 141–164.
Baron, R. M. (2007). Situating Coordination and Cooperation Between Ecological and Social Psychology. *Ecological Psychology*, 19(2): 179–199.
Bicchieri, C. (2006): *The Grammar of Society: The Nature and Dynamics of Social Norms*. Cambridge: Cambridge University Press.
Brancazio, N.; Segundo-Ortin, M. (2020): "Distal Engagement: Intentions in Perception". *Consciousness and Cognition* 79, 102897. https://doi.org/10.1016/j.concog.2020.102897
Bratman, M. (1992): "Shared Co-Operative Activity". *Philosophical Review* 101:2, 327–341.
Bratman, M. (2014): *Shared Agency: A Planning Theory of Acting Together*. Oxford: Oxford University Press.
Brownell, C. A. (2011): "Early Developments in Joint Action". *Review of Philosophy and Psychology* 2, 193–211.
Costall, A. (2012): "Canonical Affordances in Context". *Avant: Trends in Interdisciplinary Studies* 3:2, 85–93.
Costall, A. (1995): "Socializing Affordances". *Theory & Psychology* 5:4, 467–481.
Csibra, G.; Gergely, G. (2009): "Natural Pedagogy". *Trends in Cognitive Sciences* 13:4, 148–153.
Gallagher, S.; Hutto, D. (2008): "Understanding Others through Primary Interaction and Narrative Practice". In: J. Zlatev; T. Racine; C. Sinha; E. Itkonen (eds.): *The Shared Mind: Perspectives on Intersubjectivity*. New York, NY: John Benjamins, 17–38.
Gallagher, S. (2013). The Socially Extended Mind. *Cognitive Systems Research*, 25: 4–12. https://doi.org/10.1016/j.cogsys.2013.03.008
Gibson, J. J. (1950): "The Implications of Learning Theory for Social Psychology". In: J. G. Miller (ed.): *Experiments in Social Process: A Symposium on Social Psychology*. New York, NY: McGraw-Hill, 149–167.

Gibson, J. J. (1966): *The Senses Considered as Perceptual Systems*. Westport, Conn.: Greenwood Press.

Gibson, J. J. (1979[2015]): *The Ecological Approach to Visual Perception*. New York, NY: Psychology Press.

Hardecker, S.; Tomasello, M. (2017): "From Imitation to Implementation: How Two- and Three-Year-Old Children Learn to Enforce Social Norms". *British Journal of Developmental Psychology* 35:2, 237–248.

Haugeland, J. (1990): "The Intentionality All-Stars". *Philosophical Perspectives* 4, 383–427.

Heft, H. (2001): *Ecological Psychology in Context: James Gibson, Roger Barker, and the Legacy of William James's Radical Empiricism*. New York, NY: Psychology Press.

Heft, H. (2007): "The Social Constitution of Perceiver-Environment Reciprocity". *Ecological Psychology* 19:2, 85–105.

Heft, H. (2017): "Perceptual Information of 'an Entirely Different Order': The 'Cultural Environment'". The Senses Considered as Perceptual Systems". *Ecological Psychology* 29:2, 122–45.

Heft, H. (2018): "Places: Widening the Scope of an Ecological Approach to Perception–Action with an Emphasis on Child Development". *Ecological Psychology* 30:1, 99–123.

Heft, H. (2019): "Revisiting 'The Discovery of the Occluding Edge and its Implications for Perception' 40 years on". In: J. B. Wagman; J. C. Blau (eds.): *Perception as Information Detection. Reflections on Gibson's Ecological Approach to Visual Perception*. New York, NY: Routledge, 188–204.

Heras-Escribano, M. (2019): *The Philosophy of Affordances*. Cham: Palgrave Macmillan.

Hodges, B. H.; Baron, R. M. (2007): "On Making Social Psychology more Ecological and Ecological Psychology more Social". *Ecological Psychology* 19:2, 79–84.

Hutto, D. (2008): *Folk Psychological Narratives*. Cambridge, Mass.: MIT Press.

Hutto, D. D.; Myin, E. (2013): *Radicalizing Enactivism: Basic Minds without Content*. Cambridge, MA.: MIT Press.

Hutto, D. D.; Myin, E. (2017): *Evolving Enactivism: Basic Minds Meet Content*. Cambridge, MA.: MIT Press.

Hutto, D. D.; Satne, G. (2015): "The Natural Origins of Content". *Philosophia* 43:3, 521–536.

Johansson, G. (1973): "Visual Perception of Biological Motion and a Model for its Analysis". *Perception* 6:14, 201–211.

Kiverstein, J. (2015): "Empathy and the Responsiveness to Social Affordances". *Consciousness and Cognition* 36, 532–542.

Knoblich, G.; Butterfill, S.; Sebanz, N. (2011): "Psychological Research on Joint Action". In: B. H. Ross (ed.). *The Psychology of Learning and Motivation*. Burlington: Academic Press, 59–101.

Krueger, J. (2011): "Extended Cognition and the Space of Social Interaction". *Consciousness and Cognition* 20:3, 643–657.

Krueger, J. (2013): "Ontogenesis of the Socially Extended Mind". *Cognitive Systems Research* 25–6, 40–46.

Mark, L. S. (2007). Perceiving the Actions of Other People. *Ecological Psychology*, 19(2): 107–136

Marsh, K. L.; Richardson, M. J.; Baron, R. M.; Schmidt, R. C. (2006): "Contrasting Approaches to Perceiving and Acting with Others". *Ecological Psychology* 18:1, 1–38.

Marsh, K. L.; Richardson, M. J.; Schmidt, R. C. (2009a): "Social Connection through Joint Action and Interpersonal Coordination". *Topics in Cognitive Science* 1:2, 320–339.

Marsh, K. L.; Johnston, L.; Richardson, M. J.; Schmidt, R. C. (2009b): "Toward a Radically Embodied, Embedded Social Psychology". *European Journal of Social Psychology* 39:7, 1217–1225.

Menary, R. (2020). Growing Minds: Pragmatic Habits and Enculturation. In F. Caruana; I. Testa (eds.): *Habits: Pragmatist Approaches from Cognitive Science, Neuroscience, and Social Theory*. Cambridge: Cambridge University Press, 297–319.

Michael, J.; Christensen, W.; Overgaard, S. (2014): "Mindreading as Social Expertise". *Synthese* 191:5, 1–24.

Moreira de Carvalho, E. (2020): "Social Affordances". In: Vonk J.; Shackelford T. (eds.) *Encyclopedia of Animal Cognition and Behavior*. Cham: Springer, https://doi.org/10.1007/978-3-319-47829-6_1870-1.

Pacherie, E. (2013): "Intentional Joint Agency: Shared Intentions Life". *Synthese* 190:10, 1817–1839.

Reddy, V. (2015): "Self in Culture: Early Development". In: *International Encyclopedia of the Social & Behavioral Sciences*, 439–444, DOI:10.1016/B978-0-08-097086-8.23010-9

Reddy, V. (2019): "Why Engagement? A Second-Person Take on Social Cognition". In: A. Newen; L. de Bruin; S. Gallagher (eds.): *The Oxford Handbook of 4E cognition*. Oxford: Oxford University Press, 433–452.

Reddy, V.; Markova, G.; Wallot, S. (2013): "Anticipatory Adjustments to Being Picked Up in Infancy". *PLoS One* 8:2, 195–212.

Reed, E. (1996): *Encountering the World: Toward an Ecological Psychology*. Oxford: Oxford University Press.

Reed, E. S. (1993): "The Intention to Use a Specific Affordance: A Conceptual Framework for Psychology". In: R. H. Wozniak; K. W. Fischer (eds.): *Development in Context. Acting and Thinking in Specific Environments*. New York, NY: Psychology Press, 45–76.

Richardson, M. J.; Marsh, K. L.; Isenhower, R. W.; Goodman, J. R. L.; Schmidt, R. C. (2007): "Rocking Together: Dynamics of Intentional and Unintentional Interpersonal Coordination". *Human Movement Science* 26:6, 867–891.

Rietveld, E.; Kiverstein, J. (2014): "A Rich Landscape of Affordances". *Ecological Psychology* 26:4, 325–352.

Rochat, Ph. (2012): "Early Embodied Subjectivity and Inter-Subjectivity". In: N. E. Coelho; Jr. P. Salem; P. Klautau (eds.): Dimensões da Intersubjetividade, São Paulo: Escuta, 149–164.

Rochat, Ph. (2015): "Self-Conscious Roots of Human Normativity". *Phenomenology and the Cognitive Sciences* 14:4, 741–753.

Runeson, S. (1985): "Perceiving People through their Movements". In: B. D. Kirkcaldy (ed.): *Individual Differences in Movement*. Lancaster: MTP Press, 43–66.

Runeson, S.; Frykholm, G. (1983): "Kinematic Specification of Dynamics as an Informational Basis for Person-and-Action Perception: Expectation, Gender Recognition, and Deceptive Intention". *Journal of Experimental Psychology* 112, 585–615.

Satne, G. (2014): "Interaction and Self-Correction". *Frontiers in Psychology* 5, 1–11.

Satne, G. (2015): "The Social Roots of Normativity". *Phenomenology and the Cognitive Sciences* 14:4, 673–682.

Satne, G. (2016): "A Two-Step Theory of the Evolution of Human Thinking: Joint and (various) Collective Forms of Intentionality". *Journal of Social Ontology* 2:1, 105–116.

Satne, G. (2021): "Understanding Others by Doing Things Together: An Enactive Account". *Synthese* 198, 507–528.

Satne, G.; Salice, A. (2020): "Shared Intentionality and the Cooperative Evolutionary Hypothesis". In: A. Fiebich (ed.): *Minimal Cooperation and Shared Agency*, New York, NY: Springer, 71–92.

Segundo-Ortin, M.; Heras-Escribano, M.; Raja, V. (2019): "Ecological Psychology is Radical Enough: A Reply to Radical Enactivists". *Philosophical Psychology* 32:7, 1001–1023. https://doi.org/10.1080/09515089.2019.1668238

Segundo-Ortin, M. (2020): "Agency from a Radical Embodied Standpoint: An Ecological-Enactive Proposal". *Frontiers in Psychology* 11, 1319. https://doi.org/10.3389/fpsyg.2020.01319

Segundo-Ortin, M., Heras-Escribano, M. (2021). "Neither Mindful nor Mindless, but Minded: Habits, Ecological Psychology, and Skilled Performance". *Synthese*. https://doi-org.proxy.library.uu.nl/10.1007/s11229-021-03238-w

Seyfarth, R. M., Cheney, D. L., & Marler, P. (1980). "Vervet monkey alarm calls: Semantic communication in a free-ranging primate". *Animal Behaviour*, 28(4), 1070–1094. https://doi.org/10.1016/S0003-3472(80)80097-2

Tollefsen, D. (2005): "Let's Pretend! Children and Joint Action". *Philosophy of the Social Sciences* 35, 75–97.

Tomasello, M. (1999): "Emulation Learning and Cultural Learning". *Behavioral and Brain Sciences* 21:5, 703–704.

Tomasello, M. (2000): *The Cultural Origins of Human Cognition*. Harvard: Harvard University Press.

Tomasello, M. (2014): *A Natural History of Human Thinking*. Harvard: Harvard University Press.

Tomasello, M. (2019): *Becoming Human: A Theory of Ontogeny*. Harvard: Harvard University Press.

Tomasello, M.; Carpenter, M.; Call, J.; Behne, T.; Moll, H. (2005): "Understanding and Sharing Intentions: The Origins of Cultural Cognition". *Behavioral and Brain Sciences* 28:5, 675–691.

Trevarthen, C. (1979): "Communication and Cooperation in Early Infancy: A Description of Primary Intersubjectivity". In: M. Bullowa (ed.): *Before Speech: The Beginning of Human Communication*. Cambridge: Cambridge University Press, 321–347.

Vaish, A.; Missana, M.; Tomasello, M. (2011): "Three-Year-Old Children Intervene in Third-Party Moral Transgressions". *British Journal of Developmental Psychology* 29:1, 124–30.

Wagman, J. B. (2019): "A Guided Tour of Gibson's Theory of Affordances". In; J. B. Wagman; J. C. Blau (eds.): *Perception as Information Detection. Reflections on Gibson's Ecological Approach to Visual Perception*. New York, NY: Routledge, 130–148.

Short Biography

Miguel Segundo-Ortin, PhD, is a post-doctoral researcher at Utrecht University (The Netherlands). He specializes in the philosophy of the cognitive sciences, with a special emphasis on embodied and situated theories of cognition. Other research interests include the study of so-called 'minimally cognitive agents', the explanation of skilled performance, and the relationship between culture and cognitive development. Currently, Miguel is working on a research project that aims to develop an embodied and situated account of human self-control.

Glenda Satne, PhD, is a Senior Lecturer in Philosophy at the School of Liberal Arts at the University of Wollongong, Australia. She works on Philosophy of Mind and Social Ontology and has published extensively on Embodied cognition, Enactivism, Collective Intentionality, the Second-Person, Normativity, and Cultural Evolution.

Natalie Depraz
Chapter 5
Attention as Vigilant Openness

Abstract: Attention is a crucial issue of our time. The historian of society and culture tells us that from its beginnings humanity has been confronted with the changing nature of its ability to pay attention. In this contribution I will show how in history philosophers have paid little attention to attention as a systematic and thorough theme of investigation and when they did so, they either identified it with control and effort, or with a form of passive opening of the mind. By contrast, phenomenology opens the way for a renewed investigation of attention: while working out and questioning the metaphysical binary distinction between activity and passivity and autonomizing it from both perception and reflection, it provides a crucial basis for opening attention as vigilance, that is, as a receptive welcome to whatever is to come. Attention as vigilance appears then clearly as a transversal issue situated at the heart of the problems of our contemporary digital hyper-connected societies and policies.

Introduction

Attention is a crucial issue of our time. The historian of society and culture tells us that from its beginnings humanity has been confronted with the changing nature of its ability to pay attention.

In this contribution I will show how in history philosophers have paid little attention to attention as a systematic and thorough theme of investigation and when they did so, they either identified it with control and effort, or with a form of passive opening of the mind. By contrast, phenomenology opens the way for a renewed investigation of attention: while working out and questioning the metaphysical binary distinction between activity and passivity and autonomizing it from both perception and reflection, it provides a crucial basis for opening attention as vigilance, that is, as a receptive welcome to whatever is to come.

Attention as vigilance appears then clearly as a transversal issue situated at the heart of the problems of our contemporary digital hyper-connected societies and policies.

1 The Anthropological Transversality of Attention and the Bridging Role of Phenomenology

When the hunter intently watches his prey hidden behind a bush, he is in a state of continuous alertness which alone will permit him to kill the animal; when he bends his bow, he demonstrates an intense mobilization in the moment, thanks to which, alone, he will be able to hit his target. If he becomes distracted for only a moment, if he loses his concentration, it is possible that he will go without dinner and put the survival of his tribe in question. In short, attention is a state of mind, simultaneously momentary and lasting, upon which individual survival depends and which assures one's inclusion in the social sphere.

More generally, anthropologists view attention as the internal ability which is a sign of the hominization process and which defines us as humans, including our tendency to be inattentive and distracted (Jousse 1978). Although primatologists are attempting to show how the great apes have the ability to be alert and to focus while observing their fellow creatures or while grasping a desired object, this ability is only valued according to a paradigm of hominization, as is the case with respect to language ability. I would therefore suggest a distinction between alertness and wakefulness: these two forms of attention are not different by nature though, insofar as we as humans may be in a state of alertness if driven and captured by an object or an activity; and animals may develop a special wakeful ability when cultivating observational tasks.

In short, attention is an individual predisposition that crosses all human social activities and makes the world in which we are evolving a reality which matters to us, which counts for us and which acquires meaning in our eyes. The objects that surround us, the events of our life, the situations with which we are confronted, the persons with whom we have contact will change their ontological status according to the degree of attention that we give them. For example, in traditional cultures, depending on the attention that I lend them, ritual objects will appear as mere tools or as living incarnations of the divine; or again in the context of my life, according to the attention I give to it or not, an encounter will remain accidental, unimportant – peripheral – or it will acquire an existential significance which brings it to the center of my life (Piette 2007, 2009; Bidet 2010).

As an intimate dimension of our humanity, attention was un-ceaselessly cultivated with the help of various 'techniques', designed to catch it, to maintain it or, on the contrary, to relax it, as if we vaguely felt that our humanity depends on it. Does that mean that increasing our attentional quality makes us more human? It seems that it is what we vaguely feel: we build objects the function of which is not only utilitarian (for example a bag to carry my belongings) or aesthetic (a picture to decorate my office), but which are supports for drawing my attention, not to mention objects the function of which is to sustain my attention, such as totems, icons, stained glass or mandalas. The travel bag in which I put my clothes offers me a limited space where my activity is concentrated and allows me to choose the items I will bring and those I will leave behind. The picture that I hang in my bedroom has a meaning for me; it is invested with an attentional value, and, as a consequence, each time I look at it, it offers me a possible space of contemplation. Moreover, numerous athletic activities, both ancient and modern, whose explicit purpose is sometimes equally pragmatic or symbolic, secretly conceal a powerful attentional meaning: this is the case with dance, swimming, cycling and mountain-climbing, to name only a few, but also more directly with archery, martial arts, etc. Furthermore, activities involving the construction of buildings (pyramids, cathedrals, contemporary urban development), just as organizational activities of planning and city management, require a complex ability of sequential reasoning in which attention comes to order, regulate and adjust the achievement process.

It is no wonder, therefore, in view of the crucial nature of attention in our culture and for our humanity, that philosophers, along with historians and anthropologists, give so much – though scattered – importance to it. It is almost banal to state today that attention is an ancient and recurring question in philosophy: in Antiquity, and particularly for the Stoics and Epicureans, what is emphasized is the ability to concentrate on the present moment, and, as a consequence, it is attention to oneself, vigilance[1] with regards to what depends on us – what we have a grasp on – or vigilance related to what does not depend on us and awakens our desires, thus makes us suffer.[2] In similar terms Saint

---

1 I chose to translate the French word 'vigilance' with 'vigilance' and not with 'wakefulness', given the more encompassing meaning of the Latin etymology 'vigil', which also includes attention and care.
2 Cf. for example Philo of Alexandria 2001, which includes two lists of self-techniques bearing the mark of Stoicism: on the one hand, research, deep examination, reading, listening, attention (*prosochè*), self-control, indifference to indifferent things; and on the other hand, readings, meditations, therapy of the passions, memories of what is good, self-control, the accomplishment of duties. Intellectual exercises such as listening, reading and memorization prepare for meditation (*meletè*) which deepens in research and examination and eventually leads to techniques of self-control. Thus, attention or self-presence (*prosochè*) is a general orientation of self-practices and a particular technique, and meditation (*meletè*), the ultimate spiritual exercise, means in Greek, care, being concerned with someone or something and it initially designated the preparation of the orator, most often associated with memorization. In this regard, see Hadot (2002) and Foucault (1988) which makes of "souci de soi" (*epimelea*

Augustine sees in attention the present of the present and etymologically links with awaiting *attentio-attendere*, that is, how I open my mind to what is yet to come.[3]

The early modern philosophers in their turn pave the way for two central forms of attention, either following the ancient view of welcoming (Malebranche) or of will and self-control (Descartes), according to an opposition which will repeat itself in a cyclical way in later periods, along the distinctive stresses of mystics and science. Thus, with Descartes (1967, 426; 1994, art. 70)[4] and the rationalist tradition that he inaugurates (from Leibniz[5] to Fichte), attention as will is associated with judgment and effort. However, as the extracts provided in the notes below illustrate, attention comes to support perception or emotion. Its role is crucial as a help and it would not be able, so it seems, to exist alone. In the same spirit Maine de Biran will make of it an effort and a tension, Condillac a little earlier 'a strong sensation' (Condillac 2000 [1754]), whereas much later the philosopher Alain, criticizing automatic attention and favoring its persistence will exclaim: "It is a mistake to say that what one knows how to do is then done without attention [. . .]." And he will turn this into a permanent feature of human beings: "The animal is not distracted, it is absent-minded." (Alain 1941, 240)

For Malebranche, on the contrary, "attention of the mind is [. . .] a natural prayer through which we obtain that reason enlightens us" (1993 [1648], part 1, chap. V, art. 4). Attention therefore is at the same time an occupation, an exercise and an awaiting without any expectation, according to the Augustinian line of thinking, which Henri Bergson takes up again as his own at the beginning of the twentieth century:

> Every consciousness is an anticipation of the future. Consider the direction of your mind at any moment: you will find that it is occupied with what is, but above all in view of what is going to be. Attention is awaiting, and there is no consciousness without a certain

---

heautou or *cura soi*) not a state of mind but "a technique and a self-practice" involving training and exercise (*askèsis* or *exercitium*). Cf. in this regard Foucault (2004).

3 Cf. Augustine *The Confessions*, book XI, chap. 3 (Augustine 2005).

4 Cf. Descartes 1967, 426, my translation: "[The judging perception] can be imperfect and confused, as it was before, or rather clear and distinct, as it is now, according to whether my attention is focused more or less on the things in it (*the wax*) or of which it is composed." Cf. also *Passions de l'âme*, Paris, Vrin, 1994, art. 70 "Admiration is a sudden surprise of the soul which causes it to concentrate on considering with attention objects which seem to it being rare and extraordinary."

5 Cf. Leibniz (1990), chap. IX, §1, my translation: "Thought often means the operation of the mind on its own ideas when it acts and considers a thing with a certain degree of voluntary attention, but, with respect to what we call perception, the mind is ordinarily purely passive, not being able to avoid perceiving what it is currently perceiving." Cf. also §4, pp. 111–112.

attention to life. The future is there: it calls us, or, rather, it draws us to it; this uninterrupted traction drive which carries us on the path of time also causes us to act continuously. Every action is an encroachment on the future. (Bergson 2007 [1919], 8)

But it is actually Simone Weil who gives us the most acute formulation of it in *Attente de Dieu*:

> Most often, we identify attention with a muscular effort. If we say to some students: 'Now, you are going to pay attention,' we see them frown, hold their breath and contract their muscles. If, after a few minutes, we ask them: 'What did you pay attention to?' They aren't able to answer. They haven't paid attention to anything. [. . .] Attention consists in suspending one's thoughts, leaving them available, empty and penetrable to the object [. . .] the thoughts must be empty, awaiting, not looking for anything [. . .]."
> (Weil 2010, 92–93)

What more can we say in view of these many wise thoughts of ancient and modern philosophers who see in attention, alternatively, a joint effort of the will and the body or an open and receptive awaiting? They all note its role as the 'basso continuo' of human activity. What then can we claim to say that would be new or different, and, as a consequence, why is attention of interest for us today?

What is the new fact which makes a renewed reflection on attention such an emergency? For several decades our society has been experiencing a vertiginous change in methods of communication: we are well past the simple act of watching television; the multiplication of digital media *via* the development of the Internet and the downloading of TV shows, programs and films, as well as live exchanges on Facebook, Twitter, Skype, Zoom or Snap chat – which appear multidirectional and even faster than e-mail – along with the invasion of cell phones, all of that creates an unprecedented situation of multiple and simultaneous communication: everyone can contact everyone at any instant and in return waits for an immediate response. In short, an immense live communication network which keeps un-ceaselessly changing mobilizes each and every one of us and also creates a growing dependence on this 'connection' to others. What better safeguard is there against the gnawing and increasing anxiety generated by the contemporary evil of our society: the loneliness of individuals living next to each other in fragmented and atomized juxtaposition? To make up for this deficit in human relations we create an artificial connection permitting anyone at all to be connected at any moment with anyone else who is also on the 'Net'. As a consequence, we live with the illusion of being 'connected'; we feel well surrounded even if we no longer meet anyone physically. We feel good; we love feeling surrounded by multiple virtual presences who give us the impression of living at the rhythm of the immense breathing organic heart of the whole society.

This encourages a tendency of generalized 'zapping' that we can observe daily among younger people, but also that is winning over adults: The daily use of computer devices leads students to pass without transition from an information search on Wikipedia for a homework assignment to replying to an email to an online exchange on Facebook and then to an online video game. Everything is done simultaneously, and we find the same kind of behavior in the professional world of executives and politicians who for example set up a financial or diplomatic file on a computer, all the while responding to several clients or activists by e-mail and at the same time communicating with their friends on Twitter. However, whereas young people practice this multi-focused attention in a spontaneous and easy way, adults often experience this situation as a source of nervous tension and attentional confusion as if the cerebral plasticity of the young, formed from the beginning by this mobility of attention, was no longer available to older people whose neuronal flexibility is entropically deficient.

However, such a generational manichaeism remains too simple: it hides a much more dramatic reality among the young. What we designate today, thanks to the psycho-medical expression, as attention deficit disorder, describes a more widespread reality than we believe. This pathology of attention manifests itself by an inability to focus and by a fragmented and scattered attitude, resulting from a hyperactive mental state, also called 'zapping', which itself has become addictive. It results from an increasing attentional deficit, an intense internal disorder which hides behind the multiplication of sources of attention. Such a pulsional development of consumerism creates an attention reformatted by the mass media. Some will see in this a crisis of attention, even a 'destruction of the ecology of attention', which interrupts the 'deep attention' required to read a book and to develop critical thinking, to the benefit of a multitasking hyper-attention, which is fundamentally de-individualizing;[6] others, welcoming the furious race toward still newer technologies, will take note of this necessary evolution and will favor the emergence of a technologically changing humanity in a period of rapid and urgent development (Lévy 2001). If, as a 2017 study reveals, a visitor devotes no more than twenty eight seconds to each work of art in a museum (Smith 2017), is it necessary to be concerned with it and stress the importance of rediscovering contemplation in connection with an economy of attention and an ethics of presence? Or, rather, do we have to rejoice in the aptitude of our growing humanity to multiply connections and

---

6 Cf. Stiegler (2008, 2010), which focus their criticism on television and the 'capture of attention' brought about among small children, making them dependent on a consumerism deprived of critical examination.

to create interactive networks which generate novelties and produce previously unknown realities? Whatever we decide, how do we take care of the generations to come? How do we accompany the act of creation in a genuine way?

To focus our attention on attention is, in any case, of utmost importance to our contemporary humanization process. Science has certainly understood this. For more than a century, more and more work has been carried out in the fields of psychology and neurosciences that tackles this particularly complex and comprehensive function, which can be seen as an integrated network of transversal activities mobilizing memory, perception, wakefulness and emotion, but also decision-making.

And we can consider that, in the framework of this emergency to deal with attention, various ancient and contemporary meditative and spiritual techniques also have a role to play. Indeed, they are aimed less at our intellect, as is science, than at our fundamental attitude toward life. Therefore, they offer us an essential key to the practice of self-presence, or of simply being present, which is only another name for lived and living attention.

In view of the increasing impact of science and the growing visibility of meditative techniques, we can, as a consequence, legitimately wonder what role phenomenology can play here. Why exactly phenomenology? In contrast to earlier philosophical approaches, which rely on system-building, critical examination and conceptual exactitude, phenomenology has the advantage of maintaining an open view on the world: by stressing the importance of returning to experience, it is in a position to make an ally of the experimental aim of contemporary cognitive science; at the same time it has an obvious affinity with the contemplative quest of the meditative traditions because of a stepping back which it encourages through its methods of suspension of prejudices.

Therefore, phenomenology proves to be a crucial intermediary discipline for rethinking attention by considering it an experience of being open to the world more than an internal mental state, as is still the risk in science, and by providing it with a more easily shared language than that of the meditative traditions, which is often too narrowly exclusive in its expressions. It is here that phenomenology can bring something new in terms of the philosophical basis of an ethic of attention that we will refer to as 'vigilance': science, yes, but denaturalized by putting into perspective the themes of focalization and selection; philosophy, yes, but open to a lived rationality while interrupting the connection of attention with too narrow an act of self-control in relation to the will; meditation, yes, as the experimental basis of open receptivity, but described in a language and discourse more easily shared with secular society. In this way, phenomenology can be a combined methodology for overcoming disciplinary obstacles and suitable for establishing the basis of a *koinè* of the economy of attention.

2 The Autonomization of Attention from Perception and Reflection: A Condition for Attention as Vigilance

By questioning a conception of consciousness closed in on itself, understood as a solipsistic interiority or as a reflective duality, phenomenology has, since Husserl, promoted a new attitude toward the subject: intentionality. By virtue of this movement of openness directed toward things, other people and the world, intentionality amounts to being aware as being aware 'of' something. Thus, in contrast to the classical conception of the subject which encloses the conscious subject *in* herself, intentional openness offers a crucial basis both for the perceptive and, as we will see, for the proto-ethical dynamic of attention. It shows that the structure of openness is inherent in consciousness by revealing the dynamic of internal otherness. It can thus be seen as the necessary lever that turns consciousness into vigilance, this heightened quality of self-presence which is the condition for genuine presence toward others. Now, we will see that vigilance is nothing else than the key *and* the completion of the experience of attention. In short, what I would like to demonstrate is that consciousness and attention are interconnected and therefore mutually generate a new space for vigilance.

2.1 The Role of Attention in the Transformation of Consciousness from Interiority to Intentionality

The classical perspective which originates as we saw in Descartes and was developed with Leibniz's views puts to the fore being-aware as a self-apperception and sees in reflection as defined as a 'perception of perception' the formal completion of this being-aware. Husserl will be more interested in an awareness-*of* (rather than in a 'being-aware'), which he views as the turning move of the self toward the outer world. In other words, in contrast with the introspective move of the self toward itself, whereby consciousness closes up into its inner world with a tendency to navel-gazing and 'encapsulation', the founder of phenomenology chooses instead to open the subject on herself, in a double gesture of welcoming others *and* being turned toward the world. Not ensnaring oneself in the meanderings of subjectivity amounts to fully welcoming the outside world, namely the other than oneself, that which goes beyond us (i.e., transcends us). In fact, the part of strangeness which I am thus welcoming in myself is precisely what allows me to receive it fully as an irreducible other.

How does the initial move of such an intentional openness unfold itself? The formula according to which all consciousness is "consciousness of something" (Husserl 1973, §14) is well-known: it provides the formal definition of this new property of consciousness which is intentionality. Starting from such a redefinition (Depraz 1999, 69–70 and 71–79), I examine the *dynamic* of *becoming* aware, namely the genesis of the move by which the subject notices a dimension of herself that she had not seen before. Starting from the move of openness to otherness which Husserl shows in an exemplary way, I then emphasize the *genesis* of this opening through the process of *self-alteration* it contains.[7] Here emerges in contrast the objectivizing presuppositions that the model of intentionality still contains. What is therefore needed is to lead intentionality beyond itself.

Now, where is situated attention in such a shift from the inner and reflective consciousness to intentional openness? As an inner disposition, an inner ability of the subject, sometimes considered as a state or as an act, all this in a rather confused way given the tension it conveys, attention obviously shares with consciousness the property of interiority and intimacy. The question is: is this self-relation intrinsically a reflexive relation? Among the early modern philosophers, consciousness truly has these two related properties of interiority and reflexivity, the latter reinforcing interiority with the move of *turning* into oneself. Now, if interiority is a common basis for consciousness in the classical sense and attention, the latter contains a dynamic which differs from the move of reflection. There is indeed in attention an initial move of opening by which the subject welcomes that which presents itself to her. Now, the quality of this open awaiting, crucial for Malebranche as we have noted and for some of his predecessors (Philo, Augustine) is not characteristic of the Cartesian (or Leibnizian) view which Husserl both inherits and clearly modifies with his concept of intentionality. While putting aside reflexive rationality by means of his critical genealogy of consciousness, Husserl indeed ultimately produces an alternative which he names open receptivity to otherness. In short, starting from the two common properties of attention and consciousness – private interiority and the move of turning into oneself while turning away from the outside world – a distinctive property emerges as a specific difference from reflexive consciousness: the move of receptive opening. Here attention shares with consciousness its intentional property, from which emerges a third property that is not characteristic of classical reflexive consciousness, namely the opening of the self to the other.

---

7 Concerning these notions of self-otherness ('altérité à soi') and self-alteration ('altération du soi') as inherent in consciousness, cf. Depraz (1995, chap. V).

However, the situation of attention, literally 'straddling' the two conceptions of consciousness, reflexive and intentional, puts it in an uncomfortable and ambiguous position. At best, we will say that it plays the hidden role of a bridge from one conception to the other, while tracing back the meaning of intentionality from the contemplative (Augustinian-Malebranchist) view as not explicitly seen by Husserl. At worst, we will consider that attention keeps the marks of those two lines (the contemplative-mystical one and the rational-reflexive one) without becoming an autonomous concept, and remains dependent on its rational-reflexive origin, given the standard historical hermeneutics of consciousness. What is at stake at this stage of our inquiry is to identify this reflexive-rational mark, to dig up the contemplative mark and its deep connection with intentional phenomenology and, finally to rely on the mutual boundaries of attention and consciousness in order to provide a renewed understanding of attention in terms of vigilance. In other words: my contention is to consider 'vigilance' as a way to think anew while overcoming the ambiguities of its different possible understandings.

2.2 Attention, Perception and Reflection

Let us return to the beginning of this paper with the Cartesian framework which understands attention as being dependent on the reflexive conception of consciousness. Such a historical examination is necessary because it helps us to identify the places of crystallization between attention and the epistemological model of perception/reflection. Then we will see whether and, if so, how empiricism furnishes an alternative to the rationalist conception of attention. While doing so, it will be easier to understand how such identifications can still be maintained today, but also how it might be possible – and based on what arguments – to step aside and initiate another understanding of attention by freeing it from its two historical companions.

A. The Marginalization of Attention in Rationalism (Descartes, Leibniz, Fichte, Husserl)

Even if the term 'consciousness' (*conscientia*) seldom appears as a word in Descartes, it characterizes the human subject with respect to thought (*cogitatio*), mind (*mens*), reason and knowledge. The mode of the subject's consciousness depends on the mind, a faculty more "easily understood" than the body, and the relation of the subject to objects lies in their clear and distinct, namely 'true' knowledge. What is the role of attention in such a cognitive framework? If it is clear that for Descartes there is no theoretical category of attention, we see it appear regularly as a modality, a quality of the intensity of thoughts and of our perception of things:

> [. . .] but, what is worth remarking, one's perception, or rather the act by which one perceives it (*the wax*) isn't a matter of sight or touch or imagination and never has been, although it seemed to be so previously, but only an examination of the mind (*inspectio mentis*), which can be imperfect and confused, as it was before, or rather clear and distinct, as it is now, depending on whether my attention is brought to bear more or less on things that are in it (*the wax*) and of which it is composed.
> (Descartes 1967, 426; my translation)

Thus, attention is presented as a criterion of the modal truth of perception. Or rather, it modalizes the degree of intensity of the quality residing in passion, as with the notion of admiration in article 70 of the *Traité des passions*. If it is 'maintained', it provides reflection with a point of application, as felicitously formulated by the author himself regarding the methodological doubt in the first *Méditation métaphysique* (Descartes 1967, engl. translation 1993). In short, attention remains bound to the activities of the mind; it plays an auxiliary role in highlighting them, and it is soon covered up again by the judgment at stake in perception (Brown 2007). Indeed, as I perceive an object I immediately judge that what I perceive is true in such a way that a clear and distinct knowledge flows from it. For Descartes, attention, therefore, will be a simple modalization of the activity of perception, itself being structured by judgment. Moreover, the author of the *Méditations* evokes attention numerous times in connection with the notions of effort and will, as well as with laziness, which by contrast appears as the cause of distraction. Thus, it isn't so much the qualification of attention as an intensive modality of perceptive activity that is problematic (in fact, we will see that it is precisely a crucial feature); rather, what is problematic is its being concealed by judgment on the one hand, and by the effort of the will on the other.

In the framework of rational knowledge attention and reflection are linked by an oblique foundational connection: attention is an auxiliary modal quality of the perception of the object, perception itself is *judgment* of perception, insofar as it frees itself from mistaken sensation. As for reflection, it is an act of self-knowledge through which the self is apprehended as an object, and which is founded on perception as understood as an act of direct knowledge of the object, as it were. This structure of foundation situates attention both as modalizing (initiating) and subordinate: Attention to the object which modalizes the perceptive activity is the condition of self-reflection, but at the same time, reflection is an act which brings about a return to the interiority of the subject, and attention is not able to produce such a return, insofar as it is limited to supporting the aim to know the external object. In other words, the distinction between perception and reflection responds to a clear dichotomy: Perception is presented as a focalized aiming at the object which links sensation to a formal identification of the object, providing knowledge in the process, whereas reflection initiates a movement of return to oneself. In the context of such clear-cut ordering of object consciousness (perception) and self-consciousness (reflection), attention introduces a confusion which refers us back to the difficulty of assigning it a clear locus. Indeed, in connection with perception, attention contributes to the focalization on an object, but, in contrast with perception, it rather consists in a mode of presence with the object, in some form of 'sustain' or 'highlight', and takes a step back from what it targets, from the objective result. Moreover, attention doesn't only play a role in relation to the external object; it characterizes the act of the mind, engages the disposition and the attitude of the subject without as such involving a movement of *return* to oneself or a *doubling* of perception, as reflection does.

In short and at first glance, the rational construction of knowledge gives differentiated focuses to each act: the external object for attention, the internal state for reflection. In fact, such delineation reveals its limitations: On the one hand, the dichotomy interior/exterior betrays a limitation of attention to direct objectivity, whereas attention also concerns the self and describes the sphere of the subject's relation to herself, although in a different way than the reflexive return to the self; on the other hand, it shows in the sphere of internal action an omnipotence of the subject's reflexive activity, as if self-experience could only be enacted through reflection and no other act, namely through a second-order activity that is sometimes described, as we have noted, as a doubling of perception: as said earlier, Leibniz speaks of it in terms of a 'perception of perception' and therefore distinguishes perceiving and apperceiving (i.e., noticing):

> Thought often means the operation of the mind on its own ideas when it acts and considers a thing with a certain degree of voluntary attention, but in what we call perception, the mind is ordinarily purely passive, not being able to avoid apperceiving what it is currently apperceiving. (Leibniz 1990, chap. IX, §4, 111–112, my translation)

Such a distinction shows an unsatisfactory understanding of reflexive activity itself, which cannot be a simple structure repeating perceptive activity from the outside to the inside. If we consider reflection to be such an interlocking of perceptions, we quickly end up in a possible 'infinite regression' which leads to a separation from experience and the indulging in free-wheeling speculation.

The predominance of the rational model will even lead to identifying attention itself as a reflexive activity – this move starts with Leibniz when he identifies apperception with voluntary attention – i.e., to assimilating attention within reflection. This will be the later and radical option of J.G. Fichte, for whom the 'Pure I' is the main authority from which the 'Not I' takes its meaning: Self-consciousness thus founds object consciousness, which also links reflexive consciousness to the theme of freedom, i.e., to self-control. For Fichte attention has no independency and results absorbed in the reflexive activity, through which the transcendental effort allows for the subject's emancipation from the contingency of events. Thus, Fichte's concept of the Pure I systematizes the idealist theory of reflexive attention. The author presents in the *Wissenschaftstheorie* in 1804 the two moments of attention, either incomplete attention, or full and entire (and hence complete) attention, which he both situates in relation to the difference between the duality of being and thought on the one hand and their absolute synthetic unity on the other. Only complete attention is a genuine attention, and it identifies with reflexivity, itself being in its completed form. Fichte thus accomplishes the equivalence between attention and reflection and distinguishes between two stages, the first one when the pure I is complete and the second when it is not. Indeed, such an equivalence is grounded in the model of the reflexivity of consciousness and leaves no room for a specific definition of attention. So much so that we may eventually wonder if it is at all relevant to speak here of attention since it doesn't have any existence except as a reflexive activity (Maesschalck 2003, 216–233). In fact, Fichte presents here a reversal of the classical theory of consciousness, where attention to the object obliquely remains (through perception) the foundation for the reflection of the self. Fichte's radicalization produces an inversion of the conditions: attention can only be reflexive, and reflection (self-consciousness) therefore becomes the condition for object-consciousness (of the perception as such).

In short, the philosophical tradition from Descartes to Fichte builds a concept of attention which falls within the framework of knowledge of the object and therefore requires a subject characterized by judgment, self-control (will) and control of the object (reason). In this sense, being attentive is being centered on oneself and focused on the object in order to identify its properties.

Husserl situates himself partly within this rationalist-idealist heritage of reflexive philosophy and of a logical theory of perception. The founder of phenomenology shows his inheritance from the rationalist understanding of philosophy through the primacy he gives to logic. Thus, 'pure logic', as presented in chapter XI of the *Prolegomena zur reinen Logik*, is in direct connection to the Kantian theory of knowledge discussed in Chapter X and is distinguished from applied logic (cf. Husserl 2001a); earlier, it relies on Herbart and his *Psychologie als Wissenschaft*, on the one hand, and on Leibniz and his *Nouveaux Essais*, on the other, which, as we know, form a pivotal point between Cartesian idealism and Fichte's speculative idealism. Therefore, the critique of psychological logic goes hand in hand with the questioning of an empiricist conception of the theory of knowledge which considers the laws of thought as natural laws (cf. Husserl 1973). Indeed, it is the specter of 'psychologism' which justifies in the *Logischen Untersuchungen* and above all in the *Prolegomena zur reinen Logik* the exclusively rational legitimacy of attention and its claim as reflexive attention. In this regard, Husserl is initially situated within a framework where attention is considered as the activity of a knowing subject, and reflection the aptitude of the rational subject to know herself. From this critique, attention is not even thematized, but rejected in the name of the rejection of an empiricist psychological approach; only the structure of reflection is relevant, because it alone is compatible with a pure logic which highlights the ideal of the perceived experience and, as a consequence after 1913, the transcendental idealism of phenomenology.

Moreover, we find arguments in Husserl's work which stresses such an ultimately reflexive approach to attention: in the fifth *Logische Untersuchung*, he presents a theory of consciousness that joins the reflexive act with attentional experience as they both emphasize the immanent unity of the self, as it endures acts and experiences while remaining involved in them. This similar conception of the dynamic unity of the self participating and living in its acts and experiences will lead Husserl in *Erste Philosophie II* (1923–1924) to bind together in his singular way, through psychology, attention, reflection and reduction. These three modes of consciousness (attention as experience, reflection as act and reduction as method) work together to bring to light the unified and stratified life of consciousness thanks to the multidimensional presence of its immanent self. As he strives to systematize the theory of reduction, Husserl eventually bypasses attention in favor of its reflexive and then reductive completion. Once again, attentional experience remains concealed and, in this case, at best it is subordinated to the reflexive and reductive aim.

B. Empirical Attention: From Critique to a Local Emergence as a 'Function' (Locke, Hume, Husserl)

What then is the meaning of the 'empirical' aspect of attention which Husserl criticizes by putting to the fore the specter of psychologism? To what extent does an empirical meaning of attention provide a fruitful alternative to the marginalization of attention in the rationalist tradition?

In Locke's empiricism, as in Hume's, experience, along with consciousness and perception, proves to be more prominent than attention. In this regard, experience in no way has a derived and secondary importance, since it is the primordial material of all knowledge. With Locke, it is the matter of the existence of consciousness in the immediacy of feeling as felt-sensation (Locke 1997, book II, chap. XXVII; cf. Balibar 1998); with Hume, it is 'pure experience' which turns out to be safe from any *a priori* reasoning, since it is necessary to produce impressions that we have originally felt and whose ideas are copies. These impressions are all strong and noticeable and they do not admit of any ambiguity (Hume 2012, book I, part I; cf. Malherbe 1976). However, such an originality of sensation or impression, while falling within a logic of inductive derivation which makes it more than a simple fact, is not a *constitutive* component of knowledge: it remains the point of departure for a logic which proceeds by additions and associations and step by step creates unity from a progressive sum of elements.

On the other hand, turning sensation into the building block of knowledge is already to provide it with a global and structural dimension independent of any sense data. It is here that Husserl makes empiricism an ally, while implementing in it an *ideal structuring* of the felt and perceived. The critique of empiricist theories in his second *Logische Untersuchung* aims indeed at uncovering at last a genuine phenomenological empiricism deprived of any atomization process.

Husserl's main argument relies on such elementarist and atomistic conceptions of the object, in which attention has the only function of abstracting, in the literal sense of separating, of detaching from any context, i.e., of focusing on a discrete part which becomes its object. In short, object and act are shaped by each other: attention's power of abstraction corresponds to an object construed as an isolated element (cf. Husserl 2001a, 258–276). As a consequence, such a 'discretization' of the object requires a reconstruction of its identity beyond its individual factual reality, which leads to granting the association, then the generalization understood as the identification of similarities. By identifying the limits of the abstract theory of attention from the point of view of phenomenology – namely the non-distinction between object (psychological content) and act, Locke's identification of the object of attention with psychological content and the non-differentiation of the very forms of abstraction (sense data/non-sense data) – Husserl in contrast paves the way for a genuine meaning of attention, which will encompass the distinction between intuition and meaning. This means that attention is a quality of thought as much as of intuition and that I can direct my attention to any kind of acts, be they sensory, imaginary, symbolic or formal. It is in this extensive context that emerges such a transversal definition of attention, which is to become central from 1904–1905 onwards. Namely as "preference", or even as the "act of noticing" (Husserl 2001a, 269) which consists in giving a content to consciousness, in detaching it while clarifying it.

It is eventually in the fifth *Logische Untersuchung*, which presents phenomenology as a descriptive psychology of lived experiences, that Husserl identifies the respective parts of consciousness and attention; on the one hand, he examines the "varied ambiguity of the term consciousness" (Husserl 2001b, 81); on the other, he assesses "the function of attention" (Husserl 2001b, 116). Let us note for now the central and crucial role, thematic but also complicated that Husserl grants consciousness by distinguishing three of its main aspects, and by showing its complex relation to the I, insofar as conscious lived experiences and acts are not necessarily unified by the I. On the contrary, attention is reduced in one paragraph to the role of a *marker*, with the only purpose to illustrate the distinction between word and meaning, depending on the attentional stress put on the one or the other. The term 'preference', along with the term 'influence' are used here by Husserl to describe this modal function of attention, about which he clearly says that this "general fact" (of attention) is not "sufficiently elucidated" and that he does not claim here to "work out" a "'theory' of attention" (Husserl 2001b, 119).

C. The Parting of Attention from Reflection: From Static to Genetic Phenomenology

What results from the two previous steps is the weight of the reflexive act as a main component of (self)-consciousness. On the contrary, attention is most often discounted because of its mostly empiricist label, or only mentioned in terms of its functional status. Correlatively, the basic role of the perceptive act is highlighted as the understructure to which the attentional act can attach itself as an adjuvant. Such is the standard landscape of the relations between reflection, perception and attention in Husserl's phenomenology as originating from the intersection of rationalism and empiricism. Yet is this his first and last word on the subject?

I would like to show that there is more to this in Husserl's work. Even in its early stages, namely in the years around the *Logische Untersuchungen*, there is more than just this side of a critical or local understanding of attention. In the context namely of a strong discussion with his teacher Carl Stumpf about his conception of attention as a 'pleasure of noticing' [*Lust am Bemerken*], Husserl will develop an alternative understanding of attention in clear contrast with perception and, even more, with reflection (Husserl 2004). In this regard, an extract from a text entitled "Directions of attention" which emerged in the context of this discussion is rather telling:

[. . .] the consciousness in which is objectively formed what is represented is *a second order act, a reflection*. Direct representation is underlying it, that of the paper. I am perceively directed toward the paper. Then I operate the reflection. I am no longer living *as being directed toward the paper;* I no longer accomplish this directedness in a lively way; therefore, a transformation occurs by which life vanishes although the direction remains 'maintained'. And then, in a new representation and a new stance, a current direction is carried toward the events of the act which became the object and *caput mortuum* by the reflection, toward that which comes from the I, toward the direction of which what is oriented as such, toward its nature, etc. and the current objectification possesses again elements of representation and something characteristic.

But, more importantly, let us be careful about the fact that the reflection in which the modified directedness as such is captured is not any longer a mere 'turning toward', of *the apperception*, as would be this ray of orientation which is present in the act ('living' [in] the act accomplished, as it happens, and for that reason, in the living 'representation' of what is 'objective' to that which the action pertains) and confers on it precisely an orientation to that which is 'objective' for it. *The living, spontaneous act, such as it is, can only have this orientation* and, as an act of taking a stance, what is 'objective' toward what it orients itself is so with respect to what it takes a stance. As soon as the ideative reflection relating to what is directed at appears on the scene, the act transforms itself. This means that 'it' is underlying as a presupposition, but not on an unmodified mode, rather [on] a modified mode. It is no longer the act originally accomplished, but a *dead accomplishment*. The living radius of attention is withdrawn from 'it', the direction-towards is now carried elsewhere, and that means that there is a new act there, and the former direction-towards has become something changed on a non-living mode. And the same holds true for the stance taken. We now have a new living position taken, and the former one is not changed in terms of not being living.

If we understand by 'spontaneous attentional turning toward' the spontaneity of the direction-towards and of the stance taken, there is in reflection a movement of turning away provided that the attentional turning toward emerges as a new movement; talking of a reflexive turning away is justified to the extent that this kind of *vitality of the turning toward has the character of a radiating from the source point of life, from the 'Pure I', the carrier of all vitality.* (Husserl 2004, 389–390, my translation)

I wished to provide this excerpt *in extenso* because it is emblematic of the distinction of the two acts, reflection and attention – a distinction which is stated here very clearly in terms of different kinesthetic gestures: attention is a general movement by which I turn myself toward (*wende . . . zu*) completely, mind and body; I am directed toward (*richte*) and orient myself (*orientiere*) on a mode that Husserl describes as 'living' because it is 'spontaneous' and 'not modified': here, the vitality of the lived experience remains unmodified because it is not objectified in an identifying and ideative capture of the object. Reflection on the contrary is no longer a 'turning toward' (*Zuwendung*) but a 'turning away' (*Abwendung*) and a 'turning around' (*Umwendung*). In short, attention is a living and vital opening move, while reflection results as a closing up in oneself, locking down and deadly: *caput mortuum*. Beyond the semantic and kinesthetic unity of these two movements of turning (*wenden*) – Husserl will never have been any clearer –, what is at stake is the metaphorical axiology of life and death, of openness and closure, of spontaneity and modification or, again, of dynamic stance and static representation. Reflection is related to the ideality of judgment, whereas attention is the very vital source where life originates. But beyond this post-romantic axiology which presents this distinction between attention and reflection in terms of life and death, we can also notice the descriptive difference of these kinesthetic gestures: the attentional lived experience is described as a movement in which body and consciousness accompany each other in a unified rotational gesture of guided openness; by contrast, the reflexive lived experience refers to a gesture of inverted rotation, which is a turning inward into oneself. In that respect, the movement of the eyes is a striking indicator of this: my attention to someone's speech is visible in the open and quivering vitality of my gaze, which I direct toward the person speaking; when I am attentive to my companion's remarks, I am (as we say in English) 'hanging on his every word' and they resonate in me so much that I react with expressive movements of agreement (a slight opening of the mouth) or of non-understanding (a frown); the switch to reflection, by contrast, leads to a form of absence of immediate relation with the other person, an immobility in visual expression, in short, a freezing of the facial features that the other person may interpret as a rupture in the relation. Now, this absence-mindedness with respect to the other person actually betrays the intensity of my self-wakefulness and refers to an internal movement that may seem imperceptible given the seemingly external immobility.

While following the historical thread of phenomenology, it is of interest to see how this distinction evolves and is taken up again in the genetic period and, notably, in the texts devoted to passive synthesis. Indeed, in those texts the attentional turning toward emerges from an affection by the object, which initiates the dynamic of becoming-aware and results in a self-awakening of the subject. And so the theme of openness appears to describe the attentional process, whether it is called vitality and orientation as in 1904–1905, or affection, awakening and receptivity, as in 1918–1926. In any case, the dissociation between the attentional lived experience and the reflexive act in the 1920s is such that it leads to a different textual distribution. The reflexive act is analysed in the fourth Section of the 1923–1924 Course, *Erste Philosophie II*, in which the structure of the reflexivity of consciousness is presented as the accomplishment of the acts of remembering and of imagination. These structures of split-duplication of the I (*Ichspaltung*) identify the ego's different modes of self-otherness at work in the way of psychology (Depraz 1995, chap. V).

In short, we are dealing with two distinct forms of consciousness: the one, within the framework of the affective genesis of attentional turning toward, is a passive being-conscious linked to receptive openness; the other, within the context of the egoic way of psychology, is a reflexive consciousness which triggers a move towards the object's capture and identification. In the 1920s, it is clear that attention responds to the first form of consciousness and reflection to the second.

3 Conclusion

In this contribution I showed how attention was not dealt with in the history of philosophy as a systematic and thorough theme of investigation. Besides, when it came to the mind of philosophers, it was either identified with control and effort or with a form of passive opening of the mind. I analyzed how by contrast phenomenology opens the way for a renewed investigation of attention. First it works out and questions the metaphysical binary distinction between activity and passivity; second it frees it from both perception and reflection. With these two steps aside, phenomenology provides a crucial basis for opening attention as vigilance, that is, as a receptive welcome and openness to whatever is to come. It offers an alternative critical platform for questioning anew the immense problems we meet with the general disembodiment of our perceptions brought about with the increasing digitalization of our societies.

References

Alain (1941): *Éléments de philosophie*. Paris: Gallimard.
Augustine (2005): *The Confessions*. Bristol: Phœnix Press.
Balibar, É. (1998): *Identité et différence*. Paris: Seuil.
Bergson, H. (2007): *Mind-Energy. Lectures and Essays*. London: Palgrave Macmillan.
Bidet, A. (2010): "Anthropologie de la présence et de l'attention chez Albert Piette." *Sociologie du travail* 52:3, 435–438.
Brown, D. J. (2007): "Augustine and Descartes on the Function of Attention in Perceptual Awareness". In: S. Heinämaa; V. Läthennemaki; P. Remes (eds.): *Consciousness. From Perception to Reflection in the History of Philosophy*. Dordrecht: Springer.
Condillac (2000): *Treatise on the Sensations* (1754). Manchester: Clinamen Press Ltd.
Depraz, N. (1995): *Transcendance et incarnation. L'intersubjectivité comme altérité à soi chez Edmund Husserl*. Paris: Vrin.
Depraz, N. (1999): *La conscience. Approches croisées des classiques aux sciences cognitives*. Paris: A. Colin.
Descartes, R. (1993): *Meditations on First Philosophy*. Indianapolis: Hackett Classics Publishing Company.

Descartes, R. (1974): *Passions de l'âme*. Paris: Vrin.
Descartes, R. (1967): *Méditations métaphysiques, seconde méditation*. In: Œuvres philosophiques II. Paris: Garnier.
Foucault, M. (1988): *The Care of the Self*. Vintage Books: New York.
Foucault, M. (2004): *The Hermeneutics of the Subject. Lectures at the Collège de France 1981–1982*. New York: Picador.
Hadot, P. (2002): *Philosophie antique et exercices spirituels*. Paris: A. Michel.
Hume, D. (2012): *A Treatise of Human Nature*. Oxford: Clarendon Press.
Husserl, E. (1973): *Cartesian Meditations*. The Hague: Kluwer.
Husserl, E. (2001a): *Logical Investigations, Volume 1*. London: Routledge.
Husserl, E. (2001b): *Logical Investigations, Volume 2*. London: Routledge.
Husserl, E. (2004): *Wahrnehmung und Aufmerksamkeit. Texte aus dem Nachlass (1893–1912)*. In: Husserliana XXXVIII, ed. by R. Giuliani and T. Vongehr. Dordrecht: Springer.
Jousse, M. (1978): *L'Anthropologie du geste*. Paris: Gallimard.
Lévy, P. (2001): *Cyberculture*. Minneapolis: University of Minnesota Press.
Leibniz (1990): *Nouveaux Essais sur l'entendement humain*. Paris: Vrin; translation: *New Essays on Human Understanding*. Cambridge: University Press 1996.
Locke, J. (1997): *An Essay Concerning Human Understanding*. London & New York: Penguin Classics.
Maesschalck, M. (2003): "Attention et signification chez Fichte et Husserl". In: J.-C. Goddard; M. Maesschalck (eds.): *Fichte. La philosophie de la maturité (1808–1814)*. Paris: Vrin.
Malebranche, N. de, (1993): *Treatise on Ethics*. The Hague: Springer.
Malherbe, M. (1976): *La Philosophie empiriste de David Hume*. Paris: Vrin.
Piette, A (2007): *L'Être humain, une question de détails*. Marchienne-au-Pont. Socrate Editions Promarex.
Piette, A. (2009): *L'Acte d'exister. Une phénomégraphie de la présence*. Marchienne-au-Pont: Socrate Éditions Promarex.
Philo of Alexandria (2001): *On the Creation of the Cosmos According to Moses*. Leiden: Brill.
Smith, L. F.; Smith, J. K.; Tinio, P. P. L. (2017): "Time Spent Viewing Art and Reading Labels." *Psychology of Aesthetics, Creativity, and the Arts* 11:1, 77–85. https://doi.org/10.1037/aca0000049
Stiegler, B. (2008): *Économie de l'hypermatériel et psychopouvoir. Entretiens avec Philippe Petit et Vincent Bontems*. Paris: Mille et une nuits.
Stiegler, B. (2010): *Taking Care of Youth and the Generations*. Stanford: Stanford University Press.
Weil, S. (2010): *Attente de Dieu*. New York: Routledge.

Short Biography

Natalie Depraz, PhD, is a Professor of Contemporary German Philosophy and Phenomenology at the University of Rouen-Normandy, and a University Member at the Husserl-Archives (ENS-CNRS, Paris). She is a Husserl scholar (*Transcendance et incarnation. L'intersubjectivité comme altérité à soi chez Husserl,* Vrin, 1995) and she translated into French a number of Husserl's manuscripts on intersubjectivity, attention and emotion. Recently she focused on attention (*Attention et vigilance. A la croisée de la phénoménologie et des sciences cognitives,* PUF, 2014) and on surprise as philosophical issues (*La surprise du sujet. A sujet cardial,* Zeta Books, 2018). She is currently writing a second book on surprise with the aim of re-reading the history of philosophy under its light (to be published with Hermann, Paris). Parallely she published in 2019 her first novel *L'endroit* (5 sens editions, Genève), and in 2021 a second one entitled *Déni ma survie* with the same editor.

Yuko Ishihara and Olaf Witkowski
Chapter 6
Different Ways of Attending to Experience: Formalizing the Phenomenological Epoché to Translate Between Science and Philosophy

Abstract: When we reflect on our experience, our attention shifts from the objects of our experience towards the experience of the objects. This *shift of attention* can be understood in at least the following three ways: (1) an instance of *introspection* where a physical self is attending to its own experience, (2) an instance of *psychological reflection* where a psychological self is attending to its own experience in a phenomenological manner, or (3) an instance of *transcendental-phenomenological reflection* where a transcendental-phenomenological self is attending to its own experience. Misunderstandings of phenomenology often revolve around conflating phenomenological reflection with introspection or understanding it merely as a kind of psychological reflection. Such misunderstandings are detrimental not only to phenomenology alone, but also to the interdisciplinary study of experience insofar as they hinder productive exchange between disciplines. This paper presents a metareflection by a philosopher and a scientist on the nature of reflection understood as a shift of attention. We introduce a new formalism of the phenomenological method of 'bracketing', known as the phenomenological epoché, that will help clarify the aforementioned differences and prevent making logical mistakes that may arise from the ambiguity of the concept of reflection. The formalism delineates a layered model defining structural constraints on the language inherent to each discipline, with physics at the ground level, psychological phenomenology on the level above that, and transcendental phenomenology at the top. Our model, together with the notations we introduce, further illustrates how translations between the disciplines are possible and what kind of precautions one must take when undertaking them. Finally, we discuss potential implications of the model by highlighting important analogies between science and philosophy.

Introduction

I am sitting in my rocking chair inside my room looking outside at the snow falling on the ground. I then stop and attend to my experience of looking at the snow. This kind of event can occur either spontaneously or be the result of a conscious effort brought about by means of a specific method. Either way, when this happens, I am no longer attending to the snow outside of my window, but I am *reflecting* on my experience of looking at the snow. Reflection can thus be understood as a *shift of attention* from the objects of our experience towards one's own experience of the objects. But depending on what kind of reflection we have in mind, we can end up with very different ways of understanding this shift of attention. The aim of this paper is to clarify some of the ambiguity involved in the discourse on reflection and the kind of change it brings about in our attention. In the above example, the reflection on my experience of looking at the snow can have at least three meanings: (1) an instance of *introspection* where a physical self is attending to its own experience, (2) an instance of *psychological reflection*[1] where a psychological self is attending to its own experience in a phenomenological manner, or (3) an instance of *transcendental-phenomenological reflection* where a transcendental-phenomenological self is attending to its own experience. In the following, we argue that the above three meanings should be clearly distinguished. In order to help clarify the differences, we will introduce a new formalism of the phenomenological method of 'bracketing', known as the phenomenological epoché, in order to avoid logical mistakes that may arise from ambiguity of the term reflection. This will be valuable, not only for phenomenology alone, but also for the interdisciplinary study of experience. The following is a metareflection by a philosopher and a scientist on reflection understood as a shift of attention to create a metalanguage on experience that translates between science and philosophy.

But before we go any further, we must note that in interpreting reflection as involving a shift of attention, we are not thereby suggesting we understand reflection solely in terms of attention, i.e., that we reduce reflection to a form of attention. Phenomenological considerations show that whereas both attention and reflection modify our experience by accentuating and articulating what was only implicitly given, reflection involves a turning back of consciousness

---

1 'Psychological reflection' here designates *phenomenological*-psychological reflection which is a specific kind of phenomenological reflection and distinct from the kind of reflection empirical psychologists employ. We will say more about this in Section 2. Unless otherwise stated, when we say 'psychological reflection' in this paper, we mean phenomenological-psychological reflection.

onto itself, thereby introducing a new act that is founded on the reflected on experience. As Zahavi writes, following Husserl: "To pay attention to something is not to engage in two processes or activities, but to change or modify one's first-level experience or activity. Reflection, by contrast, is precisely a new (founded) act; it never occurs in isolation, but only together with the act reflected upon." (Zahavi 2005, 90) In this paper, we refrain from delving into the complex discussion regarding the exact relationship between reflection and attention.² Instead, we focus on the relatively noncontroversial fact that reflection, albeit a new founded act, involves a shift of attention. Therefore, the main question of this paper is: *What is the nature of the shift of attention involved in phenomenological reflection and how does it differ from that of other sorts of reflection?* Let us begin by introducing the phenomenological method as it was set out by Husserl.

1 Introspection vs. Phenomenological Reflection

In his influential book *Consciousness Explained* (1991) and elsewhere (1987), Daniel Dennett dismisses classical phenomenology on the grounds that its introspective method, or what he calls "introspectionist bit of mental gymnastics" (Dennett 1987, 153), does not qualify as a sound scientific method. Focusing solely on one's own inner mental life, phenomenology is essentially an 'autophenomenology' which, according to Dennett (1987, 153), by no means can yield interesting scientific results.³ One way of saving phenomenology from such criticism is by saying that such introspective reports can still be used in psychology as specific points of data.⁴ Indeed, one could even try to argue, as some have done (Gutland 2018), that phenomenological reflection is a refined form of introspection that is more scientific and systematic. But rather than trying to define phenomenological reflection in terms of introspection, we will take a more traditional approach below of trying to clarify their differences.

To begin with, phenomenology is not interested in what is going on inside the mind of a particular person at a particular point in time. This is because,

---

2 For a closer examination on this topic, see Breyer (forthcoming), Depraz (1999) and Depraz et al. (2003). Depraz et al. (2003, 24ff.) present a three phased model of the phenomenological epoché (which is a key component of phenomenological reflection) whereby attention plays a twofold function.
3 We will not be engaging with Dennett's arguments in any detail here. For a strong case against his account of phenomenology, cf. Zahavi (2007).
4 Cf. section 3: "The Role of Introspection in Scientific Psychology" of Schwitzgebel (2019).

first and foremost, phenomenology is the study of the ways in which things show themselves in our experiences and their invariant structures. It is not interested in particular mental episodes, but in understanding the essential structures of our experiences. But you may wonder how one can undertake such study without observing what is going on inside one's mind. After all, where do things show themselves but in our mind? This is where phenomenology proves its novelty: it does not buy the distinction between the 'inner mind' and 'outer reality'. But this is not to say that it denies it either. Rather, phenomenological reflection begins by acknowledging that we usually operate with this distinction. We naturally believe that the world is out there existing separately from us and that our experiences of the world take place somewhere inside our mind. Husserl (1983, §27) calls this 'the natural attitude'. But instead of remaining in this natural attitude and maintaining the distinction between the 'inner mind' and 'outer reality' as a legitimate starting point, phenomenology sets it aside. This procedure is called the *phenomenological epoché*. 'Epoché' in ancient Greek means to suspend judgment so the phenomenological epoché is the method of suspending judgment about the existence of objects and the world, which characterizes the natural attitude. To use Husserl's expressions, it is a way of 'bracketing' or 'putting out of play' our belief in the existence of objects and the world (Husserl 1983, §§30–32). And since our belief in the existence of the world is often coupled with the belief that the world is 'out there' and our mind 'in here', to suspend judgment about the existence of the world effectively means to suspend judgment on the distinction between 'outer reality' and 'inner mind'.

Before we go any further, let us stop here and clarify a typical misunderstanding of this epoché, which results in a further misunderstanding of the subject matter of phenomenology. The epoché is *not* a method of denying, negating, doubting, or even excluding the world and its existence. If it were the case, then phenomenology would deal exclusively with our experience and leave the world out of the picture. This kind of interpretation would make phenomenology susceptible to the kind of criticism Dennett and others raise. But the phenomenological epoché does none of that. Instead, it is a method of suspending our judgment about the existence of the world, which is to assume a very specific attitude (and quite an unnatural one indeed) where you neither believe nor disbelieve that the world exists. Far from leaving the world out of the picture, this method in fact allows us to see more clearly our relation to the world that was otherwise covered up in the natural attitude due to the various beliefs and theories attached to the general belief in the existence of the world. It opens us up to 'phenomena', or the ways in which things present themselves in our experience. This is what the *phenomenological reduction* accomplishes. 'Reduction' here does not mean to diminish in size, but from the meaning of the Latin word '*reducere*' (to lead back), it means that, upon the execution of the epoché, we are led back to the phenomena themselves *purely* in the way they present themselves to us. The word 'purely' is important here. No longer do we believe that these phenomena belong to the mind or that they are the result of what happens in the brain.[5] When we bracket our belief in the existence of objects and the world, we are also bracketing all beliefs and theories that are based on this universal belief. As a result of this bracketing, we are freed from the inner realm and disclosed to the field of phenomena.

Therefore, phenomenological reflection (consisting of the phenomenological epoché and reduction) is not a bending back of consciousness onto itself understood separately from the world. Rather, it involves *a change of attitude*, or a shift of attention,[6] from straightforwardly relating to objects (through our perception, imagination, thinking, etc.), to taking up a *reflective* attitude where we are attending to objects *as they relate to our experience of it*. Put differently, phenomenological reflection allows us to move away from the natural attitude to the *phenomenological attitude*; it transforms the everyday experience of being involved with objects into *phenomenological experience*, the field of experience having an intentional structure consisting of what phenomenologist call a *noetic* component (the act) and a *noematic* component (the objective correlate of the act). Husserl (1970, §46) calls this the "universal a priori correlation" between the *cogito* and the *cogitatum*. On the side of the *cogitatum*, phenomenologists work with the descriptions of the intentional object as it is intended, and these are called the noematic descriptions. On the other side of the *cogito*, we have the descriptions of the mode of consciousness, i.e., the noetic descriptions.

At this point, let us return to the example we raised in the introduction to see where our observations so far have brought us. In the example, I was sitting inside looking outside at the snow. I then reflected on this experience and said, 'I am looking at the snow'. Let us say this person is a phenomenologist. What

---

5 There is in fact another sense in which the phenomena in question are 'pure'. Once we employ the eidetic reduction through what Husserl calls the free variation in imagination, the phenomena become purified from particularities as well. Although the eidetic reduction as well as the intersubjective reduction (reduction to pure intersubjectivity) are both important components of the phenomenological method, for the purposes of this paper, we are mainly focusing on the phenomenological epoché and reduction.

6 Speaking in terms of attention rather than attitude may be more appropriate since, following Heidegger's criticism towards Husserl, our natural manner of experience is not one of taking up an attitude. When one straightforwardly relates to objects, we are not taking up any position towards those objects (which is implied in taking up an attitude), but we are attending to them (Heidegger 1985, §12).

change in attention did the phenomenological reflection bring about in our experience? Before the phenomenologist reflected on her experience, she was practically engaged with her surroundings. She may have been enjoying the peaceful scenery of the snow quietly falling to the ground. She may also have been enjoying herself in the warm room having a cup of tea with the view to the falling snow. In any case, before reflecting on this experience, there was no question that the snow that she is seeing outside the window really exists; it was merely assumed. Her own existence, namely the existence of the person that is currently sitting in the rocking chair with a specific personal history, was also taken for granted. Moreover, when we are practically engaged with our surroundings, not only do we assume the existence of the world and everything in it (including myself), but we are also *attending* to them and not our experience of them. So while I am enjoying the snow, I am attending to the snow and *not* to my experience of the snow.

But all of this changes when she decides to take up the phenomenological attitude.[7] She will first bracket her belief in the existence of the world (recall that this is the phenomenological epoché). She now abstains from believing that the snow that she is seeing is the physical snow which exists in the outside reality. This does not mean, however, that she stops looking at it or that she has denied its physical existence. Rather, she has suspended judgment about the existential status of the snow and is now purely attending to the ways in which the snow is being presented to her in her experience. Now, is this experience something that is happening in the mind? No. When she brackets her belief in the existence of the world, as we said earlier, she is effectively bracketing her belief in the distinction between the 'inner mind' and 'outer reality'. So when she attends to her experience of looking at the snow, it is not as if her attention has shifted from what is going outside to what is going on inside. In other words, it is not as if she stopped focusing on what is going on in the physical realm and instead turned her focus to what is going on inside the psychical realm. Rather, once she has bracketed the existence of the world, her attentions shifts to the experience of looking at the snow *just in the way it presents itself to her* (this is the phenomenological reduction). *This* experience is what we mean by 'phenomenological experience'. She will be surprised to discover so much

---

[7] Here we are presenting an example of how one might deliberately go about bracketing and executing the phenomenological reduction. However, it should be noted that there is an important sense in which one initially passively falls into the epoché, where the world becomes one big question mark. In his insightful article, William Jon Lenkowski (1978) has argued that the epoché presupposes the "fall into perplexity" where the world slips away, which is an event that happens to us and not something we deliberately bring about.

about this seemingly simple experience of looking at the snow. She will discover that the falling snowflake as it is perceived is always appearing to her from a specific profile while she is nonetheless somehow intending the whole snowflake. She will also discover that as she perceives the snow, there is an outer horizon that is co-given to her, such as the sound of the crackling wood in the fireplace or her childhood memory of playing in the snow. She may also discover that her perception of the snow comes with a tacit proprioceptive awareness of her bodily self, a sense of how she would be able to walk around on the snow, for example. All of this and much more will be discovered about this phenomenological experience as long as she is employing phenomenological reflection and living in the phenomenological attitude.

2 Phenomenological Psychology vs. Transcendental Phenomenology

Before we move on to the next section, we still need to clarify an important distinction that pertains to phenomenological experience. So far, we have not said anything about what kind of purpose one may have in doing phenomenology. What does the phenomenologist ultimately attempt to accomplish from studying phenomenological experiences? As a matter of fact, depending on the purpose and, accordingly, how one attends to the experience, phenomenological experience can take on different meanings. Husserl himself envisioned two ways in which one may undertake phenomenological investigations. One way is called *phenomenological psychology* (also called psychological phenomenology) and the other, *transcendental phenomenology*.[8] Phenomenological psychology is an a priori psychological discipline that studies our mind in a phenomenological manner. Just as a priori physics, such as geometry or mechanics that study the essential structures of physics, provides the foundation for the empirical natural sciences, phenomenological psychology is said to be the a priori psychology that is necessary for securing the grounds for empirical psychology. As Husserl says: "Phenomenological or pure psychology as an intrinsically primary and completely self-contained psychological discipline, which is also sharply separated from natural science, is, for very fundamental reasons,

---

[8] The following discussion of this distinction is based on one of the author's previous works (Ishihara 2016, 33–39).

not to be established as an empirical science but rather as a purely rational ('*a priori*,' 'eidetic') science." (Husserl 1997, 92)[9]

But phenomenology's role is not exhausted as an a priori psychological discipline nor does this capture the radicalness and true significance of phenomenology. This is because psychological phenomenology is not yet transcendental phenomenology and, according to Husserl, it is only as the latter that phenomenology establishes its unprecedented role in the history of philosophy. As Husserl says:

> The new phenomenology did not originally arise as pure psychology and thus was not born of a concern for establishing a radically scientific psychology; rather, it arose as *'transcendental phenomenology'* with the purpose of reforming philosophy into a strict science. Because transcendental and psychological phenomenology have fundamentally different meanings, they must be kept most rigorously distinct. (Husserl 1997, 95)

What Husserl means by the two sciences having different meanings is that they serve different purposes. The aim of psychological phenomenology is to articulate the invariant structures of our mind and, hence, is "born of a concern for establishing a radically scientific psychology". As we said above, psychological phenomenology provides the foundation for empirical psychology and thus secures the scientific rigor of psychology. Transcendental phenomenology, on the other hand, has a specifically philosophical aim of "reforming philosophy into a strict science" by articulating the meaning and validity of the world as it is constituted by the functions of consciousness. The difference can also be cashed out in terms of their scope: the transcendental problematic is much broader since its concern is not limited to a specific region, i.e., the mind, but rather extends to all possible regions. And in that sense, transcendental phenomenology is an a priori science that provides the foundation for *all* sciences, not just psychology.

Despite these differences, however, Husserl believed that the two sciences, psychological phenomenology and transcendental phenomenology, are like

---

[9] It should be noted that Husserl's use of 'a priori' does not coincide with that of Kant which is opposed to the 'a posteriori'. As he does in the quote, Husserl often used 'a priori' interchangeably with 'eidetic'; phenomenology is an 'a priori' or eidetic science (cf. Husserl 1983, xxii). Since essences, according to Husserl, are not inferred either inductively or deductively by the intellect, but directly intuited in our experience, 'a priori' and 'eidetic' are not opposed to 'a posteriori'. Understandably, he states elsewhere that he would avoid using 'a priori' and 'a posteriori' as much as possible to avoid confusion: "As already was the case in the *Logische Untersuchungen*, I avoid as much as possible the expressions 'a priori' and 'a posteriori' because of the confusing obscurities and many significations clinging to them in general use, and also because of the notorious philosophical doctrines that, as an evil heritage from the past, are combined with them." (Husserl 1983, xxii)

sisters. As he says: "[O]ne science turns into the other through a mere change in focus, such that the 'same' phenomena and eidetic insights occur in both sciences" (Husserl 1997, 95–96). And therefore, he further claims that "in a certain way purely psychological phenomenology coincides with transcendental phenomenology, proposition for proposition" (Husserl 1997, 98). Yet, this is not without an important qualification. The 'change in focus', or to use our preferred phrase, the *shift of attention*, has the effect of changing the meaning of their results fundamentally. This is to say that, while the psychologist and the transcendental phenomenologist share their findings on the intentional structure of consciousness, their *interpretations* of these insights differ substantially such that they end up with completely different understandings of the phenomenological realm that is uncovered. But how can their interpretations differ so radically? What does this 'change in focus' consist of?

It is here that we are introduced to the method called the *transcendental reduction*:

> The objectives of a transcendental philosophy require a broadened and fully universal phenomenological reduction (the transcendental reduction) that does justice to the universality of the problem and practices an 'epoché' regarding the whole world of experience and regarding all the positive cognition and sciences that rest on it, transforming them all into phenomena – transcendental phenomena. (Husserl 1997, 97)

What is interesting here is that the transcendental reduction is introduced as "the broadened and fully universal phenomenological reduction", thereby suggesting that the transcendental reduction is an extension of the phenomenological reduction and not something radically different. Indeed, this is why the phenomenological psychologist can become a transcendental phenomenologist by executing what Husserl (1997, 128) calls the "unconditioned epoché". Or, put the other way around, the transcendental phenomenologist can become a phenomenological psychologist by abstaining from taking this "unconditioned epoché" and thereby remaining transcendentally naive. But the question remains: Have we not already bracketed, through the phenomenological epoché, "the whole world of experience" and "all the positive cognition and sciences that rest on it"? In other words: *what more is there to bracket?*

Without going into too much detail, it will suffice here to say that Husserl seemed to believe that insofar as the phenomenological psychologists are interested in studying the structures of the mind, they are maintaining the existence of the mind and are effectively presupposing the existence of the world since the mind is part of a psychophysical being that exists in the world. In short, contrary to appearances, the phenomenological psychologist remains in the natural attitude. Husserl says: "Even pure psychology in the phenomenological sense, thematically delimited by the psychological-phenomenological reduction, still is and always will be a positive science: it has the world as its pre-given foundation" (Husserl 1997, 96–97). What Husserl means by "positive science" here corresponds to any science that deals with entities whose existence is posited, i.e., presupposed. Phenomenological psychology is a positive science in this sense because, whilst initially bracketing the existence of the world, the mind itself is nonetheless presupposed as existing in the world. Therefore, in order to execute the "unconditioned epoché," we have to bracket that which was left unbracketed by the phenomenological psychologist, namely the existence of the mind. What we then have as a consequence is not the mind posited as existing in the world, but a consciousness purified from all existence. Husserl calls this transcendental consciousness (or ego).[10]

Yet, here again, we should be careful not to misunderstand what the epoché achieves. To bracket the existence of both the world and the mind does not mean that we get rid of them. Rather, it means that we can now study the intentional structures of our experience independently of all beliefs and disciplines that presuppose the existence of the world or the mind. Transcendental phenomenology is therefore no longer a psychological discipline, but a philosophical discipline characterized by a higher-order reflection that seeks to understand how the meaning and validity of objects and the world are constituted in our experience based solely on the ways in which they are given to us in our experience. Borrowing Kantian terms, we can say that transcendental phenomenology is interested in disclosing the conditions of possibility of the meaning and validity of objects and the world.[11]

Let us sum up. We have said that the transcendental phenomenologist differs from the phenomenological psychologist in that, while they both attend to

---

10 This may sound like there are two egos: the empirical ego (or the mind or human subject) that exists within the world (and hence is an object among other objects) and the transcendental ego that does not exist in the world but is the condition of possibility for objects and the world. Husserl (1970, §53) called this the "paradox of human subjectivity": being a subject for the world and at the same time an object in the world. Following his way of dealing with the "paradox", we can say that it is not that there are two egos, but that they are two ways of attending to or apprehending ourselves. Namely, one and the same subject can be understood as the empirical ego or the transcendental ego depending on the kind of reflection one employs (reflection in the natural attitude or transcendental reflection).
11 One may wonder how Husserl's transcendental phenomenology differs from Kant's transcendental philosophy. While a detailed explanation requires much more space, we can very briefly describe the most important difference in the following: Kant's transcendental inquiry begins with the fact of certain scientific knowledge and moves on, in a regressive way, to the a priori subjective conditions that make such knowledge possible. Husserl did not contest this regressive method in itself (though he did not limit transcendental inquiry to scientific

our experience in a phenomenological manner, the latter remains in the natural attitude insofar as she takes the mind to be existing in the world. In order to move to the transcendental attitude, one must put into effect the "unconditioned epoché", which brackets all existence including the existence of the mind. This in turn brings us back to *transcendental-phenomenological experience*. (Here, in order to properly distinguish the method that separates transcendental phenomenology from psychological phenomenology, we can call the bracketing of *all* existence and the return to transcendental experience, 'transcendental epoché' and 'transcendental reduction' respectively.) Accordingly, attending to one's experience in a phenomenological manner can have the following two meanings: (1) attending to the intentional structures of our experience, which belongs to the mind (*psychological-phenomenological reflection*), or (2) attending to the intentional structures of our experience just in the way they present themselves to us, without any regard as to *where* this experience may be or *what* they may be apart from how they appear to us (*transcendental-phenomenological reflection*).

3 Disciplinary Layers

The above exposition has prepared the grounds for us to finally introduce the formalism of the 'bracketing' that we have promised. Any formalizing requires abstraction. So let us take a step back and make a few observations that would lead us to our formalism.

Nature arranges itself in a series of organizational levels, from its atoms near the bottom, to the biosphere near the top. The identification of such layered structure is omnipresent in philosophy and natural sciences, from the

---

knowledge or cognition but expanded it to our experience in general), but he did find it problematic that Kant did not have a way of providing sufficient evidence for the transcendental conditions. In other words, Kant sought the conditions of possibility for our cognition through transcendental arguments that bear no direct, intuitive evidence. This is why Husserl (1970, 115–116) called Kant's transcendental method "a mythically, constructively inferring [*schliessende*] method" and claimed that if he had not been bound to the naturalistic psychology of his time and had allowed himself to seek the proper intuitive method, then Kant would have discovered "a thoroughly intuitively disclosing [*erschliessende*] method, intuitive in its point of departure and in everything it discloses". To be sure, with the latter, Husserl is alluding to his own phenomenological method. Therefore, the main difference between their transcendental methods lies in Husserl's insistence on the phenomenological method and specifically, its appeal to intuitive evidence.

"layer-cake" (Oppenheim and Putnam 1958), to "levels of mechanisms" (Craver and Bechtel 2007), to the multilevel selection theory of cooperation in biology (Wilson and Wilson 2008). The reality of the world as captured by our senses also appears to divide itself into similar layers of organization. We can identify some of these layers in our example where I was sitting in my chair looking outside at the snow, and then, upon reflection, a change of events occurred where I was no longer attending to the snow itself, but to my own experience of it. There is, first of all, the existence of physical objects, like the chair and the snow, which constitute a layer. In our practical lives, we are involved with these objects in one way or another: I am *sitting in* the chair, *looking at* the snow. These experiences of the objects constitute another layer. But we can further identify yet another layer, which is the layer of reflection that allows us to distinguish our own experiences from the objects that are experienced.

In our approach, we propose that each layer should correspond to its own discipline of study, and thus be described by a particular language and structured by particular logical rules describing which relations are possible or not. In Figure 6.1, we show a matrix representing the three layers involved in the study of our experience, as well as its major studied entities. Each row corresponds to a discipline of study in the sciences or philosophy (physics, phenomenological psychology, and transcendental phenomenology), and each column is an entity (object, experience, and reflection). We will refer to this matrix as *the experience levels matrix*, or simply, the matrix. It should be noted, however, that the three layers we are introducing are by no means exhaustive of the levels of our experience. We can break down the levels into more, or more specific, categories, which would have their corresponding disciplines. However, we have limited ourselves to these three layers for the purpose of distinguishing the different kinds of reflection and the changes they bring about, and, more specifically, to clarify some of the misunderstandings and ambiguities surrounding phenomenological reflection.

We define the matrix of experience levels, as shown in Figure 6.1. To construct this matrix, we make use of a pseudo-mathematical notation, which is meant to clarify the relations among all entities (i.e., columns), when referred to in each of the languages corresponding to each layer (i.e., row). This notation allows us to write expressions referring to phenomena in a given layer, i.e., entities within the language of a discipline. An o refers to the object of an experience, whereas an x symbolizes a (subjective) experience, which can be either an experience of an object, or an experience of another experience (reflection). We introduce the notation $\circ$ to represent the composition of an experience with its object. This should be considered together with the experience, as an operator on an object o, denoted $x\circ$. From there, $x \circ o$ denotes the experience x of an object o. We also introduce the notation *[]* for the bracketing understood in the phenomenological sense of the epoché. A bracketed expression indicates the suspension of judgment over the existence of the enclosed entity. For example, *[o]* with brackets in the expression *x∘[o]* indicates that the existence of the object of experience *x* is neither assumed to exist nor assumed not to exist. Lastly, we use the apostrophe such as in *x'∘[o]* instead of *x∘[o]*, to express that experience *x'* is distinct from experience *x*. Note that all these conventions are meant to form a helpful notation and not as a fully-fledged mathematical theory.

	Object	Experience	Reflection
Physics	o Physical object	x·o Mental state	x·x·o Introspection
Phenomenological psychology	[o] Object as it is intended	x'·[o] Phenomenological experience	x'·x'·[o] Phenomenological reflection
Transcendental phenomenology	[o] Object as it is intended	[x']·[o] Pure phenomenological experience	x"·[x']·[o] Transcendental-phenomenological reflection

Figure 6.1: Experience levels matrix.
This diagram represents the matrix of levels of experience. It is structured to display a level of experience in each row, corresponding to disciplines of study, and an object of study in each column, corresponding to the studied entities: the object of an experience, the experience itself, and the reflection on the experience.

The matrix is constructed so the elements that lie on the main diagonal – from the top left to the bottom right, indicated with greyed cells in Figure 6.1, for which the index of the row equals the index of the column – are representative of the respective discipline in the sense that they are studied primarily by that discipline. Physical objects are studied primarily by physics, phenomenological experience (that has just bracketed the *o*) is studied by phenomenological psychology, and while what is studied in transcendental phenomenology is phenomenological experience purified of all existence (hence the name '*pure phenomenological experience*'), transcendental-phenomenological reflection is nonetheless representative of transcendental phenomenology in the sense that this is the unique place where the element is studied. These diagonal elements allow us to construct the rest of the matrix, as the rest of the elements of each row naturally follow from them, as will be explained later, when each layer is detailed.

The most important feature in the matrix is the function of bracketing. Bracketing accomplishes the move from one diagonal element to the next one, and more generally, from one layer to another. Importantly, each of these moves, which is accompanied by a shift of attention, is not a move within a given field of study, but is a jump operating a translation from one discipline to another, which should therefore not be treated lightly. This also means that the shift of attention, when rightly understood, leads to different disciplines. To move from the study of physical objects to the study of experiences, understood in the phenomenological sense, one brackets the existence of physical objects, leading to the middle element of the diagonal, at the center of the matrix. This bracketing of the physical object allows the phenomenologist to attend to and study the object, not as a physical object, but just as it is intended. This is the kind of move any phenomenologist would make. However, when the experience in question is understood as belonging to a mind that exists in the world, then, as we said in the previous section, this would be to stay within the realm of phenomenological psychology. Phenomenological psychologists attend to and study the intentional structures of our experience for their own sake, independently of what physicists have to say about the physical world. And as psychologists, they maintain the existence of the mind, embedded firmly against the backdrop of the world. But while their aim is to establish the foundation for empirical psychology, transcendental phenomenology has a much more ambitious, and specifically philosophical, aim of establishing the foundation for all sciences. As such, transcendental phenomenology must further bracket the existence of the mind to secure the field proper to transcendental phenomenology, rid of all presuppositions deriving from our belief in the existence of the mind and the world. This is how the move from the middle layer to the bottom layer is achieved. It should be noted that although in Figure 6.1, the bracketing of the existence of the mind corresponds to bracketing the existence of the experience $[x']$, it is more precise to say that what we are bracketing here is the existence of the mind (to which the experience is thought to belong) and not so much the existence of the experience *per se*.

With the main element fixed in each layer, we infer the other elements on the line, depending on their respective column. We start from the physical world layer, on the top row of the matrix. This layer corresponds to the discipline of physics, whose language deals with the description of the relations between physical objects, as they can be observed by subjects in the world. A physical object o can be observed by a certain subject, making for a subjective experience. Within the physical layer, we refer to such phenomenon as a mental state, studied in physical terms and which we denote $x \circ o$. When one refers to the experience of reflecting on an experience, we denote this as $x \circ x \circ o$ and refer to such phenomenon as introspection. As we are using a language describing physical phenomena, *all* references to an experience x are physical mental states, which could mean, for example, complete brain states at a given time. Note that in $x \circ x \circ o$, the reflection on experience and the experience of reflection are both denoted by x, the experience in the physical layer, or a mental state. This expression remains in the same language describing physical phenomena, and by no means is a metalanguage of a higher-order description. Therefore, we define introspection as a one-layered concept, where all elements pertain to the language of physics.[12]

The middle layer corresponds to the discipline of phenomenological psychology. The physical mental state $x \circ o$ corresponds to a subject experiencing a physical object o. But since we bracket the existence of the object when moving to the level of phenomenological psychology, at this layer, $x \circ o$ translates to a subject experiencing a bracketed object, i.e., the object as it is intended. Moreover, since the resulting experience is the phenomenologically reduced (in the technical sense of the phenomenological reduction as the method of 'going back') experience that now has the object as it is intended, rather than the physical object, as its correlate, we denote the experience in the middle layer as x' and distinguish it from the x of the physical layer. We note that this distinction is crucial for a careful translation between the layers and has the effect of safeguarding with respect to the x, which may or may not correspond to the same operation as the previous x. For example, the phenomenological experience of looking at the snow may not be equivalent to the physical mental state of the subject looking at the snow. It may well be, but it may not. This is something that a good phenomenologist would not take a stance on since to say that a phenomenological experience just is a physical mental state would be to assume a metaphysical position about the nature of our experience, which is something beyond what phenomenology claims to do. In our notation, we have followed the phenomenologists and made sure to distinguish x and x'. Therefore, phenomenological experience is denoted as $x' \circ [o]$ and phenomenological reflection, $x' \circ x' \circ [o]$.

The last layer, at the bottom of the matrix, corresponds to transcendental phenomenology. This layer is constructed by bracketing the existence of the mind to

---

[12] Physics is here intended in the broad sense, as the discipline of science studying the nature and properties of matter, its motion and behavior through space and time. We consider that empirical psychology – which defines concepts such as introspection – is part of physics taken in this broad sense, concerned with observable physical phenomena relating to mental states and human behavior.

which it was presupposed the experience belongs. This means that we translate phenomenological experience $x'\circ[o]$ in the middle layer to $[x']\circ[o]$ (which in Figure 6.1 we have called 'pure phenomenological experience' to designate that this experience is 'pure' of all existential commitments) and phenomenological reflection $x'\circ x'\circ[o]$ to the corresponding concept $x'\circ[x']\circ[o]$ in transcendental phenomenology, naturally referred to as transcendental-phenomenological reflection. However, just as the phenomenologically reduced experience x' had to be distinguished from x, i.e., the experience understood in the physical layer, here too, we must be careful to distinguish the transcendentally reduced experience from the phenomenological experience of the middle layer. Put differently, since transcendental-phenomenological reflection is freed from the belief that experience belongs to the mind (which exists in the world), we need to carefully distinguish *this* experience from the experience in the middle layer. Therefore, we distinguish x' from x'' and rename the expression to $x''\circ[x']\circ[o]$.[13]

One may note that this operation is repeatable beyond these three layers, as it is perfectly possible to create additional layers which could be relevant to more disciplines of study. As we said earlier, however, we have limited our scope to these three layers for the purpose of presenting a methodology of comparative study and clarification of the process of translation between fields related to the phenomenological study of experience. One can also note that in general, when one moves down a layer in the matrix, the preserved elements are bracketed, and the newly created elements are indicated with a prime symbol (e.g., x becomes x'). This notation is helpful in keeping track of which structures are preserved from the other layers and which are unique to the new layer, thereby enabling a smooth translation between the layers.

---

13 It should be noted that even though we bracketed the x and the o separately in our notation for the kind of experience transcendental phenomenology deals with ($[x']\circ[o]$), we can also bracket them together ($[x'\circ o]$). We opted for the separate bracket notation because, as we explained in Section 2, one can move to transcendental phenomenology by way of going through phenomenological psychology. In other words, a phenomenological psychologist could become a transcendental phenomenologist by further bracketing the o, which she had left unbracketed (this move is what Husserl called the "unconditioned epoché"). However, a transcendental phenomenologist, having a strong transcendental interest, could bracket the existence of both the mind x and the world o at once if she wanted to. In this sense, in our matrix, it is not necessary for one to go through the middle layer to get to the bottom one, though transcendental phenomenology always contains within it the possibility of becoming a psychological discipline. Furthermore, whether bracketed separately or together should make no substantial difference in terms of what the transcendental experience would offer.

4 Discussion

Let us now make a few observations based on analogies from the natural language as well as the languages of mathematics and computer science.

Every time we bracket an entity in a translation between layers, we create what a programmer may choose to call a *virtual variable* in the new layer. This signals that this newly created entity no longer answers to the same rules as the previous entity, but does keep some of its characteristics. The virtual variable ultimately lives within the logical rules of the new layer, which are distinct from those of the original layer where its existence was well defined This allows one to now look at the variable independently of the discipline to which it originally pertains.

We can also observe that this process of virtualization is reminiscent of what the practice of quoting achieves. Indeed, Søren Overgaard compares bracketing to the practice of quoting:

> By writing, Karen said: 'Snow is white' I have asserted that someone said something about the colour of snow. I have not asserted anything about the colour of snow. I have neither affirmed nor denied that snow is white, nor have I surmised, questioned, or doubted anything about the colour of snow. One may of course quote something in order to show its truth, falsity, and so on, but the point is that, in and of itself, to quote is to do none of these things. Furthermore, in order to produce an accurate quote you must focus on faithfully recounting *what the person said*, as opposed to what you might know or believe about the subject matter the person spoke about. (Overgaard 2015, 191)

Bracketing is a similar process of preserving what the experience 'says', while refraining from making any judgment regarding the various commitments made within the quoted experience. In this way, phenomenologists are "impartial reporters of our own experience" (Overgaard 2015, 191) or what Husserl calls "disinterested spectators" (1970, §45, §69). Similarly, in the translation that we perform to switch between layers in the matrix, the bracketing's virtualization preserves the reference to the corresponding element in the original layer, while freeing it from the constraints of its own layer. This operation is necessary because the language proper to a layer may imply different structural relationships from another layer, which after the translation may otherwise be subject to misinterpretation. When we bracket the existence of objects in the middle layer, for example, this gives us the freedom to investigate the object just in the way it appears to us in our experience, regardless of whether the object really exists or not.

Mathematical notations often serve as connectors between disciplinary fields, helping to expand and better translate between known theories. We may see the operation of translating from one layer to another as a *change of variables* in mathematics, a technique used to reformulate problems in which the original variables are substituted for a combination of new variables. The intent is typically that when expressed in terms of new variables, the problem may become simpler for certain purposes, or sometimes become similar to some better understood problem. The problems connected to the structures of experience are highly interdisciplinary, and similarly lend themselves to such reformulations in various disciplines of study. Since each discipline has different goals, and has developed a specific language to help address them, the hope is that the translation exercise may therefore help 'solve' the problem of experience, by taking advantage of specific disciplines offering the tools to solve specific aspects of the problem.

In the formalism we introduced above, we renamed certain notions from one layer (or row) to the other (e.g., from x to x'), and used bracketing for others (e.g., $[o]$). Both these changes add (and may even remove some) degrees of freedom to the way the notion may be handled in the other layer. While the first adds the possibility for the variable to be non-identical in the new layer's framing, the latter adds the possibility of the notion to either exist or not exist. We note that our notation deceivingly presents the phenomenological epoché and reduction as two distinct operations. However, the phenomenological epoché and reduction, which together allow for a translation from the physical layer to the phenomenological psychology layer, are one and the same operation, like two faces of the same coin. (And likewise for the transcendental epoché and reduction, which together allow for a translation from the phenomenological psychology layer to the transcendental phenomenology layer.) It would be ideal to find a notation that elegantly unifies both operations, the epoché and reduction, although we find this to be a minor caveat that should not restrict its clarifying power.

We note another mathematical analogy. Above, we chose to name our diagram a matrix, in reference to the mathematical object that contains elements from a defined set, on which a set of operations is possible, typically addition and multiplication. More generally, in mathematics, a matrix is defined over a mathematical field of elements, on which two operations are defined – not necessarily addition and multiplication as we know them, although they are typical examples if the set elements are rational numbers – and a series of properties called field axioms must be respected. One possible way of defining the field in our case can be done by considering a set of elements with the following operations: 'experience of' (which allows, for example, to move horizontally from the physical object of an experience to the mental state connected to its experience) and 'translated into the adjacent layer' (which allows, for example, to move vertically from a mental state to a phenomenological experience). We remark that the matricial notation may suggest the existence of more similarities with the mathematical object, although our intention was originally to clarify concepts between fields of study. It may be interesting to push the metaphor further to explore the fruits it may bear, in terms of new connections, relations, or conjectures the intersecting of experiential sciences with mathematics may reveal. The development of such conjectures is beyond the scope of this paper, but we intend to explore them further in the future.

Moving on from mathematics to computer science, we observe that the structure of the Internet may offer a complementary perspective to this model. In the context of computer networks, the so-called Open Systems Interconnection (OSI) model (cf. Figure 6.2) characterizes all communications that occur, from the moment an individual sends the message 'Hello!' via an application on their smartphone, to the moment it reaches its destination on another smartphone. Both the source and the destination are able to read the message in English. In between, however, the message will be *encapsulated* and *decapsulated*, at least partially, several times, in the process of translation into the language or protocol proper to each layer (e.g., the application layer uses a protocol called HTTP, while the transport layer uses TCP). At each step downwards, before reaching the next layer, the message is being quoted so to speak, by adding keywords around it such that it can be preserved in the underlying layer. This is done from the top layer in Figure 6.2, which corresponds to the level of a message written in English on a smartphone application, all the way down to the physical layer, which consists of signals exchanged in binary over electrical wires and wireless networks. Each step upwards then decapsulates the message by removing the headers and trailers that surround it, thus effectively translating it back to the language of the layer above, until it is back to the English language. This computational view of a layered model with multiple interfaces of translation is similar to the one we have brought up in this paper in the context of the study of experiences. Here too, we can see a comparable operation to the bracketing, preserving a message in a different layer, which relies on a distinct set of logical rules, mechanisms, and specific language. And here too, we can see the importance of a careful distinction between the literal and the bracketed messages not being actionable using their original rules outside the context of their original layer, but nevertheless being usable by the virtue of the bracketing and clearly distinguishing functions between layers.

This computational perspective also highlights the importance of studying the transfer of information while describing the relation between elements involved in subjective experience. Information not only lives in the metalanguage – the language one layer downstream in the matrix – but also directly affects the layer itself. This is also true for the physical layer, where information is already instantiated and can influence the dynamics of physical systems (Deutsch 2013). When looking at transfers between layers, one notices similar effects where the same information is being reframed and takes on a new interpretation. For example, let us consider two models describing the dynamics of liquid water, the first one based on the behavior of its molecules, and the second in terms of equations of fluid mechanics. While it is possible to transfer the information from one level of description to the other, the variables considered in the first model (e.g., position, velocity, and angular velocity of each molecule) will be drastically different from the second (e.g., viscosity, temperature, and pressure of the fluid). In every layer, the key to understanding the emergent rules resides in reconstituting the complete picture where each part of the system has access to appropriate information in its own context (Wiener 1961). It is therefore crucial to keep track of which bits of information are preserved or modified through translations between layers. One possible application is that by doing so, one may learn about computational complexity constraints involved in moving through a series of embedded layers (Lloyd 2012), which may in turn be more informative than proofs of uncomputability (Aaronson 2012). An advantage of such information perspective is in the way it can capture contingencies between elements of different layers.

This type of computational model is based on a series of consecutive layers building step by step to different perspectives. Each layer corresponds to a different level of description that forms a complete model of reality. From a layer, one may move directly to a neighboring layer, using a well-defined translation process on the elements and operators of the layer's language, to the neighboring layer's homologues. In order to move to a more remote layer, it is necessary to successively compose such translations together, back to back.

Depending on how such neighbor-to-neighbor translations are defined, any intrinsic constraint raised at any layer of the hierarchy, will automatically bring corresponding constraints into existence in all other layers of the model. This suggests the existence of a network of causal relations, which connects with the discussion about levels of organizations we mentioned at the beginning of Section 3. Close to such considerations live the controversial discussions about the 'explanatory gap' (Levine 1983) between the phenomenal aspects of our experience and the underlying physical states, and downward causation (Campbell 1974), where the question is whether higher-level entities or properties can exert causal influence on lower-level ones. The unidirectionality of the downward causation arrow is not evident. Craver and Bechtel (2007, 559–560) discuss the illusion of causation in an organism at the top level of a hierarchy, which in spite of seemingly having an effect on its constitutive parts at the bottom chemical level, turns out to be due to other bottom-level effects in the first place. There would thus just

Chapter 6 Different Ways of Attending to Experience — 155

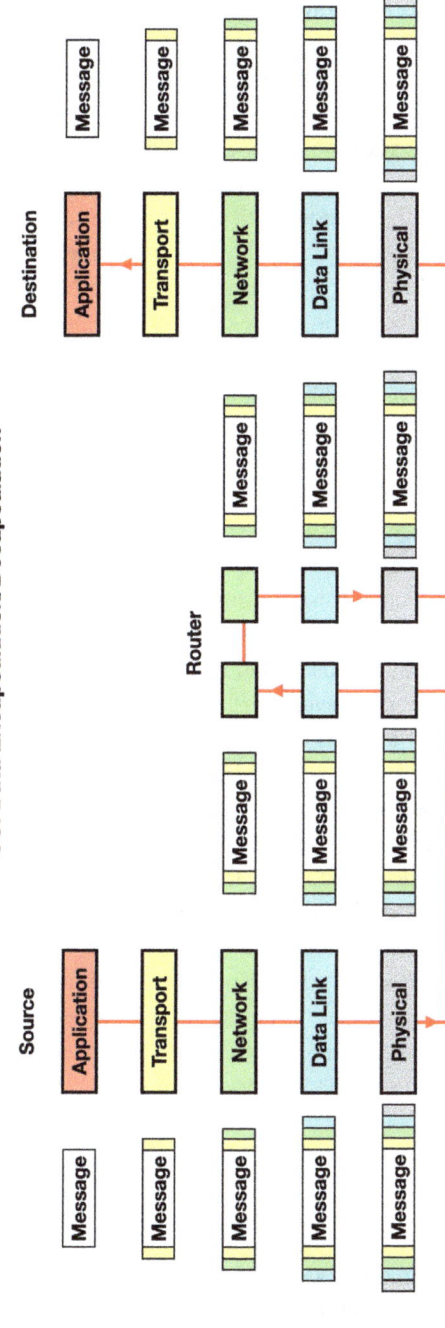

Figure 6.2: OSI Data Encapsulation/Decapsulation.
This diagram shows the Open Systems Interconnection model of the Internet, exemplifying the process of sending a message that will be successively encapsulated and decapsulated. The key concepts of this computational model, as well as several others imported from mathematics, shed light on the importance of bracketing and distinguishing mechanisms between layers.

be regular same-level causal relationships that have "mechanistically mediated" effects downwards, due to constraints defined on relationships between elements in different levels. A different way to address such downward causation is to consider that processes at the lower levels of a hierarchy are constrained by and act in conformity to the laws of the higher level (Campbell 1974, 180). For example, in DNA, the specific way that amino acids code for functions and the way the proteins they form fold at the molecular level both strongly depend on environmental conditions which wound up selecting for the specific coding system in the context of the units of selection, and external factors in the cell, such as the presence of molecular chaperones that facilitate folding. It should be noted that although we did not deal with these issues in this paper, there has been much discussion about how phenomenology may contribute to these discussions (Roy et al. 1999; Thompson 2007; Kirchhoff and Hutto 2016).

In the model we propose relating physics, psychology, and transcendental phenomenology, there is no preferential direction for a causation arrow. The model may be constructed starting from any of the three layers, to move to the other ones, with no sense of precedence or supervenience of either of them over the others. This means that one may reason starting from their experience of the experience of objects (reflection), as well as from objects of such experiences, as the neighboring layers may perfectly well be reconstructed from any starting point. Furthermore, there is no reason why starting in reverse order, from transcendental experience to physics, should not feel just as natural as the opposite way around. The only way to find out about our experience is to turn our attention towards it, which can only be done by using a proper metalanguage to describe specific relations among experiences. Next, one may use the language of phenomenological psychology to describe subjective experiences of objects. The key point is that certain constraints may only appear from the level of description of subjective experiences to the interactions between fundamental material objects of our universe. This should not come as a surprise, as our knowledge of physics in the first place comes from constructing a shared language based on our subjective experience of the world. Physics, in that sense, is intrinsically intersubjective, and always implicitly incorporates the structures of experiences in its theory.

5 Conclusion

We started this paper by pointing out three ways of understanding what it may mean to reflect on one's experience of looking at the snow: (1) an instance of *introspection* where a physical self is attending to its own experience, (2) an instance of *psychological reflection* where a psychological self is attending to its own experience in a phenomenological manner, and (3) an instance of *transcendental-phenomenological reflection* where a transcendental-phenomenological self is attending to its own experience. The notation that we introduced in the experience levels matrix clarifies the distinction very clearly. Reflection is understood as a form of introspection – one attends to one's experience introspectively – when both the reflected on experience and reflection are understood in physical terms, thereby remaining in the single layer of physics ($x \circ x \circ o$). Phenomenological reflection of the psychological sort operates over two layers between the phenomenological experience and the physical object, and they can do so legitimately by bracketing the existence of the physical object. Therefore, the phenomenological psychologist attends to the intentional structures of our experience, studying both the modes of consciousness and the objects as they are intended. The distinction between this kind of reflection *($x' \circ x' \circ [o]$)* and transcendental-phenomenological reflection *($x'' \circ [x'] \circ [o]$)* becomes evident when we see that transcendental-phenomenological reflection, unlike psychological-phenomenological reflection, no longer operates with the identity of the reflected on experience with itself as the mind that exists in the world. Having bracketed the existence of the mind presupposed by phenomenological psychologists, the transcendental phenomenologist operates with three layers, the transcendental, the psychological, and the physical, and legitimately so by bracketing both the existence of the mind (psychology) and the world (physics).[14] In this way, the transcendental phenomenologist attends to the intentional structures of our experience with a view of understanding how the meaning and validity of objects and the world are constituted in our experience. Besides other suggestions based on analogies from mathematics and computer science, the matrix also clarifies the conditions in which variables and operators require careful translation, which is a key process in a multidisciplinary field such as the study of one's experience. We hope that our formalism of the bracketing, together with the matrix, will open up constructive conversations between disciplines and contribute towards a more comprehensive understanding of the structures of our experience.

---

[14] One could argue that transcendental-phenomenological reflection only operates with two layers, the transcendental and the psychophysical. This is justifiable when one bypasses the psychological layer to move to the transcendental layer, as noted in footnote 13. We have here defined transcendental-phenomenological reflection as operating with three layers for the purpose of distinguishing it from psychological-phenomenological reflection.

References

Aaronson, S. (2012): "Why Philosophers Should Care About Computational Complexity". In: B.J. Copeland; C. Posy; O. Shagrir (eds.): *Computability: Gödel, Turing, Church, and Beyond*. Cambridge, MA: MIT Press, 1–50.

Breyer, T. (forthcoming): "Reflection, Attention, and the Practice of Phenomenology". In: O. Davis; A. Fuentes; D. Schilling (eds.): *FaberSocius: Integrating Humanity*. Cambridge, MA: Cambridge University Press.

Campbell, D. (1974): "Downward Causation in Hierarchically Organised Biological Systems". In: F.J. Ayala; T. Dobzhansky (eds.): *Studies in the Philosophy of Biology: Reduction and Related Problems*. London, UK: Macmillan, 179–186.

Craver, C.F.; Bechtel, W. (2007): "Top-Down Causation Without Top-Down Causes". *Biology and Philosophy* 22:4, 547–563.

Dennett, D.C. (1987): *The Intentional Stance*. Cambridge, MA: MIT Press.

Dennett, D.C. (1991): *Consciousness Explained*. Boston, MA: Little, Brown and Company.

Depraz, N. (1999): "The Phenomenological Reduction As Praxis". *Journal of Consciousness Studies* 6:2–3, 95–110.

Depraz, N.; Varela, F.J.; Vermersch, P. (2003): *On Becoming Aware: A Pragmatics of Experiencing*. Amsterdam: John Benjamins Publishing Company.

Deutsch, D. (2013): "Constructor Theory". *Synthese* 190:18, 4331–4359.

Gutland, C. (2018): "Husserlian Phenomenology as a Kind of Introspection". *Frontiers in Psychology* 9, 896.

Heidegger, M. (1985): *History of the Concept of Time: Prolegomena*, trans. by T. Kisiel. Bloomington: Indiana University Press.

Husserl, E. (1970): *The Crisis of European Sciences and Transcendental Phenomenology: An Introduction to Phenomenological Philosophy*, trans. by D. Carr. Evanston: Northwestern University Press.

Husserl, E. (1983): *Ideas Pertaining to a Pure Phenomenology and to a Phenomenological Philosophy: First Book*, trans. by F. Kersten. The Hague: Martinus Nijhoff.

Husserl, E. (1997): *Psychological and Transcendental Phenomenology and the Confrontation with Heidegger (1927–1931)*, trans. by T. Sheehan; R.E. Palmer. Dordrecht: Kluwer Academic Publishers.

Ishihara, Y. (2016): *Transcendental Philosophy and Its Transformations: Heidegger and Nishida's Critical Engagements with Transcendental Philosophy in the Later 1920s*. PhD dissertation: University of Copenhagen. http://curis.ku.dk/ws/files/170161777/Ph.d._af handling_2016_Ishihara.pdf

Kirchhoff, M.; Hutto, D. (2016): "Never Mind the Gap: Neurophenomenology, Radical Enactivism and the Hard Problem of Consciousness". *Constructivist Foundations* 11, 346–353.

Lenkowski, W.J. (1978): "What is Husserl's Epoche?: The Problem of the Beginning of Philosophy in a Husserlian Context". *Man and World* 11, 299–323.

Levine, J. (1983): "Materialism and Qualia: The Explanatory Gap". *Pacific Philosophical Quarterly* 64, 354–361.

Lloyd, S. (2012): "A Turing Test for Free Will". *Philosophical Transactions of the Royal Society A: Mathematical, Physical and Engineering Sciences* 370:1971, 3597–3610.

Oppenheim, P.; Putnam, H. (1958): "Unity of Science as a Working Hypothesis". In: H. Feigl; M. Scriven; G. Maxwell (eds.): *Concepts, Theories, and the Mind-Body Problem*. Minneapolis: University of Minnesota Press, 3–36.

Overgaard, S. (2015): "How to Do Things with Brackets: The Epoché Explained". *Continental Philosophical Review* 48, 179–195.

Roy, J-M.; Petitot, J.; Pachoud, B.; Varela, F.J. (1999): "Beyond the Gap: An Introduction to Naturalizing Phenomenology". In: J. Petitot; F.J. Varela; J. Pachoud; J-M. Roy (eds.): *Naturalizing Phenomenology: Issues in Contemporary Phenomenology and Cognitive Science*. Stanford: Stanford University Press, 1–80.

Schwitzgebel, E. (2019): "Introspection". In: E.N. Zalta (ed.): *The Stanford Encyclopedia of Philosophy* (Winter 2019 Edition). https://plato.stanford.edu/archives/win2019/entries/introspection/

Thompson, E. (2007): *Mind in Life: Biology, Phenomenology and the Sciences of Mind*. Cambridge,MA/London: Harvard University Press.

Wiener, N. (1961): *Cybernetics or Control and Communication in the Animal and the Machine*. Vol. 25. Cambridge, MA: MIT Press.

Wilson, D.S.; Wilson, E.O. (2008): "Evolution 'for the Good of the Group': The Process Known as Group Selection Was Once Accepted Unthinkingly, Then Was Widely Discredited; It's Time for a More Discriminating Assessment". *American Scientist* 96:5, 380–389.

Zahavi, D. (2005): *Subjectivity and Selfhood. Investigating the First-Person Perspective*. Cambridge, MA: MIT Press.

Zahavi, D. (2007): "Killing the Straw Man: Dennett and Phenomenology". *Phenomenology and the Cognitive Sciences* 6:1–2, 21–43.

Zahavi D. (2015): "Phenomenology of Reflection". In: A. Staiti (ed.): *Commentary on Husserl's Ideas I*. Berlin/ Boston,MA: De Gruyter, 177–194.

Short Biography

Yuko Ishihara, PhD, is an associate professor at the College of Global Liberal Arts at Ritsumeikan University in Osaka, Japan. Her research areas include the Kyoto School philosophy and classical phenomenology, with a specific interest in transcendental philosophy. She completed her PhD dissertation in 2016 on a comparative study of Heidegger's and Nishida Kitaro's critical engagements with transcendental philosophy in the late 1920s. Her recent research focuses on the topic of play and how modern philosophers, both eastern and western, have turned to the notion of play to overcome the metaphysics of subjectivity.

Olaf Witkowski, PhD, is the director of research at Cross Labs in Kyoto, Japan, a research institute in machine intelligence, cognitive science, and artificial life. He received his PhD in computer science from the University of Tokyo in 2015, co-founded various ventures in science and technology on three continents, and is an elected member of the board of directors of the International Society for Artificial Life. He is also a research scientist at the Tokyo Institute of Technology, a lecturer at the University of Tokyo, and a regular visitor at the Institute for Advanced Study in Princeton. His research specializes in collective intelligence and open-ended evolution.

Part 2: **Attention and Mediation**

Elizaveta Solomonova and Michelle Carr
Chapter 7
The Role of Attention and Intention in Dreams

Abstract: In this chapter, we review the different ways that attention works in relation *to* dreams and how it may function *in* dreams, and apply the framework of attention, proposed in this volume – as a means of accessing and mediating interactions with the world – to the dreaming world. We first review prior work on the role of attention as 1) access to dreams, e.g., how practices of recording and sharing dreams act as enabling factors for improving dream recall and enhancing richness of dream experience; and as 2) a mediator of dreams, e.g., how incubation, imagery rehearsal, and ultimately lucidity can be cultivated as cognitive skills enabling agency in the dream experience. We propose that attention functions as a constitutive factor in dream experience and that it is a trainable, developmental cognitive skill. We argue that dreams are not simply experiences that happen to the dreamer, rather, through employing attentional techniques in various ways, the dreamer may cultivate different degrees of agency in the dream.

Introduction

> *I was walking along the seafront. It was snowing lightly. The sea was frozen over. I decided to skate on the frozen sea and went down to the beach. I was suddenly wearing ice skates. I started skating out over the sea in a straight line away from the coast line. I reached an island and took a rest. It was very warm but the sea was still frozen. I laid down on the white sand in the sun.* (Carr et al. 2020)

Traditionally, dreams have been seen as experiences that one cannot control, as something that happens *to* the dreamer (at times through involvement of supernatural powers), without the dreamer's permission, volition or agency. This view was famously challenged in the advent of psychoanalysis: in his *Interpretation of Dreams*, Freud proposed that while we may not be consciously in control of our dreams, our unconscious mind is actively constructing dream content, and that dream content is symptomatic of our repressed, accumulated neuroses (Freud 2010 [1900]). This shift in perspective signaled that dreams may in fact be subject to individual experience and to one's mental state (which is something that is

https://doi.org/10.1515/9783110647242-008

possible to change, or even to control), therefore bringing, at least partially, responsibility for dreams to the dreamers themselves. Emerging neuroscience of dreams, in 1970s, however, adopted a more conservative behaviorist position, and for a relatively long time the dominant view has been one of neuro-reductionism, where dreams were seen as random hallucinatory products of the activity of the sleeping brain (Hobson and McCarley 1977). In recent years, a more nuanced picture of dreams is gradually emerging. Research from psychology, philosophy and anthropology converges on the idea that dreams may be individual or even collective practices rather than uncontrolled brain events. Developmentally and temporally, dreaming can be recognized as cognitive achievement (alongside other cognitive abilities, such as memory, perception, attention, etc.) (Foulkes 2014). And dream qualities, including what is possible in the dream state, how rich the dream experience is, and how well the dream will be remembered, may change as a result of attentional practices during waking hours. Research on dream incubation, dream sharing and lucid dreams shows that the dreamer is an active participant and co-creator of their dream life, and that the dreamer's agency, awareness and degrees of control are all dynamic, continuous and potentially trainable skills. Further, in line with work on 4E cognition (Menary, 2010) and following evidence from sensory incorporation studies (Nielsen 1993, 2017; Sauvageau et al. 1998), it has been proposed that dreams are not simply experiences of virtual reality confined in the sleeping brain, but rather can be conceptualized as processes of *embodied imagination* (Thompson 2014; Solomonova and Sha 2016; Solomonova 2017), rooted in lived sensorimotor experience and responsive to sensory information from the outside world.

This chapter is separated into two sections, treating attention in the context of dreaming: 1) attention *to* dreams as a means of access to dream state; and 2) attention *in* dreams as a mediator of dream experience. We argue that attention is a cross-state cognitive skill that allows for creation and modulation of dream experience, and that attention helps elucidate aspects of agency in dreaming. We argue that dreams are not simply something that passively happens to the dreamer, but rather that dreams are processes of creative embodied and performative engagement with the oneiric world.

1 Attention: Access to the Dream State

In this section we will discuss how attention facilitates and improves access to the dream state. First, we will show how attention *to* dreams (or to memory of dreams) during waking state shapes our capacity to have vivid dreams and our ability to report and describe the dream experience. Second, we will review work on attention *in* dreams during sleep, and will suggest that attentional processes during sleep reveal quasi-perceptual creative processes of dream imagination.

1.1 Paying Attention to Dreams During Waking Life

In researching dreams one of the fundamental methodological difficulties lies in the fact that we do not have direct access to the dream state; instead, we rely on the dreamer's memory for the dream experience and on the dreamer's capacity to report on their experience. A number of things must happen before dream researchers have access to a dream narrative, and all of them rely on various forms of attention and memory. First, the dreamer must be able to remember the dream as it happens during sleep. To do so, the dreamer's attentional skills during sleep must be sufficiently active in order to be able to be engaged in the dream scenario and to commit the dream experience to memory. In the majority of dreams these attentional skills do not require recognizing that the dream is a dream. The dreamer then largely makes use of habitual patterns of attending to the world, applying them to the dream experience. The dream state, however, has its own rules, and attention in dreams must adapt to the altered neurobiological state of sleep. We will discuss some of the unique phenomenology of the dream experience further in this chapter. Second, the dreamer must change neurobiological state and transition from sleep into wakefulness. During this process there is risk that the dream may be forgotten or distorted due to interference from other stimuli. Third, the dreamer must be able to narrate the dream bearing in mind the dream's temporal structure, emotional tone, visual, auditory, somatosensory and other aspects. This requires sustained awake attention to the fragile and quickly dissipating memory of the dream. Little work to date has focused on attention in context of dreams, likely due to the limitation that it is nearly impossible (unless one is in a lucid dream, which is discussed in Section 2 of the present chapter) to study attention experimentally in the dreaming subject.

In relation to neurocognitive bases of dream recall, some evidence points to the notion that remembering dreams may be a function of an individual attentional profile. For instance, Conduit and colleagues (Conduit et al. 2000), proposed that dream recall is not a function simply of having had or not having had an experience during sleep, but is dependent on activation of attentional circuits during dreams, which, then, enables the dream to be attended to, encoded in memory and then remembered upon awakening. Similarly, work in Ruby's group revealed that individuals who claim to have higher frequency of dream recall, exhibit specific traits in brain activity and brain architecture in areas associated with attentional, motivational and memory processes (higher white matter density and cerebral blood flow in medial prefrontal cortex, higher cerebral blood flow in temporo-parietal junction) (Eichenlaub et al. 2014b; Vallat et al. 2018), as well as larger attentional-orienting responses (evoked potentials, brain signatures signaling that an experimental stimulus was attended to) during wake (Eichenlaub et al. 2014a).

Empirical research on dream recall, as well as a wealth of popular wisdom, shows that attending to dreams (whether by keeping a dream journal or by cultivating positive attitudes towards dreams) improves dream recall (Aspy 2016; Beaulieu-Prevost et al. 2004, 2007; Schredl 2002). In other words, paying attention *to* dreams (or to memory of dreams) while awake increases one's capacity to remember future dreams, possibly improving ability to pay attention (and to engage memory-encoding processes) while *in* a dream. The most common advice given to those who wish to remember more dreams is to get into the habit of recording dreams first thing in the morning. In addition to this, dreamers are also often encouraged to recall and mentally rehearse any previous dream prior to going to bed, and to set intention, to remind themselves, that they must remember their dreams upon awakening. These practices are very efficient, but raise two important questions: by changing the way one attends to their dreams, does one reveal a dream life that was already happening but was forgotten? Or does one effectively change dreams and create a richer and more memorable dream life by virtue of *intending* to attend to the dream experience, and thus learning to more effectively deploy attentional and memory-building abilities while in the dream state?

An important methodological difficulty in dream research is related to the fact that dreams are not restricted to the stage of the rapid eye movement (REM) sleep, as previously thought, and may happen at any moment and in any sleep stage. This led some scholars to hypothesize that perhaps we always dream, and only when we are able to pay attention to the dream and therefore to engage memory encoding mechanisms, that we remember a dream. On the other end of the spectrum, Dennett's 'cassette theory' of dreams (Dennett 1976) suggests that dreams are not happening in real time at all and are constructed retroactively by the dreamer upon awakening. This latter framework presupposes that during sleep there is nothing to pay attention to. One example of the question of whether or not dreams are happening somehow undetected by the reflective consciousness unless one learns to pay attention to them, comes from research on what is sometimes termed 'white dreams'. White dreams are dream reports that contain only a feeling of having had a dream, but no memory of sensory experience or narrative. Traditionally, white dreams have been considered to be dreams that are forgotten, i.e., the assumption is that if one has a memory of 'something', it must have been a full dream, which is now forgotten due to interference of other stimuli upon awakening or insufficient attention to the dream. This reflects a bias in the literature and popular view of the dreams which represents 'real' dreams as fully immersive spatio-temporal and audio-visual experiences. A recent paper (Fazekas et al. 2019) challenges this assumption and offers an alternative explanation: white dreams are full experiences of low perceptual content (less vivid, less clear and less stable than most waking and dream experiences) which are then remembered as such upon awakening. Similarly, while 'deep' sleep (NREM stage 3 sleep) has traditionally been seen as 'dreamless' sleep, more careful empirical examination of cognitive phenomena associated with this stage of sleep has revealed a variety of diminished forms of dreaming (Windt et al. 2016). These include, on the one hand, dramatic sleepwalking or night terror experiences, and, on the other hand, minimal states of awareness and potentially contentless experiences (different from unconsciousness) of 'having slept', what Windt referred to as 'pure temporality' (Windt 2015b). These examples highlight a tension between the opposing possibilities that either a) we dream all the time but sometimes (as in deep sleep) our attentional capacities are not 'on' enough therefore we do not remember our dreams; or b) dream reports reflect exactly what the experience was, and when we do not remember, in fact, we do not dream.

Another context in which attention to dream reports matters is in practices of dream interpretation and dream sharing. In this case, attending to memory of dreams and constructing and rehearsing a dream narrative serves not only as a future dream-enhancing cognitive tool, but also plays an important social role. By bringing dreams into the shared space of narrative and interpretation, it was suggested (Blagrove et al. 2019) that attending to and sharing dreams with others has a social bonding function and potentially increases empathic response towards the dreamer. Similarly, attending to dreams in context of psychotherapy has been shown to increase connection and trust between therapists and patients (Hill and Knox 2010; Hill et al. 2014; Pesant and Zadra 2004).

Lastly, there is a growing body of evidence that attention training during wakefulness may change attentional processes in dreams, in particular with regards to increasing qualities of lucid dreams. For example, Baird and colleagues (Baird et al. 2019) found that long-term meditators (but not novices) had higher frequency of lucid dreams. Similarly, dispositional mindfulness (a trait predisposing an individual to perceive the world with increased awareness) was associated with an increase in lucid dreaming (Stumbrys et al. 2015). Developing lucidity during sleep is also an integral part of the Tibetan Buddhist training. The practice of dream yoga (Holecek 2016; Norbu 2002; Wallace and

Hodel 2012) is aimed at cultivating awareness throughout the night. Specifically, the practitioners are learning to use attentional skills acquired during waking meditation practice while dreaming in order to gain insight into the nature of the mind and to advance on contemplative path.

1.2 Paying Attention While Dreaming

Much work has been done in psychology and phenomenology of perception with regards to how an individual may learn to attend to aspects of the world. Elements of the environment that may attract attention may be salient, surprising, motivating, engaging. The enactive approach in cognitive science and philosophy (Gallagher et al. 2013; Noë 2004; Stewart et al. 2010; Varela et al. 1992) teaches us that the act of perceiving the world is not a passive input-output information processing activity, and that there seems to be a clear relationship between one's experience and the way that one is oriented towards or, in Merleau-Ponty's terms, *geared* towards the world (Merleau-Ponty 1945). In other words, we see what we have learned to see, and we engage with the world in the way in which we have habituated to meet it. But what about dreams? How does a world, brought forth by the dreamer, anchor and mobilize attentional capacities? How does the dream or the dreamer determine what to pay attention to and which elements of the experience will be remembered? Some debates, both in philosophy and in cognitive science of dreams, have focused on the question whether dreams are more like perception or more like imagination (for a review of the literature see (Windt 2015a).

In waking visual perception, our eyes are constantly making rapid jerky movements, known as saccades (Hutton 2008). These movements correspond to what we are attending and orienting to in visually rich environments. During the Rapid Eye Movement (REM) stage of sleep, associated with the highest rates of immersive, vivid and complex dream experiences (Nielsen 2000), even though the body is paralyzed, the eyes are rapidly moving (Aserinsky 1965), resembling the activity during waking perceptual saccades. One idea aiming to explain the reason for such movements is known as the "scanning hypothesis" (Herman et al. 1983), which suggests that these eye movements represent quasi-perceptual activity during dreams mapping on to what the dreamer is looking at. Indeed, some research (Arnulf 2011) suggests that eye movements during sleep may, in fact, correspond directly to reports of the direction where the dreamer was likely looking during the dream. In other words, eye movements represent an idiomorphic relationship between attentional acts during wake and during dreaming, suggesting that both the experiential aspects and the dreamer's orienting and attending behavior during dreaming are very similar to that during waking. Recent work by LaBerge and colleagues (LaBerge et al. 2018) aimed at clarifying some aspects of the debate around whether dreams were more like perception or like imagination. Experienced lucid dreamers (dreamers who learned to become aware of the dream state while in a dream) were asked to trace a shape with their finger and to follow the finger with their eyes. In waking cognition, attending to an external stimulus is reflected in smooth eye movements, while imagining that same stimulus produces jerky eye movements. Lucid dreamers, when following a dreamt stimulus with their eyes, showed a smooth pattern of eye movements, thus suggesting that the dreamt environment, even if endogenously brought forth, still has experiential qualities and attentional response similar to that of the external world.

While there are clearly some similarities in the way in which we attend to the world around us while awake and while dreaming, there are also some important nuances in how attention and perception function during dreaming which are not quite so similar to waking life. In fact, in waking life, whether we are paying attention or not, the world makes demands on our perceptual system and even stimuli that are not consciously attended to are often parts of a more general sense of awareness of the environment. In dreams, however, it appears as if the entire dream reality directly depends on attention. In Thompson's account of dreams as processes of enactive imagination, he writes: "If I don't notice trees in my flying dream, is that because I fail to notice their presence or because I'm not generating any sensory imagery of trees? It strikes me as more phenomenologically accurate to say that the reason I don't notice them is that I'm not visually imagining them at the moment" (Thompson 2014, 182).

Further, paying attention to the dream world during a non-lucid dream (a dream in which the dreamer is not aware of the fact that they are dreaming) often does not result in the realization that the dream content is implausible/bizarre (as compared to waking life). Indeed, relatively high prevalence of plot discontinuities, instability of characters and objects, scene changes, and narrative jumps in dreams (Mamelak and Hobson 1989; Rittenhouse, Stickgold and Hobson 1994) all attest to the idea that it is quite difficult to sustain attention on a stimulus or a task during a dream. This fact has often been used to suggest that 'failure' to realize that a dream is bizarre represents a form of cognitive deficiency, akin to psychosis (Dresler et al. 2014; Hobson and Voss 2011). An alternative view would imply that in dreams we are perhaps pre-reflexively aware (even if not explicitly lucid or rational) of the fact that we are dreaming. Thus, our dreamworlds are familiar to us, and are organized differently with different kinds of affordances (e.g., we can fly or spend time with deceased loved ones without being delusional). Dreaming is one of the earliest forms of cognitive life, likely starting at the same time as waking cognition. Some evidence for this notion comes the developmental trajectory of dreaming-rich REM sleep, forms of which develop in already in utero and are abundant in newborns (Roffwarg et al. 1966). It is unclear, therefore, why dream life would have to be subject to the same criteria for rationality, realness or plausibility as waking life. In our waking life, too, much time is quite productively spent on 'implausible' or 'bizarre' pursuits, ranging from private experiences of daydreaming or imagining to public and collective practices of art, filmmaking and literature. Dreams, we argue, are better understood as imaginary lives and performative processes with their own fields of affordances and with their own personalized experiential vocabulary for what is possible, it does not and nor does it need to map directly onto waking life.

Another way in which the dreamer's attention may be hijacked or reoriented in the dream, is through incorporation of outside stimuli into dreams. Studies of sensory incorporation into dreams show that the dreaming mind reacts to the physical world in a variety of ways (Nielsen 1993, 2017; Sauvageau et al. 1998; Schredl et al. 2009; Solomonova and Carr 2019). Interestingly, stimuli only rarely appear directly in the dream, rather, this new intruding information is often transformed by the dream into something else, which may more or less fit within the ongoing dream scenario. Furthermore, stimuli may change their modality upon entering the dream world: somatosensory stimulation may produce a change in visual or auditory experience. In the dream examples below, the dreamers were undergoing a somatosensory stimulation protocol, which consisted of inflating a blood pressure cuff on the participant's ankle during sleep, followed by a dream interview (Solomonova 2017).

> *Liza was there to wake me up. She turned on the lights and asked me about my dreams. I was answering her. I could feel the pressure pump on my leg. She asked me what does it feel like, I said it feels like a hug. She said 'Doesn't it feel like someone pulling on your leg?'*
>
> *[. . .] I was in my parents' car [. . .] and a large marine animal approached me as if he knew me. It wanted to play with my purse and it bit me! I didn't want (it to continue) and I was saying Oh No!**

These excerpts show two different dreamt responses to the intrusion of the somatosensory stimulus (pressure on an ankle). In the first one, the dreamer immediately associates the sensation of the pressure cuff with the ongoing protocol in the lab (they are expecting this to happen), and then dreams of the experimenter, imagines giving a dream report and describes the phenomenology of feeling the pressure cuff inflate on their leg, in a somewhat humorous manner. In the second example, the pressure cuff intensified the ongoing dream. It produced a dramatic scene shift (from a parents' car to encountering a large animal), a confrontation with a threatening animal, and a physical sensation of being bitten. In both cases, the external stimulation was processed by the brain and transformed by the dreamer into something that was part of the ongoing dream experience. Thus, the dreamer's attention was shifted from what was going on in the dream prior to the stimulation and was now focused on integrating this new and demanding stimulus into the dream. This suggests that the stimuli we attend to in the dream are processed in a multisensory (Nielsen 2017) and perhaps even synesthetic manner, and are in line with contemporary theories of dream formation that suggest that dreams function in an associative (Hartmann 1996; Horton and Malinowski 2015) manner, bringing together disparate elements from memory. Moreover, examples from somatosensory dream incorporation highlight the role of the body in dream formation, lending further support to the idea that dreams are not happening exclusively 'in the head' but are indeed, as the enactive view of the mind would suggest, forms of embodied imagination (Thompson 2014).

In this section we discussed the ways in which attention to dreams during wake and attention in dreams during sleep facilitates access to the dream world. Attentional practices in wake and dreams also have potential for insight in modulating the dreamer's agency. In the following section we will discuss how attention and intention function as mediators of dream content.

2 Mediation: Intention, and Agency in the Dream

2.1 Agency Over Dreams

It has been proposed that attention and intention play a role in skilled agency (Wu 2016). However, fundamental questions of agency over dreams remain largely unanswered. Does the dreamer direct the dream or does the dream direct the dreamer? On the one hand, the dreamer is the source of the dream, i.e., everything that unfolds in the dream is formed by the dreamer's mind. On the other hand, the dreamer is often left at the mercy of the dream with little or no control over the dream scenario. In Section 1 we discussed how attention in dreams functions as a way to bring forth the dream environment. In this section we will discuss the different aspects of how a dreamer's self may be instrumental in the oneiric narrative through practices of intention and skilled attention to and in dreams.

Traditionally, dreaming is thought of as an experience that happens *to* the dreamer, and over which the dreamer has very little agency or control. In support of this view, to some extent, are the cognitive and neurophysiological evidence, suggesting that during dreaming one has relatively less cognitive control than during waking, due to relative inhibition of frontal cortical activation during REM sleep (Cicogna and Bosinelli 2001; Rechtschaffen 1997). In line with this, dream reports demonstrate relatively less cognitive agency compared to waking or even daydream reports. In this context, cognitive agency refers to whether one's thoughts seem to be controlled or directed by the dreamer, also termed 'reflective' or 'metacognitive' thoughts. To illustrate, below is an example of a daydream report, which while imaginative still demonstrates the intact cognitive agency more typical of waking thought.

> *I was thinking about being calm and the image of a small forest came to mind. I thought about walking through that forest with my closest friends and what we would talk about and how safe and happy I feel with them. Spending time with them motivates me to work on my art projects. I then started thinking about all the projects I have going and what I'm going to work on later today.* (Carr and Nielsen 2015)

In general, cognitive agency is lower in dream reports, whereas immersive and embodied imagery is clearly higher in dreams than in either daydream or waking reports. Speth and Speth (2018) suggest that in dreaming, 'to think is to do', meaning rather than reflecting on or directing thoughts and actions, the dreamer simply reacts to and interacts with the dream imagery in a relatively impulsive and non-reflexive way.

Many dream researchers have claimed that such a temporary suspension of reflective thought (akin to Coleridge's 'suspension of disbelief'), e.g., a general inability to recognize that one is in a dream, or acceptance of dream scenario despite its implausibility or bizarreness, allows us to become fully immersed in dream imagery, and may enable functional outcomes associated with dreaming, including emotion regulation (Cartwright 2011), memory consolidation/integration (Horton and Malinowski 2015; Nielsen and Stenstrom 2005; Wamsley and Stickgold 2010), insight (Edwards et al. 2013), among others. For instance, Freud (1900) famously suggested that the fact that dreams are bizarre and therefore 'unreal' provides the dreamer with a 'safe space' to enact their repressed anxieties and traumas, thus allowing for a cathartic emotional release, while being non-threatening to the ego, since the manifest content of the dream has undergone many transformations and distortions. These distortions, according to the psychoanalytic view, hide the real, latent, source of the dream.

From another angle, dreaming has been related to a form of involuntary autobiographical memory, which likewise occurs during daydreaming and mind-wandering (Rasmussen and Berntsen 2011). Involuntary autobiographical memory is triggered by associative and context-dependent retrieval and requires little executive control (Singer 1966; Smallwood and Schooler 2006).

Involuntary autobiographical recall tends to occur during periods of unfocused attention, such as daydreaming and dreaming, compared to voluntary recall, which is more frequent during goal-directed cognition. Support for a link between involuntary autobiographical memory and dreaming is shown first by evidence that more than 80% of dreams contain autobiographical memory features (Malinowski and Horton 2014), and second by a phenomena termed the 'dream rebound', that suppressing thoughts before sleep prompts their involuntary incorporation in dream content (Wegner et al. 2004). These findings support that there are some important 'involuntary' processes underlying dream generation. Nevertheless, there is also evidence that intentional and voluntary pressures influence dream content, and that levels of agency within dreams vary along a continuum.

Co-creative dream theory acknowledges, on the one hand, contributions of a somewhat autonomous or involuntary process of dream imagery generation, and, on the other hand, a role for dreamer agency in dream content creation (Sparrow 2014). Specifically, the theory suggests that even in non-lucid dreams, where the dreamer does not appear to have explicit awareness of or control over the dream, the dreamer is actively co-creating the dream by making specific choices or taking specific actions within the dream. Co-creative theory regards the dream as an interactive process between the dreamer as an agent and the unfolding dream narrative, thus the dreamer is an active and determinant participant in the process of dreaming. To some extent, this theory is supported by other research that acknowledge the presence of a range of levels of awareness and metacognition in ordinary, nonlucid dreams, in which the dreamer can reflect on and determine how to engage with the dreamworld, contrary to claims that reflective awareness is completely absent in dreaming (e.g., the Metacognitive, Affective, Cognitive Experiences) (Kahan and LaBerge 2011; Kahan and Sullivan 2012; Kozmová and Wolman 2006). Such approaches acknowledge the varied levels of directed attention and controlled behaviors a dreamer can have, guiding the ongoing dream narrative through choices, and actions. The dreamer is capable of directing the dream's development, in a dynamic and reciprocal process of dream creation (Sparrow and Thurston 2010).

Another recent approach suggests that simply 'feeling', the private mental experience of emotions (Damasio et al. 2000), informs attention and behavior in the world (Hoeksma et al. 2004). For example, if undesired anger arises, specific behaviors such as gaze aversion (looking away from the object of unpleasant emotions) or physically changing the situation (i.e., leaving a room) are common responses enacted to regulate emotion, particularly in social interactions (Nummenmaa and Calder 2009). If the entire dream experience depends on where attention is directed, the affective tone of the dream, then, mediates attentional quality. In dreams, gaze aversion or gaze shifts from what one was initially paying attention to often occur as novel elements appear in the dream, or as narrative jumps that commonly occur as changes in dream scenes or situations. These shifts and the continual injection of novel elements into the dream may to some extent allow the dreamer to witness and interact with emotionally laden imagery without becoming overwhelmed. Several theorists (Nielsen and Levin 2007; Stickgold 2005; Stickgold and Walker 2013) support this as a function of dreaming, to replay emotional memory in combination with novel associative elements, as a means of integrating salient experiences into broader autobiographical memory. However, this process is often interrupted by intense emotion, as evident in the case of nightmares (Nielsen and Levin 2007; Nielsen and Carr 2017). The nightmare seems to demonstrate a process of maladaptive emotion regulation strategies, particularly attentional narrowing, i.e., maintaining attention on an aversive stimulus. In such a case, the focus on a threat in a nightmare takes on a form of perseverative loop, where it appears that difficulty in voluntarily redirecting attention somewhere else leads to an amplified persistence of threatening or disturbing imagery. In this situation, the dreamer's attention and therefore their entire experience is 'locked' onto the dysphoric scenario.

> *I was walking along the beach and it was raining lightly. The sun was setting over the ocean and I thought this was weird, as it was setting in the south and as I for some reason felt it was the middle of the day still. I watched the sunset until the sun was gone completely. Suddenly it was very dark, there were no visible stars, I couldn't see the moon, and there weren't any street lights. I didn't know which direction to walk in to avoid ending up in the sea. I started walking but slipped over and ended up in the water. I tried to swim to the surface but I couldn't figure out which way to go so just went what I thought was upwards. My lungs were burning and I couldn't reach the surface, everything was still dark and I felt dizzy and felt my eyes close. I'm fairly certain I died.* (Carr et al. 2020)

While negative thoughts and attention can result in increasingly negative, narrowed and repetitive content, the complementary position may be true for intense positive emotion. That is, it broadens mindsets, increases openness to others, encourages exploration, and results in vivid and elaborate dreams (Carr and Nielsen 2017). In contrast to 'locked' qualities of dream attention in nightmares and bad dreams, archetypal, or 'Big', dreams tend to describe euphoric, integrative, altered state experience that demonstrate the expanding influence of positive emotion within a dream (Spadafora and Hunt 1990).

> *My family and I went to this amazing sight seeing place. It was on the top of a massive mountain so when you looked out it looked like it was the edge of the earth. You could see the stars in the darkness above the clouds and the clouds were pink and orange and they were falling off the mountain like a water fall. It was so stunning. We then went on a helicopter ride and it took off over the edge of the mountain and then there was nothing beneath*

you. I got the same feeling you get on a scary rollercoaster and like my stomach was in my mouth. I felt happy when I woke up. (Carr er al. 2020)

In these examples, the dreamer's emotional appraisals and attentional focus seem to direct the developing dream narrative, leading to the manifestation of anxiety in the case of the first dreamer who is fearful of falling into the ocean and then being unable to surface, yet leading to an aesthetically vivid and positive experience in the case of the latter dreamer who is filled with awe. Thus, the quality of attention, thoughts and habitual reactions of the dreamer partly shape the dream narrative. Through changing patterns of thinking or reacting in waking life, e.g., cultivating positive emotion, practicing adaptive emotion regulation strategies, etc., we can influence how and where attention is applied during dreaming. More intentional practices of attention can also be used to direct and exert varying degrees of control over dream content.

2.2 Intentional Practices Using Attention to Change Dreams

Dream incubation is the practice of setting an intention prior to sleep in order to influence dream content. While dream incubation has been practiced for millennia as a means of guiding dreams to provide insight around social, cultural or personal problems, experimental research also shows that reflecting on current concerns prior to sleep can effectively incubate relevant dream content and even provide novel solutions. Barrett (1993) asked college students to incubate a specific, personally relevant problem of their choosing. After simply thinking about the problem for 15 minutes before sleep, participants rated 49% of their dreams as related to the problem and 34% of them as containing a solution. Saredi and colleagues (1997) also report that for a small sample of participants in a sleep laboratory study, reflecting on a question related to a significant current concern prior to sleep increased the frequency of dream references to the concern.

Imagery rehearsal (a form of dream incubation) and other types of dreamwork (working with dreams while awake through different intentional practice such as visualization or contemplation of remembered dreams) can introduce elements of change into recurring dream patterns, with some element of agency in the dreamer but not full lucidity. Imagery rehearsal therapy (IRT) is a promising treatment for frequent, distressing, or recurring nightmares (Aurora et al. 2010). This treatment focuses on rewriting a nightmare scenario during wake, typically writing down an alternate, more positive ending to the nightmare, and then mentally rehearsing the modified version of the dream several times prior to sleep. The principle of IRT is the idea that individual dream patterns, including nightmares, are examples of learned cognitive behavior, which have the potential of being unlearned and changed. Thus, working with and visualizing alternative nightmare narratives during wake helps 'decondition' the dreamer and allows for new possibilities for the otherwise locked negative dream scenario. This simple technique is effective at treating nightmares and reduces both the frequency of nightmare experience and mental distress associated with nightmares (Krakow and Zadra 2006).

Dream Report

I'm put in a [prison] cell. [. . .] The cell has big windows that look down into a kitchen area that's dark and quiet [. . .] I'm looking through the window and I can see two kitchen drawers opening by themselves, as if it's by a ghost [. . .] It's a feeling of dread, I'm really frightened, my hairs go up and I'm cold . . . I just start crying, and I panic then.

Re-scripting the Dream

I think if there were other people in the room, I'd feel better. [. . .] I feel like I could hold their hands, and I wouldn't be on my own. Just feels safer with other people. [. . .] the drawers are going back on their own now, we're not pushing them, we're just kind of standing there – like holding hands . . . and the drawers are going back . . . .and it feels a bit like when a room stretches, goes longer [. . .] it feels like, that's quite powerful . . . it's like winning . . . .and I can see the floor, the tiles, dark tile, light tile . . . checkered . . . I can see more of that . . . [. . .] As it's going away . . . .the fear is going away . . . .and it's replaced with like "wow . . . this is working" [. . .]I feel calmer . . . and like lighter in the sense of weight, but also light – as in I would be a different shade of, it's gone from grey to shades of blue . . . I'm like a pale blue now. (Carr et al. 2020)

The re-scripting exercise allows the dreamer to imagine a more positive resolution to the dream even in the waking state. Thus, intentional practices harnessing attention in waking life can influence dream imagery and elevate the dreamer's agency in shaping the narrative. Specifically, it appears that the ability to recall, bring to attention and manipulate imagery in space of one's own imagination while awake, increases one's capacity for responding in a different way when facing spontaneous and uncontrolled imagery during dreams. Beyond this, it is possible to experience wake-like agency in a dream, through the practice of lucid dreaming, that is, becoming aware of the fact that you are dreaming while still asleep.

2.2.1 Forms of Attention and Agency in Lucid Dreams

Lucid dreams are unique in that the dreamer does have awareness of the fact that they are dreaming, enabling the dreamer to exert volition over the dreaming self and at times over the dreamworld. Lucid dreamers can choose how to attend to and respond to dream imagery moment-to-moment. Lucid dreaming offers a means of directing dreams from within the sleep state. Although there are individual differences in spontaneous lucid dreaming frequency, there are techniques individuals can use to induce lucid dreams and increase lucid dreaming frequency (La Berge 1980). A few attentional techniques can be practiced during the day and just prior to sleep to increase dream lucidity. One technique, 'reality checking,' involves asking yourself repeatedly throughout the day whether you are currently dreaming, and this habit can then spill over into sleep and lead to the realization that one is dreaming. Another method involves setting the strong intention to become lucid by repeating an affirmation in mind prior to sleep such as, "The next time I am dreaming I will remember that I'm dreaming". This can be combined with visualizing a recent dream and imagining becoming lucid within the dream. These techniques are thus similar to dream incubation and imagery rehearsal methods, in that they utilize waking attention to influence dream content, but the goal is to then become lucid and able to further control the dream while asleep.

As discussed above, comparisons of dreaming with waking attention have typically focused on claims that the dreaming mind lacks the self-reflective awareness and metacognition that are assumed in wake. However, even in waking we often wander around in a state of semi-attention, simply perceiving and reacting to experience without affording much reflective attention to the present. Mindfulness practices are often aimed at improving a capacity to maintain attention to the present moment (Bishop et al. 2004), and to diminish habits of mind-wandering. By improving attention to the present moment in waking life, mindfulness may similarly increase the likelihood and frequency of lucid dream experiences. For example, one study (Stumbrys et al. 2015) showed that participants who reported having prior meditation experience also reported higher lucid dream frequency, with approximately 4.28 lucid dreams per month (vs. 2.55 in non-meditating study participants). In addition, participants who had prior meditation experience showed a significant correlation between mindfulness and lucid dreaming frequency. Thus, cultivating certain forms of focused attention in waking life may similarly elevate lucidity in dreaming.

Further, lucidity is not a discrete phenomenon, and levels of lucidity seem to vary across a continuum. Dreams can be considered semi-lucid when showing some levels of insight into the dream-state (Mallett et al. 2021). For instance, in Table 1, dreamers can demonstrate moderate levels of lucidity if they begin

Table 1: Semi-lucidity interrogation. Representative defenses of why a participant picked each DLQ-1 response (i.e., lucidity level). (from Mallet et al., 2021, with permission).

I was aware that I was dreaming.	Why did you rate your awareness the way you did?
Not at all	I believed it was reality when I was dreaming and had no doubt that it wasn't.
Just a little	I had no idea I was asleep. I had no idea I was dreaming, I only knew the situation was odd. I was only partially aware that it was a dream, and only towards the end of the dream.
Moderately	The dream at moments felt real. Because the things occurring in my dream were too bizarre to be real life. I was aware that what was happening in the dream made no sense…
Pretty much	… I was able to partially control it. Because I realized I was dreaming. Because I was aware that I was dreaming, but … it took me a while to realize I was asleep. I was mostly aware that I was dreaming but at times things felt more real. … I was aware that I was dreaming but could not completely grasp the idea of what I was doing …
Very much	I was fully aware I was dreaming.

to notice, e.g., 'the things occurring in my dream were too bizarre to be real life' or become 'aware that what was happening in the dream made no sense.' In such cases, there is an element of sustained attention which seems to enable the dreamer to notice how their experience moment-to-moment differs from normal waking life, and which precedes the full insight that they are experiencing a dream. This is in contrast to completely nonlucid dreams, in which the dreamer fully accepts without question the dream, for example, one participant states, "I believed it was reality when I was dreaming and had no doubt". There is a lack of critical awareness or reflection on the dream experience. As in Bishop et al. (2004), lucidity as a form of mindfulness requires sustained attention, without which the moment-to-moment experience, however bizarre or unusual, is not questioned and the dreamer is not even capable of questioning it.

In addition to variations in degree of awareness within the dream, lucid dreamers have variation in degree of control over the dream environment, which seems to rely on attentional control. In one recent study (Mallett 2020), participants were invited to the laboratory and asked to study an experimental room that contained a variety of objects and were then asked to attempt to recreate this scene while in a lucid dream. Several lucid dreamers were able to call to mind and accurately remember elements of the experimental scene while in the dream, but they experienced varying degrees of success when attempting to transmute remembered objects into forms in the dream-world. Various techniques of attentional focus and intention could be applied to attempt to create dream objects. One participant closed their eyes in the dream, thought of an object, and then opened their eyes, and the object would appear. This lends further support to the idea that stability of the dream environment is dependent on what the dreamer is attending to, and therefore, whatever the dreamer is *not* attending to becomes labile. In this example, the dreamer makes use of this quality of attention-based perception in dreams and decides to 'reset' the dream environment in order to make manifest an imagined object. In another example, a participant reported: "I saw the rattlesnake, it was mostly orange and black. The first time I saw the tail it had three black bands but when I looked again there were five, then seven and finally thirteen". Thus, again, objects that are not being attended to appear to be unstable and often mutate in transitions between attention.

Thus levels of insight, attention and control vary in lucid dreams; the control of the actual dream environment is somewhat unique compared to waking attention and is sometimes termed 'supernatural control', being able to transform and mutate aspects of the dream-world at will, a power we cannot exercise in the real waking (external) world. While degrees of control over the self, one's thoughts, and where one directs attention seem relatively common in lucid dreams, this 'supernatural control' likely requires more skillful levels of sustained attention, focusing one's intention to project imagined content into the dream-world. It seems that practices of waking visualization may to some extent enable 'supernatural control' in dreaming, perhaps most commonly observed when applying lucid dreaming as a treatment for nightmares.

Lucid dream therapy is in fact one of the American Academy of Sleep Medicine's recommended treatments for nightmares, in which the aim is to re-write bad dreams from within the dream state as they are occurring, In contrast to IRT, discussed above, where memory for dreams is re-written during visualization and rehearsal practices in waking state. The goal of lucid dream therapy is to become lucid within a recurring nightmare, using the recurring features of the nightmare as signs to trigger insight (similar to examples above in which bizarre or unusual dream features can be 'noticed' in nonlucid dreams if/when the dreamer is paying attention, and this increases levels of lucidity). Once lucid, the dreamer first has the benefit of realizing that the ongoing nightmare is not 'real', it is just a dream. This initial shift in perspective allows the dreamer to feel less overwhelmed, decreasing negative emotion, and providing the time and space for the dreamer to re-orient towards the dream with an open or curious or empowered attitude, and use the increased cognitive capacity inherent in lucidity to reflect on actions rather than impulsively reacting to the dream. In general, the practice of using lucid dreams to modify nightmares seems to enable dreamers to engage in constructive and adaptive patterns of attention to and meditation of the dream world, in a manner more consistent with waking skillful attention. Below is an example in which the dreamer realizes they are dreaming through the act of noticing, attending to, a recurring dream scenario, and then decides to shift the dream to get out of a stressful situation.

> *I was learning to drive with my previous instructor, and there was also another instructor in the car. I remember feeling frustrated because my feet could not move to press the pedals and the car started to slow down. At this point I realized I was dreaming because this is a recurring dream I have had before and recognized it. After realizing it was a dream I stopped the car and got out. The dream ended with my getting out of the car.*
>
> (Mallett et al. 2021)

In general, practices of lucid dreaming engage varying levels of attention, first through simply noticing one is dreaming, and further through directing attention, controlling behavior, and finally, leveraging the malleability of the dream world to one's advantage.

3 Summary and Conclusions

In Section 1 we discussed the many ways in which attention to dreams and attention in dreams functions as a means of *access* to the dreamworld. Attention to dreams *while awake* is a strong enabling factor for remembering dreams. The ability to remember dreams may also depend on individual traits, such as a patterns of brain activation of the attentional circuits during both sleep and wakefulness, or mindfulness, either acquired through long-term meditation practice, or as a general psychological disposition. Practices such as recording, rehearsing and sharing dreams play an important role in shaping future dreams. In the following section we will discuss how attention works while dreaming. Attention in dreams *while asleep* has qualities of both perception and of imagination. The way the dreamworld appears to us depends directly on what is attended to. We argue that dream life does not have to correspond to the criteria for plausibility or reality, because we conceptualize dreaming as a performative, imaginative, creative activity, and not a mere rehearsal of memories or of waking life. The dreamer's attention is also dependent on stimuli from the outside world: the dream body reacts to the experiences of the sleeping body, and the dreaming mind incorporates outside stimuli into the dream narrative, albeit often in a distorted, transformed manner.

In Section 2 we have discussed how attention works as a *mediator* of the dream experience through changes in agency and awareness in both lucid and non-lucid dreams. Co-creative theory of dreaming, for example, places the dreamer as an active agent in the dream, able to reason within the dream scenario and to make choices. Emotional valence of dreams, on the other hand, may demand and direct attentional resources of the dreamer, by either narrowing (in case of nightmares) or expanding (in case of positive, euphoric dreams) the attentional capacity. Dream incubation practices in waking life allow the dreamer to take charge of the mental imagery of the dream by either focusing on desirable dream aspects or by actively rescripting dream narrative. Lastly, awareness and agency in lucid dreams opens up possibilities for controlling or modulating the dream, including an ability to rescript nightmare scenarios as they unfold.

William James famously wrote of the constitutive power of attention and agency for experience: "[. . .] each of us literally chooses, by his way of attending to things, what sort of universe he shall appear to himself to inhabit" (James 1902, 424). This poetic intuition appears to be even more *literally* applicable to dreams. If the enactive view of phenomenology of dreaming is right, then a dream is dynamically composed by the dreamer by bringing forth imagery, environments, characters and objects as the dream unfolds. Even though most dreams happen (or at least appear to happen) without any obvious volition or intention, and are not entirely controllable by the dreamer, it is possible to change, enrich or dedramatize the dream by engaging in different attentional practices both in wake and during the dream. Attention to dreams during wake, in form of paying attention to memory of dreams, by recording, rehearsing and sharing dreams functions as a means of access to and mediates potential qualities of future dreams. Active engagement with dream imagery, whereas the dreamer learns to own their experience, to recognize its different qualities and to modify aspects of it in order to re-encode and re-appropriate the dream, has a power to change how future dreams will unfold. Finally, learning to be aware of the dream during the dream, and to deploy and practice attentional skills in order to become the author of the dream narrative, provides the dreamer with cognitive tools for experiencing the dream as an expansive and positive mental space.

References

Arnulf, I. (2011): "The 'Scanning Hypothesis' of Rapid Eye Movements During REM Sleep: A Review of the Evidence". *Archives Italiennes de Biologie, 149*: 4, 367–382. doi:10.4449/aib.v149i4.1246

Aserinsky, E. (1965): "Periodic Respiratory Pattern Occurring in Conjunction with Eye Movements During Sleep. *Science 150*: 3697, 763–766. Retrieved from http://www.ncbi.nlm.nih.gov/pubmed/5844080

Aspy, D. J. (2016): "Is Dream Recall Underestimated by Retrospective Measures and Enhanced by Keeping a Logbook? An Empirical Investigation". *Consciousness and Cognition 42*, 181–203. doi:10.1016/j.concog.2016.03.015

Aurora, R. N.; Zak, R. S.; Auerbach, S. H.; Casey, K. R.; Chowdhuri, S.; Karippot, A.; . . . American Academy of Sleep (2010): "Best Practice Guide for the Treatment of Nightmare Disorder in Adults". *J Clin Sleep Med, 6*: 4, 389–401. Retrieved from https://www.ncbi.nlm.nih.gov/pubmed/20726290

Baird, B.; Riedner, B. A.; Boly, M.; Davidson, R. J.; Tononi, G. (2019). "Increased Lucid Dream Frequency in Long-Term Meditators but not Following MBSR Training". *Psychology of Consciousness (Washington, D. C.) 6* :1, 40–54. doi:10.1037/cns0000176

Barrett, D. (1993). "The 'Committee of Sleep': A Study of Dream Incubation for Problem Solving". *Dreaming*, 3(2), 115.

Beaulieu-Prevost, D.; Zadra, A. (2004): "How Dream Recall Frequency Shapes People's Beliefs About the Content of their Dreams". *North American Journal of Psychology 7*: 2, 253–264.

Beaulieu-Prevost, D.; Zadra, A. (2007): "Absorption, Psychological Boundaries and Attitude Towards Dreams as Correlates of Dream Recall: Two Decades of Research Seen Through a Meta-Analysis". *Journal of Sleep Research 16*: 1, 51–59. doi:10.1111/j.1365-2869.2007.00572.x

Bishop, S. R.; Lau, M.; Shapiro, S.; Carlson, L.; Anderson, N. D.; Carmody, J.; Segal, Z. V. (2004): "Mindfulness: A Proposed Operational Definition". *Clinical Psychology: Science and Practice 11*: 3, 230–241.

Blagrove, M.; Hale, S.; Lockheart, J.; Carr, M.; Jones, A.; Valli, K. (2019): "Testing the Empathy Theory of Dreaming: The Relationships Between Dream Sharing and Trait and State Empathy". *Frontiers in Psychology 10*, 1351. doi:10.3389/fpsyg.2019.01351

Carr, M.; Nielsen, T. (2015): "Daydreams and Nap Dreams: Content Comparisons. *Consciousness and Cognition* 36, 196–205.

Carr, M.; Nielsen, T. (2017): "A Novel Differential Susceptibility Framework for the Study of Nightmares: Evidence for Trait Sensory Processing Sensitivity. *Clinical Psychology Review* 58, 86–96. doi:10.1016/j.cpr.2017.10.002

Carr, M.; Summers, R.; Bradshaw, C.; Newton, C.; Ellis, L., Johnson; E.; Blagrove, M. (2020): "Frontal Brain Activity and Subjective Arousal During Emotional Picture Viewing in Nightmare Sufferers. *Frontiers in Neuroscience* 14, 1032.

Cartwright, R. (2011): *The Twenty-Four Hour Mind: the Role and Sleep and Dreaming in Our Emotional Lives*. Oxford: Oxford University Press.

Cicogna, P. C.; Bosinelli, M. (2001): "Consciousness During Dreams". *Consciousness and Cognition, 10*: 1, 26–41. doi:10.1006/ccog.2000.0471

Conduit, R.; Crewther, S. G.; Coleman, G. (2000): "Shedding Old Assumptions and Consolidating What We Know: Toward an Attention-Based Model of Dreaming". *Behavioral and Brain Sciences 23*: 6, 924–928.

Damasio, A. R.; Grabowski, T. J.; Bechara, A.; Damasio, H.; Ponto, L. L.; Parvizi, J.; Hichwa, R. D. (2000): "Subcortical and Cortical Brain Activity During the Feeling of Self-Generated Emotions". *Nature Neuroscience 3*: 10, 1049–1056. doi:10.1038/79871

Dennett, D. C. (1976): Are Dreams Experiences? *The Philosophical Review 85*: 2, 151–171.

Dresler, M.; Wehrle, R.; Spoormaker, V. I.; Steiger, A.; Holsboer, F.; Czisch, M.; Hobson, A. (2014): „Neural Correlates of Insight in Dreaming and Psychosis". *Sleep Medicine Reviews*, In press. doi:DOI: 10.1016/j.smrv.2014.06.004

Edwards, C. L.; Ruby, P. M.; Malinowski, J. E.; Bennett, P. D.; Blagrove, M. T. (2013): "Dreaming and Insight". *Frontiers in Psychology 4*, 979. doi:10.3389/fpsyg.2013.00979

Eichenlaub, J. B.; Bertrand, O.; Morlet, D.; Ruby, P. (2014a): "Brain Reactivity Differentiates Subjects With High and Low Dream Recall Frequencies During Both Sleep and Wakefulness". *Cerebral Cortex 24*: 5, 1206–1215. doi:10.1093/cercor/bhs388

Eichenlaub, J. B.; Nicolas, A.; Daltrozzo, J.; Redoute, J.; Costes, N.; Ruby, P. (2014b): "Resting Brain Activity Varies With Dream Recall Frequency Between Subjects". *Neuropsychopharmacology 39*: 7, 1594–1602. doi:10.1038/npp.2014.6

Fazekas, P.; Nemeth, G.; Overgaard, M. (2019): "White Dreams are Made of Colours: What Studying Contentless Dreams Can Teach About the Neural Basis of Dreaming and Conscious Experiences". *Sleep Medicine Review 43*, 84–31. doi:https://doi.org/10.1016/j.smrv.2018.10.005

Foulkes, D. (2014): *Dreaming: A Cognitive-Psychological Analysis*. New York: Routledge.

Freud, S. (2010[1900]). *The Interpretation of Dreams*, trans. by J. Strachey. New York: Basic Books.

Gallagher, S.; Hutto, D. D.; Slaby, J.; Cole, J. (2013): "The Brain as Part of an Enactive System". *Behavioral and Brain Sciences 36*: 4, 421–422. doi:10.1017/S0140525X12002105

Hartmann, E. (1996): "Outline for a Theory on the Nature and Functions of Dreaming". *Dreaming 6*: 2, 147–170.

Herman, J. H.; Barker, D. R.; Roffwarg, H. P. (1983): "Similarity of Eye Movement Characteristics in REM Sleep and the Awake State. *Psychophysiology 20*: 5, 537–543. doi:10.1111/j.1469-8986.1983.tb03008.x

Hill, C. E.; Knox, S. (2010): "The Use of Dreams in Modern Psychotherapy. *International Review of Neurobiology 92*, 291–317. doi:10.1016/S0074-7742(10)92013-8

Hill, C. E.; Knox, S.; Crook-Lyon, R. E.; Hess, S. A.; Miles, J.; Spangler, P. T.; Pudasaini, S. (2014): "Dreaming of You: Client and Therapist Dreams About Each Other During Psychodynamic Psychotherapy". *Psychotheraphy Research*, doi:10.1080/10503307.2013.867461

Hobson, J. A.; McCarley, R. W. (1977): "The Brain as a Dream State Generator: An Activation-Synthesis Hypothesis of the Dream Process. *American Journal of Psychiatry 134*: 12, 1335–1348. Retrieved from http://www.ncbi.nlm.nih.gov/pubmed/21570

Hobson, J. A.; Voss, U. (2011): "A Mind to Go Out of: Reflections on Primary and Secondary Consciousness. *Consciousness and Cognition 20*: 4, 993–997. doi:10.1016/j.concog.2010.09.018

Hoeksma, J. B.; Oosterlaan, J.; Schipper, E. M. (2004): "Emotion Regulation and the Dynamics of Feelings: A Conceptual and Methodological Framework. *Child Development 75*: 2, 354–360. doi:10.1111/j.1467-8624.2004.00677.x

Holecek, A. (2016): *Dream Yoga: Illuminating Your Life Through Lucid Dreaming and Tibetan Yogas od Sleep*. Louisville: Sounds True.

Holzinger, B.; Klösch, G.; Saletu, B. (2015): "Studies with Lucid Dreaming as Add-on Therapy to Gestalt Therapy". *Acta Neurologica Scandinavica 131*: 6, 355–363.

Horton, C. L.; Malinowski, J. E. (2015): "Autobiographical Memory and Hyperassociativity in the Dreaming Brain: Implications for Memory Consolidation in Sleep. *Frontiers in Psychology 6*, 874. doi:10.3389/fpsyg.2015.00874

Hutton, S. B. (2008): "Cognitive Control of Saccadic Eye Movements". *Brain and Cognition 68*: 3, 327–340. doi:10.1016/j.bandc.2008.08.021

James, W. (1902): *The Principles of Psychology*. Volume 1. New York: H. Holt and Company.

Kahan, T. L.; LaBerge, S. P. (2011): "Dreaming and Waking: Similarities and Differences Revisited". *Consciousness and Cognition 20*: 3, 494–514. doi:10.1016/j.concog.2010.09.002

Kahan, T. L.; Sullivan, K. T. (2012): "Assessing Metacognitive Skills in Waking and Sleep: A Psychometric Analysis of the Metacognitive, Affective, Cognitive Experience (MACE) Questionnaire". *Consciousness and Cognition 21*: 1, 340–352. doi:10.1016/j.concog.2011.11.005

Kozmová, M.; Wolman, R. N. (2006): "Self-Awareness in Dreaming. *Dreaming 16*: 3, 196–214.

Krakow, B.; Zadra, A. (2006): "Clinical Management of Chronic Nightmares: Imagery Rehearsal Therapy. *Behavioral Sleep Medicine 4*: 1, 45–70. doi:10.1207/s15402010bsm0401_4

La Berge, S. P. (1980): "Lucid Dreaming as a Learnable Skill: A Case Study". *Perceptual and Motor Skills 51*: 3), 1039–1042.

LaBerge, S.; Baird, B.; Zimbardo, P. G. (2018): "Smooth Tracking of Visual Targets Distinguishes Lucid REM Sleep Dreaming and Waking Perception from Imagination". *Nature Communications 9*: 1, 3298. doi:10.1038/s41467-018-05547-0

Malinowski, J. E.; Horton, C. L. (2014): "Memory Sources of Dreams: The Incorporation of Autobiographical Rather Than Episodic Experiences". *Journal of Sleep Research*. doi:10.1111/jsr.12134

Mallett, R. (2020): "Partial Memory Reinstatement While (Lucid) Dreaming to Change the Dream Environment". *Consciousness and Cognition 83*, 102974. doi:10.1016/j.concog.2020.102974

Mallett, R.; Carr, M.; Freegard, M.; Konkoly, K.; Bradshaw, C.; Schredl, M. (2021): "Exploring the Range of Dream Lucidity". *Philosophy and the Mind Sciences 2*, 1–23. doi: https://doi.org/10.33735/phimisci.2021.63

Mamelak, A. N.; Hobson, J. A. (1989): "Dream Bizarreness as the Cognitive Correlate of Altered Neuronal Behavior in REM Sleep". *Journal of Cognitive Neuroscience* 1:3, 201–222. doi:10.1162/jocn.1989.1.3.201

Menary, R. (2010): "Introduction to the Special Issue on 4E Cognition". *Phenomenology and the Cognitive Sciences 9*: 4, 459–463.

Merleau-Ponty, M. (1945): *Phénoménologie de la perception*. Saint-Amand: Gallimard.

Nielsen, T. (1993): "Changes in the Kinesthetic Content of Dreams Following Somatosensory Stimulation of Leg Muscles During REM Sleep". *Dreaming 3*: 2, 99–113.

Nielsen, T. (2000): "A Review of Mentation in REM and NREM Sleep: 'Covert' REM Sleep as a Possible Reconciliation of Two Opposing Models". *Behavioral and Brain Sciences 23*: 6, 851–866; discussion 904–1121. Retrieved from http://www.ncbi.nlm.nih.gov/pubmed/11515145

Nielsen, T. (2017): "Microdream Neurophenomenology". *Neuroscience of Consciousness 3*: 1, 1–17. doi: https://doi.org/10.1093/nc/nix001

Nielsen, T.; Levin, R. (2007): "Nightmares: A New Neurocognitive Model". *Sleep Medicine Review 11*: 4, 295–310. doi:10.1016/j.smrv.2007.03.004

Nielsen, T.; Stenstrom, P. (2005): "What Are the Memory Sources of Dreaming?" *Nature 437*: 7063, 1286–1289. doi:10.1038/nature04288

Nielsen, T. A.; Carr, M. (2017): "Nightmares and Nightmare Function". In: M. H. Kryger; T. Roth; W. Dement (eds.): *Principles and Practices of Sleep Medicine, 6e* (pp. 546–554): Elsevier.

Noë, A. (2004): *Action in Perception*. Cambridge, MA: The MIT Press.

Norbu, N. (2002): *Dream Yoga and the Practice of Natural Light, Revised*. Ithaca, New York: Snow Lion Publications.

Nummenmaa, L.; Calder, A. J. (2009): "Neural Mechanisms of Social Attention". *Trends in Cognitive Sciences 13*: 3, 135–143. doi:10.1016/j.tics.2008.12.006

Pesant, N.; Zadra, A. (2004): "Working With Dreams in Therapy: What Do We Know and What Should We Do?". *Clinical Psychology Review 24*: 5, 489–512. doi:10.1016/j.cpr.2004.05.002

Rasmussen, A. S.; Berntsen, D. (2011): "The Unpredictable Past: Spontaneous Autobiographical Memories Outnumber Autobiographical Memories Retrieved Strategically". *Consciousness and Cognition 20*: 4, 1842–1846. doi:10.1016/j.concog.2011.07.010

Rechtschaffen, A. (1997): "The Single-Mindedness and Isolation of Dreams". In: M. S. Myslobodsky (ed.): *The Mythomanias: The Nature of Deception and Self-Deception*. Mahwah, NJ: Lawrence Erlbaum, 203–224.

Rittenhouse, C.; Stickgold, R.; Hobson, A. (1994): "Constraint on the Transformation of Characters, Objects, and Settings in Dream Reports. *Consciousness and Cognition 3*: 1, 100–113.

Roffwarg, H. P.; Muzio, J. N.; Dement, W. C. (1966): "Ontogenetic Development of the Human Sleep-Dream Cycle". *Science 152*: 3722, 604–619. doi:10.1126/science.152.3722.604

Saredi, R.; Baylor; G. W., Meier, B.; Strauch, I. (1997): "Current Concerns and REM-Dreams: A Laboratory Study of Dream Incubation". *Dreaming 7*, 195–208.

Sauvageau, A.; Nielsen, T.; Montplaisir, J. (1998): "Effects of Somatosensory Stimulation on Dream Content in Gymnasts and Control Participants: Evidence of Vestibulomotor Adaptation in REM Sleep. *Dreaming 8*, 125–134.

Schredl, M. (2002): "Questionnaires and Diaries as Research Instruments in Dream Research: Methodological Issues. *Dreaming 12*: 1, 17.

Schredl, M.; Atanasova, D.; Hormann, K.; Maurer, J. T.; Hummel, T.; Stuck, B. A. (2009): "Information Processing During Sleep: The Effect of Olfactory Stimuli on Dream Content and Dream Emotions". *Journal of Sleep Research 18*:3, 285–290. doi:10.1111/j.1365-2869.2009.00737.x

Singer, J. (1966): *Daydreaming: An Introduction to the Experimental Study of Inner Experience*. New York: Random House.

Smallwood, J.; Schooler, J. W. (2006): "The Restless Mind". *Psychological Bulletin 132*: 6, 946–958. doi:10.1037/0033-2909.132.6.946

Solomonova, E.; Sha, X.W. (2016): "Exploring the Depth of Dream Experience: The Enactive Framework and Methods for Neurophenomenological Research". *Constructivist Foundation 11*:2, 407–416

Solomonova, E. (2017): *Embodied Mind in Sleep and Dreaming: A theoretical Framework and an Empirical Study of Sleep, Dreams and Memory in Meditators and Controls*. (PhD). Université de Montréal, Montreal.

Solomonova, E.; Carr, M. (2019): "Incorporation of External Stimuli into Dream Content". In: K. Valli; R. Hoss (eds.): *Dreams: Understanding Biology, Psychology, and Culture* (Vol. 1). Santa Barbara, CA, USA: ABC-CLIO.

Spadafora, A.; Hunt, H. T. (1990): "The Multiplicity of Dreams: Cognitive-Affective Correlates of Lucid, Archetypal, and Nightmare Dreaming". *Perceptual Motor Skills, 71*: 2, 627–644. doi:10.2466/pms.1990.71.2.627

Sparrow, G. S., & Thurston, M. (2010). The Five Star Method: A Relational Dream Work Methodology. *Journal of Creativity in Mental Health*, 5(2), 204–215.

Sparrow, G. S. (2014): "Analyzing Chronic Dream Ego Responses in Co-Creative Dream Analysis". *International Journal of Dream Research 7*: 1, 33–38.

Speth, C.; Speth, J. (2018): "A New Measure of Hallucinatory States and a Discussion of REM Sleep Dreaming as a Virtual Laboratory for the Rehearsal of Embodied Cognition". *Cognitive Science 42*: 1, 311–333. doi:10.1111/cogs.12491

Stewart, J.; Gapenne, O.; Di Paolo, E. (eds.) (2010): *Enaction: Toward a New Paradigm for Cognitive Science*. Boston: MIT Press.

Stickgold, R. (2005): "Why We Dream". In: M. H. Kryger; T. Roth; W. Dement (eds.): *Principles and Practices of Sleep Medicine*. Philadelphia: Elsevier, 579–587.

Stickgold, R.; Walker, M. P. (2013): "Sleep-Dependent Memory Triage: Evolving Generalization Through Selective Processing". *Nature Neuroscience 16*: 2, 139–145. doi:10.1038/nn.3303

Stumbrys, T.; Erlacher, D.; Malinowski, P. (2015): "Meta-Awareness During Day and Night: The Relationship Between Mindfulness and Lucid Dreaming". *Imagination, Cognition and Personality 34*: 4, 415–433.

Thompson, E. (2014): *Waking, Dreaming, Being: Self and Consciousness in Neuroscience, Meditation, and Philosophy*. Columbia: Columbia University Press.

Vallat, R.: Eichenlaub, J. B.; Nicolas, A.; Ruby, P. (2018): "Dream Recall Frequency Is Associated With Medial Prefrontal Cortex White-Matter Density". *Frontiers in Psychology 9*, 1856. doi:10.3389/fpsyg.2018.01856

Varela, F. J.; Thompson, E.; Rosch, E. (1992): *The Embodied Mind: Cognitive Science and Human Experience*. Cambridge: MIT Press.

Wallace, B. A.; Hodel, B. (2012): *Dreaming Yourself Awake Lucid Dreaming and Tibetan Dream Yoga for Insight and Transformation*. Boston, MA: Shambhala Publications.

Wamsley, E. J.; Stickgold, R. (2010): "Dreaming and Offline Memory Processing". *Current Biology 20*: 23, R1010–1013. doi:10.1016/j.cub.2010.10.045

Wegner, D. M.; Wenzlaff, R. M.; Kozak, M. (2004): "Dream Rebound The Return of Suppressed Thoughts in Dreams". *Psychological Science 15*: 4, 232–236.

Windt, J. M. (2015a): *Dreaming: A Conceptual Framework for Philosophy of Mind and Empirical Research*. Cambridge: MIT Press.

Windt, J. M. (2015b): "Just in Time – Dreamless Sleep Experience as Pure Subjective Temporality". In: J. Windt; T. Metzinger (eds.): *Open MIND*. Frankfurt am Main: MIND Group.

Windt, J. M.; Nielsen, T.; Thompson, E. (2016): "Does Consciousness Disappear in Dreamless Sleep?". *Trends in Cognitive Sciences 20*: 12, 871–882. doi:10.1016/j.tics.2016.09.006

Wu, W. (2016): "Experts and Deviants: The Story of Agentive Control". *Philosophy and Phenomenological Research 93*: 1, 101–126.

Short Biography

Elizaveta Solomonova, PhD, is a researcher at the Neurophilosophy Lab in the Culture, Mind and Brain research group of the Division of Social and Transcultural Psychiatry at McGill University in Montreal, Canada. Elizaveta received her interdisciplinary PhD from the University of Montreal in Psychiatry and Philosophy, working in the Dream and Nightmare Laboratory (University of Montreal) and in the Topological Media Lab (Concordia University). Elizaveta's work is centered around the intersection between enactive and embodied cognition, sleep neurophenomenology and socio-cultural aspects of consciousness. Her research has focused on sleep and memory, nightmares, sleep paralysis, sleep and meditation, perinatal mental health and delusional ideation. She is interested in the complex ways in which waking and dreaming mind are interconnected.

Michelle Carr, PhD, is a researcher at the University of Rochester Sleep and Neurophysiology Laboratory. She previously conducted research at the Swansea University Sleep Laboratory in the UK, and holds a PhD in Biomedical Sciences from the University of Montreal, where she conducted research at the Dream and Nightmare Laboratory. Her work focuses on the role of REM sleep and dreams in emotional memory, along with comparative studies of nightmare disorder and PTSD. Her other research interests include sleep paralysis, lucid dreaming, consciousness studies, and the use of dreamwork in psychology. Dr. Carr is currently the president of the International Association for the Study of Dreams.

Luis R. Sandoval and Betzamel López
Chapter 8
Improving Attention in Psychosis With Digital Tools

Abstract: Schizophrenia is one of the most debilitating psychiatric illnesses, and is accompanied by significant personal, social and economic burden to both the individual and the family. Schizophrenia is characterized by positive (i.e., visual and auditory hallucinations) and negative symptoms (i.e., anhedonia, apathy) as well as by cognitive decline. This chapter presents a clinical case of an individual who suffers from schizophrenia with persistent auditory hallucinations and with significant cognitive decline, in particular in attention and vigilance. The treatment approach targeted these neurocognitive deficits by combining interpersonal therapy with a computerized neurocognitive treatment, which allowed the patient to improve his attention. This clinical case illustrates that improvements in attention and vigilance can serve as cognitive mediators to control one's experiences, helping the patient to encapsulate his auditory hallucinations (i.e., psychosis) and expand his social skills.

Introduction

Schizophrenia is one of the most debilitating psychiatric illnesses. It is often accompanied by significant personal, social, and economic burden to both the individual and the family. Schizophrenia is characterized by positive symptoms (i.e., visual and auditory hallucinations) and negative symptoms (i.e., anhedonia and apathy), as well as cognitive decline, including difficulties with attention. This chapter presents a clinical case of an individual who suffers from schizophrenia with persistent auditory hallucinations and with significant cognitive deterioration, in particular in attention and vigilance. The treatment approach aimed to target these neurocognitive deficits by combining interpersonal therapy with a computerized neurocognitive treatment, which allowed the patient to improve his cognitive abilities. This clinical case illustrates that improvements in attention and vigilance can serve as cognitive mediators, in this case helping the patient to encapsulate his auditory hallucinations (i.e., psychosis) and expand his social skills.

This chapter contains six sections: first, we present an overall description of what schizophrenia is and the cognitive deficits that are characteristic in this chronic psychiatric illness. We describe the implications of the attentional difficulties in schizophrenia. Further, we discuss how auditory hallucinations impact attention. Additionally, we give an overall description of Cognitive Remediation Therapy (CRT). We present different theoretical approaches and describe the most common clinical techniques used in CRT to treat neurocognitive deficits associated with schizophrenia. Moreover, we present current evidence of CRT in improving neurocognition and other symptoms that are part of this illness. In the third section, we present the clinical case of a patient living with schizophrenia, Peter, to illustrate the impact of this disease. We first cover his early education years, his interpersonal relationships with his friends and parents, and the life goals he had before presenting symptoms of psychosis. Then, we describe the onset of his psychiatric disorder, followed by the progression of the illness and the aftermath in his life. In the fourth section, we describe the computerized neurocognitive tool used for this clinical case, the clinical procedures applied in the treatment, and how these two approaches worked in tandem to improve neurocognition. Finally, in the last section, we present the results of Peter's clinical progression in attention and concentration. To conclude, we discuss the clinical implication of improving attention as a mediator mechanism to improve the symptoms of schizophrenia; in Peter's case, improving his attention was the mechanism that helped him "encapsulate" his auditory hallucinations, ultimately allowing him to have improved social interactions and reintegrate into society.

1 Theoretical Foundations: Background on Schizophrenia and CRT

1.1 Schizophrenia

Schizophrenia is a severe psychiatric illness estimated to affect approximately 21 million individuals globally. It is one of the most disabling disorders, accounting for 13.4 million disability-adjusted life years (DALYs) worldwide (Charlson et al. 2016). It significantly alters the life of patients who suffer from this illness, as well their families and caregivers. Some of the defining features of this illness include (1) *positive symptoms*, which are experiences or beliefs that are not usually present, such as hallucinations and delusions; (2) *negative symptoms*, characterized by the deficiency in abilities that were previously present, such as lack of motivation and social withdrawal; and 3) *cognitive impairments*, such as deficits in attention, memory, and problem-solving skills. The diagnosis of this illness, as defined by the *Diagnostic and Statistical Manual*, fifth edition (DSM-5® 2013), relies on the presence of positive and negative symptoms for at least 6 months, resulting in functional impairment to the individual with the disease.

1.2 Cognitive Profile in Schizophrenia

Cognition is commonly impaired in schizophrenia. The cognitive deficits in this condition were not recognized until the early nineteenth century when Emil Kraepelin described the attention impairments he noted in patients with the illness. He first described a deficiency in active attention, whereby individuals were not able to keep their attention fixed for any amount of time, as well as a deficit in passive attention, explained as an irresistible attraction to external stimuli, i.e., easy distractibility (Kraepelin 1919).

It is now well known that the cognitive domains that are typically affected in schizophrenia include not only attention, but also working memory, verbal memory, and processing speed, among others (Cella and Wykes 2013; Mohn and Torgalsbøen 2018; Ichinose and Park 2019; Gur and Gur 2013). As Kraepelin initially noted, individuals with schizophrenia often have difficulty attaining focus and maintaining sustained effort towards cognitive activities, such as engaging in a task or having a conversation. These attentional difficulties, in turn, affect higher-level cognitive functions that require active attention. In other words, in order to successfully perform any cognitive task, the individual first needs to put attention into said task.

Furthermore, social cognition is also impaired; individuals with schizophrenia present difficulties identifying and processing emotions, connecting with others around them, and responding appropriately to social cues (Green et al. 2015). It has been shown that in cognitive tasks people with schizophrenia perform at one standard deviation below the mean for the general population (Fatouros-Bergman et al. 2014; Heinrichs and Zakzanis 1998; Savla et al. 2013; Schaefer et al. 2013). While the positive and negative symptoms certainly contribute to the suffering experienced by these patients, the cognitive defects further impact patients' ability to communicate and interact with the world around them. Cognitive impairments not only affect patients' social interactions, but also their engagement in higher-level activities such as school or work, therefore limiting their integration into meaningful aspects of society.

Unfortunately, almost all patients with schizophrenia report cognitive decline as compared with their prior level of function. The cognitive deficits in schizophrenia tend not to improve over the course of the illness, and some patients show signs of cognitive decline even prior to their first symptoms of schizophrenia (Jones et al. 1994). Correspondingly, the attentional deficits are usually present years before the first psychotic episode; by the time the diagnosis is made, the impairments in attention can be profound (Bowie and Harvey 2006). Interestingly, the duration of the illness is not correlated to the presence of cognitive deterioration, as both patients with chronic and recently diagnosed schizophrenia experience these deficits (Czepielewski et al. 2015). The cognitive deficits in schizophrenia, including the decreased attentional capacities, have been attributed to decreased brain volume as a result of gray matter loss (Delisi 2008). However, the exact mechanisms by which cognitive dysfunction persists are not fully understood, although it has been hypothesized that people with schizophrenia present defective neuroplasticity of neural circuits associated with cognition (Keshavan et al. 2015). Unfortunately, to date, there are no medications that assist with cognitive decline (Green 2016), although there are therapies that help slow its progression or improve cognition in some areas, as will be discussed in this chapter.

1.3 Attention in Schizophrenia

This chapter focuses specifically on the attentional deficits found in schizophrenia. In the field of psychology, attention is defined as "a state in which cognitive resources are focused on certain aspects of the environment rather than on others and the central nervous system is in a state of readiness to respond to stimuli" (American Psychological Association 2013). Vigilance, in turn, can be defined as sustained attention or tonic alertness, which includes both the degree of wakefulness and arousal and, more importantly to this discussion, the level of cognitive performance as it pertains to attention (Oken et al. 2006).

The attentional difficulties in schizophrenia have been described as having three aspects – inability to set attention (selective attention), inability to maintain attention (sustained attention or vigilance), and inability to shift attention (shift attention) (Hagh-Shenas et al. 2002). Given the severe attention deficits, schizophrenia has been associated with symptoms of hyper-distractibility and lack of an attentional filter (Robbins 2005). In cognitive tests assessing attention function, individuals with the disease have shown to have limited capacity to suppress distractions and restrict attention to the task at hand. Researchers have postulated that the attentional difficulties specifically arise from the impaired ability to control attention, therefore leading to the rest of the attentional symptoms typically seen in these patients (Gold et al. 2007); thus, impairments in perceptual processes such as attention represent a significant aspect of the cognitive dysfunction present in the disease. Nevertheless, it is crucial to emphasize that the attentional difficulties are not under explicit voluntary control of patients, but a well-established symptom brought about by the neurobiological aspects of the disease.

1.4 Auditory Hallucinations and its Relationship with Attention in Schizophrenia

Positive symptoms in schizophrenia – in particular, auditory hallucinations (i.e., hearing voices that are not there) – are perception-like experiences that occur without an external stimulus. These are vivid and clear, with the full force and impact of normal perception and are involuntary to control (DSM-5 2013). Auditory hallucinations are one of the most commonly experienced and distressing symptoms of schizophrenia (Rayner et al. 2015), occurring to up to 70% of those with the diagnosis (Turkington et al. 2016).

Auditory hallucinations are involuntary, intrusive, and unwanted and are perceived either as coming from the outside or from a specific area inside the head. Auditory hallucinations have great inner variability (Beck and Rector 2003) and can take the form of familiar forms (i.e., friends or family members) or unfamiliar forms (i.e., strangers, demons, God, celebrities). They are perceived as a distinct input from the individual's own thoughts (DSM-5 2013), making it difficult for patients to direct their attention between their inner thoughts and auditory hallucinations.

There are three competing theories of the etiology of auditory hallucinations and why they are present in schizophrenia. The first theory suggests that these hallucinations are the result of abnormally enhanced sensitivity of the auditory regions of the brain to external auditory stimuli (Rayner et al. 2015). The second theory postulates that the hallucinations are the result of impaired self-monitoring (Allen at al. 2007). Self-monitoring is the ability to monitor self-willed intentions and actions, and such inability would bring hallucinators to mistake inner speech for an external event, and thus misattribute their inner voices to an external source (Cardela and Gallemi 2019).

The third theory, derived from a phenomenological approach, argues that schizophrenia is a disorder of consciousness, self-awareness, and self-experience underlined by disturbances in hyper-reflexivity and diminished self-affection. In this context, hyper-reflexivity refers to an exaggerated self-consciousness, while diminishment of self-affection is a sense of basic self-presence that is manifested in a way that involves no distinction between a subject and an object, meaning individuals cannot distinguish between their own self and the reality around them (Sass and Parnas 2003). In this theoretical approach, self-affection is a necessary condition for consciousness, otherwise individuals will lose the sense of inhabiting their own actions, thoughts, feelings, impulses, bodily sensations, or perceptions, making them feel that these processes are in the possession or under the control of some alien being or force. Following the phenomenological approach, auditory hallucinations of verbal material involve unrecognized perceptions of one's own 'inner speech' (Johnson 1978; Cahill and Frith 1996; David 1999), due to an apparent deficit in self-monitoring (Frith 1987), creating an inability to recognize the source of one's auditory hallucinations.

Despite the competing theories regarding the etiology of auditory hallucinations, it is universally known that it is paramount to treat auditory hallucinations in individuals with schizophrenia in order to enhance attention skills. However, patients often face a challenging choice between paying attention to internal stimuli in the form of hallucinations vs. external stimuli (i.e., their own thoughts, emotions, conversations with friends and family), which makes it exceedingly difficult to navigate daily inter- and intra-personal interactions.

1.5 Cognitive Remediation Therapy in Schizophrenia

The deficits in attention and other cognitive functions have been targeted through Cognitive Remediation Therapy (CRT). CRT is a treatment modality that seeks to improve cognition via cognitive exercises that train patients in skills such as attention, memory, and problem-solving, which can then be applied to real-world situations. While this technique was first implemented to treat WWI veterans who suffered severe brain trauma (Boake 1991), its success in rehabilitating cognitive function in that population motivated psychiatrists to integrate CRT into mental health research in the early 1970s (Ben-Yishay 1979). One of the first records of the use of CRT in schizophrenia is by Meichenbaum and Cameron (1973), who used verbal commands to improve attention in this patient population. Several higher-level cognitive tasks (such as memory and executive function) were integrated into later forms of CRT, including the Wisconsin Card Sort test (Goldberg et al. 1987; Stratta et al. 1994), the Integrated Psychological Therapy (Brenner et al. 1994), and the Cognitive Enhancement Therapy (Hogarty and Flesher 1999a, 1999b).

In 2010, the Congress of Cognitive Remediation for Schizophrenia defined CRT as "a behavioral training-based intervention that aims to improve cognitive processes (attention, memory, executive function, social cognition or metacognition) with the goal of durability and generalization" (Wykes et al. 2011). Today, CRT is an indispensable tool for the cognitive rehabilitation and treatment of schizophrenia and many other psychiatric conditions. Several modalities of CRT have emerged, which offer different techniques and approaches to cognitive remediation and improvement.

1.6 CRT Theoretical Approaches

CRT can generally be classified into two types of approaches: perceptual skills ('bottom-up') and executive skills ('top-down'). 'Bottom-up' approaches are based upon the assumption that initial training focused on basic cognitive processes (i.e., attention, vigilance, and processing speed) facilitates improvement in higher-level cognition and executive functions (Hogarty et al. 2004).

'Top-down' approaches assume that the requirements for different tasks are variable. Top-down approaches are designed to train higher-level cognitive processes with the aim of transferring gains to more basic cognitive domains, including attention capabilities, in turn improving community outcomes, such as integration into society and enhanced capacity of communication and social interaction (Eack et al. 2010). These top-down programs are effective at improving attention and problem-solving skills for individuals with schizophrenia and other psychotic disorders, and these improvements persist at durability assessments (Medalia et al. 2000; Medalia et al. 2002).

1.7 CRT Clinical Techniques

CRT programs are based on two main techniques: (1) drill and practice and (2) strategic training. Drill and practice consist of repetitively practicing a problem or exercise in order to reach a peak level of performance. It is based on the notion that repeated practice leads to improvement. Strategic training teaches mental strategies to boost cognitive function, such as using a story as a memorization technique by incorporating the different items to be memorized into a narrative. By practicing these strategies through CRT, individuals with schizophrenia can inherently target their attentional difficulties through both bottom-up and top-down approaches. With exercises that target attention specifically, they improve their attentional capacities by repeated practice and by incorporating pertinent strategies. Exercises that instead focus on higher-level cognitive aspects, such as working memory and executive function, indirectly boost attention capacity through a bottom-down approach.

1.8 Evidence of CRT's Impact on Attention

Given that attentional difficulty is a hallmark of schizophrenia, it is one of the most commonly targeted cognitive domains in cognitive remediation therapy (CRT) programs. Attention training through CRT showed to enhance attention capacity, improve vigilance, and minimize distractibility in patients with chronic schizophrenia (Medalia et al. 1998). These improvements, in turn, translated into clinical progress as measured by the Brief Psychiatric Rating Scale (BPRS), a validated tool to assess psychiatric symptoms. Specifically, attention training improved somatization, emotional withdrawal, as well as hallucinatory behavior, which are clinically meaningful disease parameters that can have a significant impact on outcomes.

In order to optimize outcomes, it is imperative to implement a multimodal approach to attention training, including vigilance, response inhibition, and multiple attention processing (Bell et al. 2009). It is important to note, however, that despite CRT's vast success at treating attentional impairments in schizophrenia, the attentional capacity of these patients tend not to match that of healthy controls even after cognitive training (Mohn and Torgalsbøen 2018).

1.9 Evidence of CRT's Impact on Cognition and Functioning in Schizophrenia

Several studies have sought to assess the efficacy of CRT at treating overall cognition in schizophrenia and have found benefits in cognition and improvements in individual domains, including attention and vigilance, social cognition, verbal memory, working memory, and processing speed (McGurk et al. 2007; Wykes et al. 2011). The effects of CRT on cognition have been shown to persist up to three years after the treatment has ceased. In a study evaluating patients with both early-course and chronic schizophrenia, the treatment effects persisted one year after therapy, with enduring improvements in neurocognitive and social cognitive domains (Eack et al. 2009; Hogarty et al. 2006). Another study found cognitive gains and significant improvement in patients' quality of life at one-year follow-up; importantly, the researchers also assessed outcomes two and three years after the treatment and found that patients had a reduction in the number of acute psychiatric episodes requiring hospitalization (Garrido et al. 2017).

CRT has not been found to improve positive symptoms directly, since these are often managed with antipsychotic medications. However, CRT has been found to impact negative symptoms favorably (Cella et al. 2014; Eack et al.

2013; Farreny et al. 2013; Sanchez et al. 2014), showing small to moderate effects on negative symptoms (Cella et al. 2017). The encouraging impact of CRT on negative symptoms could potentially be explained by a partial overlap of negative symptoms with cognitive impairments. This is of utmost significance given that negative symptoms have traditionally been thought of as not having many treatment options, pharmacological or otherwise (Fusar-Poli et al. 2015), and CRT serves as a step in the right direction to help treat the negative symptoms in schizophrenia.

Cognitive performance is associated with functional outcomes in schizophrenia, with higher performance correlated with improved ability to live independently, attend work or school, contribute to society, and develop meaningful social and professional relationships (Green et al. 2000). Furthermore, cognitive enhancement has even been associated with better quality of life (Fujii et al. 2004). While individual cognitive domains, such as attentional domains, have been found to impact the course of the illness positively, overall cognitive function has been more highly related to functional outcome than individual domains have.

2 The Case of a Patient with Schizophrenia

The following patient narrative illustrates how cognitive impairments can affect the daily life and social functioning of an individual who developed schizophrenia. To preserve the patient's anonymity, he will be referred to as Peter.

Early Development

Peter was born a healthy boy and met all his developmental milestones on time. According to Peter's teacher, he was a 'sweet, empathetic, smart, and healthy kid'. In elementary school, Peter expanded his social network. He played soccer, practiced gymnastics, and was on the swimming team. When he was in 3rd grade, his little brother was born, and he proudly became a big brother. In 6th grade, Peter's dad found employment in a different city, and Peter transferred to a new school. According to Peter's father, even though Peter was a well-adjusted kid, he had difficulties adapting to his new city, new school, and creating new friendships and relationships. To ameliorate the situation, Peter's parents stayed in communication with some of Peter's old friends via phone calls. This seemed to help him stay connected to the life he had in the previous city.

Middle School and High School

Like many children who transition into adolescence, Peter's interests changed. He started playing video games, listening to music, and being more social. He continued swimming.

During his high school years, he developed a special interest in computers. He opened an email account so that he could be socially active with his friends. His grades were average. In 10th grade, he decided to join the debate club and became the president of the gamers club. In the latter group, he formed close friendships and met his first girlfriend. During his junior and senior years he started to show great passion for music. He formed a small jazz band, which he named "The 3.0". In the last semester of his senior year, before leaving for college, Peter decided to teach music to children and assist in coaching first-graders in their swimming practice. Peter attended prom with his girlfriend, and he was nominated to be prom king; it was there where they decided to go to the same university after graduation.

The Summer before College

The summer before entering college, he decided to take a 7-day road trip with his five closest friends, who were also members of his jazz band. The trip was great, according to all, except for one incident. Peter reported that one day, while he was driving, he felt that something "bad was going to happen to them". When his friends asked him why he was worried and looking at the rear-view mirror constantly, he said, "The police and the FBI are after us; they want something from us (. . .). So, if I see the police, I will not stop; otherwise, we will go to jail". His friends were confused about what he said, and they laughed at the idea of being chased by the FBI. Peter felt confused at first but laughed about it himself because it sounded like a 'conspiracy'.

First Year of College

Two weeks into the first semester, Peter felt lethargic, sad, nostalgic, and had difficulty concentrating. He did not have enough energy to play music, and he felt nervous and suspicious every time he visited his girlfriend's dorm. When he talked about these thoughts with his girlfriend, she told him that he was probably feeling homesick. Peter believed that explanation. However, a week after that incident, Peter felt more anxious and began to experience suicidal ideation. He left campus and returned home where he was seen by his PCP. He was prescribed an anxiety medication and referred to therapy, which he refused. The explanation that he received was that it is common to feel sad and anxious during the first weeks of college and that what he was feeling was normal. He went back to college and discontinued medication after a week because he was feeling better. Shortly after discontinuing medication, he began to feel paranoid (that people were watching him) and started seeing lights of different colors and believing that each represented a particular message (e.g., red indicated not to trust people, blue was neutral, yellow meant 'they' were trying to steal his ideas). He also began to dress differently, including painting his nails and wearing make-up, as he believed that by doing that he would be able to protect his ideas from other people. He attempted telling his girlfriend about what he was experiencing, but she did not take him seriously because she thought he was using recreational drugs. That same night, Peter went from one bar to another until the next morning. He was wandering around the streets when the police found him. The police asked what he was doing, and he responded that he was trying to hide from the vampires. He appeared disoriented and disheveled; thus, he was taken to an emergency room (ER) for evaluation, where his toxicology screen for alcohol and other drugs was negative, meaning his behavior was not attributable to an intentional ingestion of alcohol or drugs.

During the first 24 hours under observation in the ER, Peter developed difficulty sleeping and poor appetite, and there was a significant decline in his overall cognitive functioning. He did not remember who he was, where he was, nor which day it was. He also had difficulty understanding simple tasks and reported having foggy thoughts. That night in the ER, Peter presented with psychotic symptoms: paranoid thoughts and visual and auditory hallucinations. He was stabilized with medication and referred to an inpatient unit after a few hours. After two weeks in the inpatient psychiatric unit, he was discharged with a diagnosis of psychotic episode. Peter was 18 years old at the time and that incident was the beginning of a long history of chronic psychosis and several psychiatric hospitalizations.

Three weeks after his first psychiatric hospitalization, he was found throwing rocks at several windows of a hotel and was arrested for damage to private property. When asked what he was doing, he said that the hotel had a yellow-colored aura, which meant that there were "spies" using that hotel as their "hideout". With the understanding that his behavior was related to mental illness, he was thus released from jail and taken to an inpatient unit where he was stabilized on the medications risperidone and lorazepam.

Peter's Life after his Diagnosis

Social
Peter discontinued his university studies and returned to live with his parents and younger brother. He gradually began losing his social connections and his ability to maintain and sustain conversations. Some relations were lost because his friends did not know how to connect with him, and they struggled to listen and understand Peter's stories about being followed by spies and FBI agents and other 'conspiracy theories'. Other relationships were lost as Peter rejected them based solely on the color-aura system he saw around things (i.e., red, yellow, or blue auras). At 19, one year after his first incident, Peter only had three main social relationships: his younger brother, his mom, and his dad. Although they were his family, they also had difficulties engaging and managing Peter.

Work
Over the years, Peter has had multiple jobs. He has tried working in retail stores, coffee shops, and restaurants; he has experimented with yard work and has worked as an office assistant. He even tried to be in a music band. However, he has not been able to continue any form of employment for a significant amount of time. He is usually fired or he resigns, mostly because his active visual and auditory hallucinations interfere with his performance. For example, Peter stated, "When I was looking for a job, I saw the ad in the newspaper, and the voices told me that it was a blue-like aura job, so I applied. Two days after I got the job, I noticed that the aura of the restaurant changed color to red and yellow, which meant that it was not safe there and I needed to quit".

School
Peter has tried to re-engage in his university studies, but his overall cognition has declined significantly, which makes any school task extremely difficult, if not impossible, to complete. Peter has failed in his attempts of returning to school three times and the only courses that he has taken have been through video (i.e., YouTube).

Treatments

Over the past 22 years, Peter has been on and off psychiatric and psychological treatments, achieving a partial recovery at best. Several antipsychotic trials have reduced the frequency and intensity of his visual and auditory hallucinations. Cognitive Behavioral Therapy has helped him develop adaptive coping mechanisms to deal with his chronic illness, while interpersonal therapy has improved his emotional management and his inter- and intra-personal relationships. However, although these treatments are recommended to treat psychosis, his cognition has not improved and has been declining over the years.

Peter is now 41 years old and has had at least 20 psychiatric hospitalizations ranging from 3 days to 9 months.

3 Using Computerized Neurocognitive Exercises with a Cognitive Remediation Approach

3.1 Sagacity

Sagacity™ is an interactive computerized neurocognitive training program designed by the software company FishBuffalo. The Sagacity™ suite consists of twenty modules that are specifically designed to improve neurocognitive areas including, but not limited to, attention (verbal, non-verbal, auditory, and visual), memory, concentration, vigilance, mathematical thinking, reaction time, problem-solving, and processing speed. Each module has a wide range of settings and levels of difficulty, allowing the clinician to individualize a comprehensive training path for each patient. The program is designed to help clinicians treat individuals who have neuro- and social cognitive declines due to medical, developmental, learning, and/or psychiatric conditions.

3.2 Combining Computerized Training with Cognitive Remediation Treatment Session

These cognitive remediation sessions are structured in five phases: (1) brief psychotherapy session, (2) computerized neurocognitive training, (3) free discussion, (4) computerized neurocognitive training, and (5) real-life application.

Phase 1. Procedure of the Psychotherapy Session

In the first phase of the session, the clinician checks in with the patient about current and/or past physical or emotional symptoms, thought distortions, and cognitive concerns. Determining the patient's current emotional and physical states are important to discuss and assess before doing any neurocognitive exercises as these states take priority over cognitive tasks. During this phase, the clinician works with the patient on addressing emotional discomfort as well as creating and implementing adaptive coping mechanisms. At the end of this phase, the patient can begin training.

Phase 2. Computerized Neurocognitive Training

In phase two, the clinician encourages the patient to work on five components before starting the exercise: (1) cognitive anchors, (2) cognitive strategies, (3) emotional strategies, (4) thought strategies, and (5) mechanical/environmental strategies.

Cognitive anchors: Cognitive anchors are an important part of this cognitive remediation treatment approach. They are any visual and/or auditory pieces of information that are present in the task that help the patient have a point of reference when approaching a simple or difficult cognitive task. These anchors are gathered by having a thorough descriptive observation, and not inference observation, of the task presented. Some questions during this phase are: "Describe what you see on the computer screen". "Are there any forms, shapes, words, colors that you see? If so, describe them".

During this phase, the patient is highly encouraged to say out loud what he or she is seeing and thinking (i.e., protocol analysis), so both the clinician and the patient can understand the approach to the given task.

Cognitive strategies: Cognitive strategies are mental maps that can help the patient approach cognitive tasks. To create cognitive strategies, the patient is asked to infer about the goal of the exercise while using previous cognitive anchors to create a strategy. Some of the questions to use here are: "What is the goal of the task?" "Based on your previous description, what can you use to better understand the task?" "What do you see that you can use to create an initial strategy?" "What is your initial strategy?"

Once the patient has a preliminary cognitive strategy, they can begin to tackle the neurocognitive exercise. Thoughts and cognitive strategies are addressed and shaped by using Self-Instructional Training and Protocol Analysis principles. Self-Instructional Training helps patients perform cognitive tasks by self-controlling thoughts between a stimulus situation, including cognitive tasks, and their response to those situations (Meichenbaum 1977), while teaching them to interpolate adaptive mechanisms to improve cognitive performance. Protocol Analysis is used to help the patient figure out, in-vivo, how they are tackling certain exercises (Lundgrén-Laine 2010). It also allows the clinician to modify and challenge patient strategies in order to improve cognitive tasks.

Emotional strategies: Feeling anxious, nervous, unmotivated, and/or approaching neurocognitive tasks with low self-esteem is very common. These feelings arise due to the cognitive limitations attached to having a severe and chronic psychiatric condition such as schizophrenia. Interpersonal Therapy principles and Cognitive Behavioral Therapy (CBT) for psychosis are used to address these issues. Creating adaptive coping mechanisms for emotional regulation before, during, and after each cognitive exercise is an important part of this step. Some of the coping mechanisms used here can include breathing exercises, behavioral activation, and mindfulness.

Thought strategies: The patient is trained to be aware of the different types of thoughts that can be present during the neurocognitive training. In patients with schizophrenia, the most frequently seen thoughts are tangential, disorganized, intrusive, negative, and obsessive thoughts, as well as negative self-criticism. For that reason, it is important to create adaptive coping mechanisms that can be easily deployed if these thoughts become a distraction and/or trigger undesirable emotions. Thought distortions are managed with Self-Control Therapy principles, in particular, self-monitoring strategies to attend to negative events in order to avoid the negative consequences of these thoughts. Another technique used here is CBT for psychosis when active auditory and visual hallucinations are also present.

Mechanical/environmental strategies: The patient is taught how to create strategies to improve performance by incorporating cues for their immediate environment or contextual settings during the training. Some examples of these strategies include paying attention to their body position (e.g., hands close to the keyboard or touch screen, sitting in front of the computer or tablet) and blocking out visual and auditory distractions (an open window or door, a person in another room, chair or desk squeaking when they move, etc.).

Phase 3. Free Discussion

In the third phase, the patient is encouraged to talk freely about any topic for 10 minutes. The goal of this phase is to allocate some time for the patient to debrief about the cognitive tasks or to take a brief break. During this phase, the clinician will intervene if the patient brings up a difficult topic that is moving towards eliciting clinical vulnerability, such as severe trauma. If this happens, the clinician should address those concerns before moving forward with the second stage of cognitive remediation training.

Phase 4. Neurocognitive Training

The fourth phase begins immediately following the third phase and follows the same structure as the second phase.

Phase 5. Real-Life Application

In the fifth (and final) phase, the clinician and patient discuss current cognitive anchors (emotional and thought anchors) that can be applied in real-life scenarios. The patient's disability is demonstrated through concrete experiences rather than with abstract discussion. Errors, strategies, and accurate decisions are discussed during this phase too. In other words, this last phase is important to translate any past and newly acquired skills into the patient's daily activities.

4 Intervention treatment with Sagacity

On initial evaluation, Peter was found to have significant impairments in concentration and attention. For this reason, he started cognitive rehabilitation therapy using a computerized intervention to treat these cognitive deficits.

At first, Peter was defensive and did not want to interact with the computer. Upon further exploration, he reported that he did not have a lot of experience with computers and that he felt ashamed because he knew he had problems with attention. Moreover, Peter reported that his "voices" prohibited him from interacting with the computer because the software was designed to "steal his thoughts". He stated that the voices also called him names and threatened to tell each of his family members about his "secret spy duties" and "deviated sexual identity". For these reasons, Peter was afraid and avoidant to start cognitive remediation therapy. Nevertheless, after a couple of weeks of talking to him about the benefits of this treatment, Peter accepted to continue.

He began the cognitive treatment using the Reaction Time module. This module proposes to train the patient to react to an external stimulus (auditory and visual) by aiming to gradually decrease the time it takes for the patient to respond to said stimulus. Thus, it aims to improve attention, alertness, and reaction time by fostering the ability to rely on internal cues to respond to external stimuli. This, in turn, translates to improvements in patients' reactions to environmental, everyday stimuli they might encounter in their social interactions. The task consists of pressing the spacebar as soon as the patient sees and hears an audiovisual cue prompting a reaction. Each attempt lasts no longer than 5 seconds, and 10 attempts are usually done per training segment. The module has different levels of difficulty that range from requiring the patient to react within 500 milliseconds of a stimulus to having to react within 175 milliseconds of said stimulus, with decreasing support from visual and auditory cueing aids.

Based on Peter's low attention and concentration, his treatment plan started with the easiest level of difficulty (500 milliseconds with all the audio and visual aids). The end goal was for him to react within 175 milliseconds with no aids at all, which is a clinically favorable outcome even for individuals with no cognitive deficits.

Initially, Peter's reaction time was extremely slow (cf. Table 8.1). When he was asked about his performance, he responded, "It is hard for me to pay attention to the sounds made by the computer because there is a lot of noise [. . .] [the] voices are talking to me while I try to do the exercises".

Table 8.1 shows Peter's reaction times in milliseconds. Peter underwent 7 training sessions, and although the number of sessions and trials are too few to show a statistically significant difference between sessions, clinically speaking, these numbers show extraordinary progress in his attention, reaction time and, most importantly, his ability to control his cognition. To put these numbers into context, we divided the scores into color code ranges to illustrate his progression in reaction time over the course of the training program. A short reaction time is a favorable outcome because it means that Peter reacted appropriately quickly to the cue: a reaction time from 0.0–0.50 seconds (or 0–500 milliseconds) was the target reaction time; 0.51–1 seconds showed mild impairment; 1.01 to 1.5 seconds meant moderate impairment; 1.51 to 2.0 seconds was high impairment; lastly, 2.0+ seconds showed extreme impairment.

Table 8.1: Reaction Times at Each Training Session Tab. 1. Reaction Times at Each Training Session.

Session	Trial 1	Trial 2	Trial 3	Trial 4	Trial 5	Trial 6	Trial 7	Trial 8	Trial 9	Trial 10
Session 1	29000.47	22000.51	N/A	N/A	N/A	N/A	N/A	N/A	N/A	N/A
Session 2	-3825	495	705	1335	1620	1155	-4680	-4365	1035	960
Session 3	1290	1845	1380	1410	1755	1830	1665	1335	1350	735
Session 4	1500	1740	1665	1830	1530	1590	1815	1215	1770	1170
Session 5	1755	1500	1365	360	1170	855	1485	645	1020	1770
	1065	1185	1545	1575	1215	1140	1215	1785	-3585	1770
Session 6	795	1620	420	600	705	825	705	900	1155	465
	1875	1125	1500	1275	1485	1485	1305	1830	1485	1470
Session 7	675	510	1095	540	765	1485	1485	795	1035	900
	375	630	915	1065	1215	1125	495	1395	780	1500
	660	1020	1350	810	1530	795	630	585	1260	1020

0.0 to 0.50 seconds	On Target
0.51 to 1.0 seconds	Mild Impairment
1.01 to 1.5 seconds	Moderate Impairment
1.51 to 2.0 seconds	High Impairment
> 2.01 seconds	Extreme impairment

4.1 Attention and Clinical Progression

Session 1: In his first training session, Peter only tolerated 2 out of 10 attempts at pressing the spacebar when he saw and heard the cueing stimulus. This module requires the patient to react within a time frame of 0.0 to 0.5 seconds. His reaction times for these trials were 29.47 seconds and 22.51 seconds for the first and second trials, respectively. He discontinued training because he reported having "too many auditory hallucinations". He also stated, "the voices are too frequent and too loud, and it is impossible to do something about it, I can't, and I don't want to do it anymore". He was anxious and frustrated because the auditory hallucinations did not go away. These scores show that his attention and concentration skills were extremely impaired at baseline, given the 4300% delay in his response compared to target reaction time.

Session 2: Peter reported feeling afraid the voices would talk back to him or threaten him for doing the exercises, especially since, according to him, during that time the voices were too loud. Given the intense emotions Peter was experiencing, the previously described therapeutic principles were implemented in this session, with successful results. First, Peter was able to tolerate ten trials. Second, his reaction time improved significantly since the scores now ranged from −4.68 seconds (meaning that he reacted 4.68 seconds *before* the target time, an impulsive response) to a maximum of 1.33 seconds (which shows a delayed response). Third, 30% of his scores were in the 2.0+ seconds range; 30% landed in the 1.0 to 1.5 seconds range; 20% were in the 0.5 to 1.0 seconds score range; 10% were in the 1.5 to 2.0 seconds range; and the last 10% were in the correct reaction time range of 0.0 to 0.5 seconds. These results show not only that Peter was trying to react on time but that he was motivated to work on his cognitive performance and improve his attention.

Session 3: By the third session, Peter once again engaged in the cognitive task and was able to complete ten trials. His scores were as follows: 40% of his reaction times were in the 1.5 to 2.0 seconds range, 50% were in the 1.0 to 1.5 seconds range, and 10% stayed in the 0.5 to 1.0 seconds range. These numbers show that his reaction times were improving and that he did not have any impulsive attempts (as no reactions happened before target time), showing more control of his impulses. However, Peter reported that the voices were coming and going, and it made him feel anxious. In addition to anxiety, Peter felt confused because he did not know what was happening to him nor to his auditory hallucinations. In further exploration, he initially thought that the reason the voices were absent for some time was because they were playing games and were plotting to harm him.

Session 4: In this session, as expected in any cognitive training, the numbers did not change drastically compared to the previous session. This time, 70% of his scores landed in the 1.5 to 2.0 seconds range, while the remainder 30% of the scores were in the 1.0 to 1.5 seconds range. Although these numbers were far from close to the correct and expected target time, Peter continued not having impulsive reactions during the exercise. Emotionally speaking, Peter continued with a sense of ambivalence and confusion because of the frequency of the voices and their content. Sometimes he was distracted by the voices and became upset by the things for which he was blamed and felt responsible. For example, one of the voices said, "You are a failure because you can't focus on things". At this time, Peter started to understand that his anxiety was a key player in this cognitive training, and he noticed that when his anxiety was under better control, he had better chances of sustaining attention to other things that were not his auditory hallucinations.

Session 5: This session was particularly interesting. Peter was able to complete twenty trials, and his overall scores improved. In this session, 50% of his scores were in the 1.0 to 1.5 seconds range; 30% were in the 1.5 to 2.0 seconds range; 10% were in the 0.5 to 1.0 seconds range; 5% of the scores were higher than 2 seconds, and 5% were in the correct target time of 0.0 to 0.5 seconds. Among the twenty trials, just 5% of his attempts showed impulsivity. The clinical interpretation of this session indicates that, little by little, Peter was gaining more control of his attention and impulses, while also building more cognitive stamina and endurance. It also showed that, even though he continued experiencing auditory hallucinations, Peter was able to silence, or put on hold, at least for a few milliseconds, the internal auditory stimuli, thus allowing him to focus his attention on the cognitive training (external stimuli) rather than on his internal processes.

Session 6: By this session, Peter was able to complete twenty trials without having mental fatigue. His scores, as shown in the table, improved significantly. Forty-five percent of his scores landed in the 1.0 to 1.5 seconds range; 30% in the 0.5 to 1.0 seconds range; 15% in the 1.5 to 2.0 seconds range; and 10% landed in the correct target time. In this session, there were no impulsive attempts and there was more control of his sustained attention. Most importantly, Peter expressed, "It is the first time in 10 years that I have been able to pay attention to one thing at a time and for a long, long time [. . .] I feel dizzy, but I also noticed that I can hear the noise made by the computer, and for the first time in years, I did not hear my voices". He also replied, "I feel relieved because I didn't know I have some control over my auditory hallucinations, but I am also afraid because I don't know if this is normal". He even expressed concerns about this situation because he did not know what to do if he does not hear the voices anymore.

Session 7: This session was exceptional. Peter was now able to complete 30 trials without feeling anxious, tired, or inattentive. In this session, 43.3% of his scores were in the 1.0 to 1.5 seconds range; 46.6% of his scores landed in the 0.5 to 1.0 seconds range; 3.3% of his scores were in the 1.5 to 2.0 seconds range; and 6.6% were in the correct target reaction time. These results are impressive, given that they show that Peter was able to respond more precisely to the external stimuli, rather than spending time on his internal stimuli (i.e., auditory hallucinations), as was the case in previous sessions. By this session, in Peter's words, "I am able to hear the noise from the computer more than the noises from the [sic. auditory hallucination] voices [. . .] I can do this now". He added, "I know the voices are there but I feel like I can now encapsulate the voices, so I do not need to completely devote my attention to them [. . .]. It is nice to hear also the voices of my doctors, friends, and family members as well". This was the last neurocognitive training session for Peter, and it became evident how much his auditory hallucinations decreased between the first and last sessions. They were still present, but to a much lesser degree and causing significantly less impairment, as they were during the first session.

An essential factor that contributed to Peter's success was the clinician's ability to integrate Peter's symptoms, emotions, and training requirements into these sessions and adapt the neurocognitive training and level of difficulty to a level that was challenging but accessible to him. Neurocognitive training is not meant to be implemented as a standardized, structured program. A properly trained clinician ought to incorporate different aspects of psychotherapy and neurocognitive training to tailor the therapy session according to the patient's presenting symptoms and cognitive deficits, as was done with Peter. This multifaceted approach to cognitive remediation is imperative to maximize clinical outcomes in schizophrenia.

5 Discussion

The clinical case discussed in this chapter illustrates the typical clinical picture of an individual who was diagnosed with one of the most debilitating psychiatric illnesses and describes how his symptoms manifested throughout the course of his illness. Despite the common belief that traditional psychological and pharmaceutical agents treat and manage this psychiatric illness, the typical interventions are far from having a significant impact on repairing neuro- and social cognitive skills, which leaves patients with a major obstacle to recovery. Despite a significant and growing body of empirical literature on different and dissociable aspects of cognition in schizophrenia, the majority of treatment interventions tend to approach cognitive deficits as a package.

This clinical case also demonstrates that combining cognitive remediation principles with computerized neurocognitive training can improve positive and negative symptoms in schizophrenia, and most importantly, restore essential neurocognitive functions such as attention that are the pillar to improving social functioning and community integration.

Peter's case depicts that improvements in attention and vigilance served as mediators to 'encapsulate' his auditory hallucinations. In other words, this treatment approach allowed Peter to switch and sustain attention by choice to external stimuli rather than to his internal stimuli, which in turn allowed him to interact more with his external environment. Furthermore, this clinical case highlights three fundamental aspects of attention in the context of schizophrenia: (1) how does one learn to pay attention and understand internal stimuli, including hallucinations, (2) how does one deploy attention strategically towards the external world, and lastly, (3) what does it take to be able to skilfully and appropriately switch the attentional focus back and forth between internal and external stimuli. These aspects may be relevant not only to clinical practice but also to rethinking some core issues of attention as a way of interacting with both internal and external worlds, outside of the domain of psychiatry.

5.1 Attention to Internal and External Worlds and Switching between the Two

In human cognition, sources of stimuli can come both from external and from internal environments. We can argue that when Peter is interacting with his auditory hallucinations (i.e., internal stimuli) while excluding other stimuli (i.e., external conversations), he is showing an active attentional mechanism, since he is capable of engaging with his voices for minutes and sometimes hours. This suggests that he may not have attentional impairments in terms of basic attentional capacity to attend to specific internal stimuli. However, schizophrenia is typically characterized by diminished capacity to sustain attention on the external, not internal, tasks (Hagh-Shenas et al. 2002). Therefore, from a clinical point of view, his attention is impaired given that he cannot voluntarily select nor sustain his attention by choice, for instance, from an internal (e.g., voices) to an external stimulus (e.g., other people, external environment, etc.). This implies that his specific challenge lies in the ability to switch back and forth and sustain attention on either external or internal cues, depending on the demands of specific contexts and situations.

6 Conclusions: Attention as a Mediator Mechanism

This clinical case illustrates that improvements in attention and vigilance can serve as cognitive mediators, in this case helping the patient to encapsulate his auditory hallucinations (i.e., psychosis) and expand his social skills.

The cognitive domain of attention is of utmost importance. Attention is the first cognitive domain that needs to be mastered before more complex tasks can be performed. For instance, in order to encode information into working (and subsequently long-term) memory, individuals first need to attend to said information. In a similar fashion, without attending to a task, it becomes impossible to fulfill it or complete it successfully. Attention must first be addressed prior to advancing to any further cognitive functions, such as verbal and working memory, processing speed, problem solving, and executive function. This suggests that a bottom-up cognitive remediation training approach, which focuses on the most basic attentional skills, may strengthen the foundation to allow for more complex attentional capacities and other higher-order cognitive competencies, including social skills.

Overall, when Peter's distress was attenuated and his basic attentional skills were improved, Peter was able to 'encapsulate' (diminish or silence) his auditory hallucinations. This can be understood by understanding the relationship between the patient's neurocognitive state and symptoms of psychosis. When emotional vulnerability is high, the patient's attention is 'hijacked' by the internal stimuli (voices). This consumes the majority of the attentional bandwidth available to the patient at any given moment. Therefore, attenuating the emotional response liberates a significant number of attentional resources, which become suitable to engage skilfully with the external world. CRT then functions as an anchor by which the brain is retrained to volitionally switch and sustain attention when appropriate. Previous work showed that attention training plays an important role in mediation of auditory hallucinations (Ensum and Morrison 2003). Thus, targeting attentional skills in this clinical case revealed an important mediator role for attention in managing both cognitive skills necessary for interacting with the world and in encapsulating distressing auditory hallucinations.

In sum, the reduction of the voices and associated distress helped Peter engage in regular day-to-day activities and had a dramatic impact on his general level of function. Needless to say, Peter was still affected by a chronic form of schizophrenia and continued to experience cognitive problems associated with this disorder; however, through the tandem of CRT and technology he re-learned to cope with the voices, gained control over the intrusiveness of his symptoms (internal stimuli), and altogether experienced a reduction in the emotional distress caused by his disease. This shows that Peter was able to develop new and more functional self-regulatory processing pathways and increased his cognitive reservoirs, which, in turn, led to an improvement in perceived control over his symptoms and to a significant amelioration of his daily functioning and overall health and wellbeing.

References

Allen, P.; Aleman, A.; Mcguire, P.K. (2007): "Inner Speech Models of Auditory Verbal Hallucinations: Evidence from Behavioural and Neuroimaging Studies". *International Review of Psychiatry* 19:4, 407–415.

American Psychiatric Association (2013): *Diagnostic and Statistical Manual of Mental Disorders* (DSM-5®). Arlington: American Psychiatric Publishing.

Beck, A.T.; Rector, N.A. (2003): "A Cognitive Model of Hallucinations". *Cognitive Therapy and Research* 27:1, 19–52.

Bell, M.D.; Fiszdon, J.M.; Bryson, G. (2009): "Attention Training in Schizophrenia: Differing Responses to Similar Tasks". *Journal of Psychiatric Research* 43:4, 490–496.

Ben-Yishay, Y.; Diller, L.; Rattok, J.; Ross, B.; Schaier, A.; Scherger, P. (1979): *Working Approaches to Remediation of Cognitive Deficits in Brain Damaged Persons*. Supplement to the Seventh Annual Workshop for Rehabilitation Professionals Department of Behavioral Science. New York: New York University Medical Center, Institute for Rehabilitation Medicine.

Boake, C. (1991): "History of Cognitive Rehabilitation following Head Injury." In: J. S. Kreutzer; P. H. Wehman (eds): *Cognitive Rehabilitation for Persons with Traumatic Brain Injury: A Functional Approach*. Baltimore: Paul H. Brookes Publishing, 3–12.

Bowie, C.R.; Harvey, P.D. (2006): "Cognitive Deficits and Functional Outcome in Schizophrenia". *Neuropsychiatric Disease and Treatment* 2:4, 531–536.

Brenner, H.D.; Roder, V.; Hodel, B.; Kienzle, N.; Reed, D.; Liberman, R.P. (1994): *Integrated Psychological Therapy for Schizophrenic Patients (IPT)*. Göttingen: Hogrefe & Huber Publishers.

Cangas, A.J.; Errasti, J.M.; García-Montes, J.M.; Álvarez, R.; Ruiz, R. (2006): "Metacognitive Factors and Alterations of Attention Related to Predisposition to Hallucinations". *Personality and Individual Differences* 40:3, 487–496.

Cahill, C.; Frith, C.D. (1996): "A Cognitive Basis for the Signs and Symptoms of Schizophrenia". In: C. Pantelis; H.E. Nelson; T.R.E. Barnes (eds.): *Schizophrenia: A Neuropsychological Perspective*. New York, NY: John Wiley and Sons, 373–395.

Cardella, V., & Gangemi, A. (2019). "From the Categorical to the Dimensional Approach in Psychopathology: The Case of Auditory Hallucinations". *Mediterranean Journal of Clinical Psychology*, 7(3).

Cella, M.; Preti, A.; Edwards, C.; Dow, T.; Wykes, T. (2017): "Cognitive Remediation for Negative Symptoms of Schizophrenia: a Network Meta-Analysis". *Clinical Psychology Review* 52, 43–51.

Cella, M.; Reeder, C.; Wykes, T. (2014): "It is all in the Factors: Effects of Cognitive Remediation on Symptom Dimensions". *Schizophrenia Research* 156:1, 60–62.

Cella, M.; Wykes, T. (2013): "Understanding Processing Speed – its Subcomponents and their Relationship to Characteristics of People with Schizophrenia". *Cognitive Neuropsychiatry* 18:5, 437–451.

Charlson, F.J.; Ferrari, A.J.; Santomauro, D.F.; Diminic, S.; Stockings, E.; Scott, J.G.; McGrath, J.J.; Whiteford, H.A. (2018): "Global Epidemiology and Burden of Schizophrenia: Findings from the Global Burden of Disease Study 2016". *Schizophrenia Bulletin* 44:6, 1195–1203.

Czepielewski, L.S.; Massuda, R.; Goi, P.; Sulzbach-Vianna, M.; Reckziegel, R.; Costanzi, M.; Kapczinski, F.; Rosa, A.R.; Gama, C.S. (2015): "Verbal Episodic Memory along the Course of Schizophrenia and Bipolar Disorder: A New Perspective". *European Neuropsychopharmacology* 25:2, 169–175.

David, A. (1999): "Auditory Hallucinations: Phenomenology, Neuropsychology, and Neuroimaging Update". *Ada Psychiatrica Scandinavica Supplement* 395, 95–104.

DeLisi, L.E. (2008): "The Concept of Progressive Brain Change in Schizophrenia: Implications for Understanding Schizophrenia". *Schizophrenia Bulletin* 34:2, 312–321.

Eack, S.M.; Greenwald, D.P.; Hogarty, S.S.; Cooley, S.J.; DiBarry, A.L.; Montrose, D.M.; Keshavan, M.S. (2009): "Cognitive Enhancement Therapy for Early-Course Schizophrenia: Effects of a two-year Randomized Controlled Trial". *Psychiatric Services* 60:11, 1468–1476.

Eack, S.M.; Hogarty, G.E.; Cho, R.Y.; Prasad, K.M.; Greenwald, D.P.; Hogarty, S.S.; Keshavan, M.S. (2010): "Neuroprotective Effects of Cognitive Enhancement Therapy against Gray Matter Loss in Early Schizophrenia: Results from a 2-year Randomized Controlled Trial". *Archives of General Psychiatry* 67:7, 674–682.

Eack, S.M.; Mesholam-Gately, R.I.; Greenwald, D.P.; Hogarty, S.S.; Keshavan, M.S. (2013): "Negative Symptom Improvement during Cognitive Rehabilitation: Results from a 2-year Trial of Cognitive Enhancement Therapy". *Psychiatry Research* 209:1, 21–26.

Ensum, I.; Morrison, A.P. (2003): "The Effects of Focus of Attention on Attributional Bias in Patients Experiencing Auditory Hallucinations". *Behaviour Research and Therapy* 41:8, 895–907.

Farreny, A.; Aguado, J.; Ochoa, S.; Haro, J.M.; Usall, J. (2013): "The Role of Negative Symptoms in the Context of Cognitive Remediation for Schizophrenia". *Schizophrenia Research* 150:1, 58–63.

Fatouros-Bergman, H.; Cervenka, S.; Flyckt, L.; Edman, G.; Farde, L. (2014): "Meta-Analysis of Cognitive Performance in Drug-Naïve Patients with Schizophrenia". *Schizophrenia Research* 158:1–3, 156–162.

Frith, C.D. (1987): "The Positive and Negative Symptoms of Schizophrenia Reflect Impairments in Perception and Initiation of Action". *Psychological Medicine* 17, 631–648.

Frith, C.D. (1992): *The Cognitive Neuropsychology of Schizophrenia*. Hove, U.K.: Lawrence Erlbaum.

Fujii, D.E.; Wylie, A.M.; Nathan, J.H. (2004): "Neurocognition and Long-Term Prediction of Quality of Life in Outpatients with Severe and Persistent Mental Illness". *Schizophrenia Research* 69:1, 67–73.

Fusar-Poli, P.; Papanastasiou, E.; Stahl, D.; Rocchetti, M.; Carpenter, W.; Shergill, S.; McGuire, P. (2015): "Treatments of Negative Symptoms in Schizophrenia: Meta-Analysis of 168 Randomized Placebo-Controlled Trials". *Schizophrenia Bulletin* 41:4, 892–899.
Garrido, G.; Penadés, R.; Barrios, M.; Aragay, N.; Ramos, I.; Vallès, V.; Faixa, C.; Vendrell, J.M. (2017): "Computer-Assisted Cognitive Remediation Therapy in Schizophrenia: Durability of the Effects and Cost-Utility Analysis". *Psychiatry Research* 254, 198–204.
Gold, J.M.; Fuller, R.L.; Robinson, B.M.; Braun, E.L.; Luck, S.J. (2007): "Impaired Top-Down Control of Visual Search in Schizophrenia". *Schizophrenia Research* 94:1–3, 148–155.
Goldberg, T.E.; Weinberger, D.R.; Berman, K.F.; Pliskin, N.H.; Podd, M.H. (1987): "Further Evidence for Dementia of the Prefrontal Type in Schizophrenia? A controlled Study of Teaching the Wisconsin Card Sorting Test". *Archives of General Psychiatry* 44:11, 1008–1014.
Green, M.F.; Kern, R.S.; Braff, D.L.; Mintz, J. (2000): "Neurocognitive Deficits and Functional Outcome in Schizophrenia: Are we Measuring the 'Right Stuff'?". *Schizophrenia Bulletin* 26:1, 119–136.
Green, M.F. (2016): "Impact of Cognitive and Social Cognitive Impairment on Functional Outcomes in Patients with Schizophrenia". *The Journal of Clinical Psychiatry* 77 Suppl. 2, 8–11.
Green, M. F.; Horan, W. P.; Lee, J. (2015): "Social Cognition in Schizophrenia". *Nature Reviews Neuroscience* 16:10, 620–631.
Gur, R.C.; Gur, R.E. (2013): "Memory in Health and in Schizophrenia". *Dialogues in Clinical Neuroscience* 15:4, 399–410.
Hagh-Shenas, H.; Toobai, S.; Makaremi, A. (2002): "Selective, Sustained, and Shift in Attention in Patients with Diagnoses of Schizophrenia". *Perceptual and Motor Skills* 95:3 Pt. 2, 1087–1095.
Heinrichs, R.W.; Zakzanis, K.K. (1998): "Neurocognitive Deficit in Schizophrenia: A Quantitative Review of the Evidence". *Neuropsychology* 12:3, 426–445.
Hogarty, G.E.; Flesher, S.; Ulrich, R.; Carter, M.; Greenwald, D.; Pogue-Geile, M.; Kechavan, M.; Cooley, S.; DiBarry, A.L.; Garrett, A.; Parepally, H. (2004): "Cognitive Enhancement Therapy for Schizophrenia: Effects of a 2-year Randomized Trial on Cognition and Behavior". *Archives of General Psychiatry* 61:9, 866–876.
Hogarty, G.E.; Flesher, S. (1999a): "Developmental Theory for a Cognitive Enhancement Therapy of Schizophrenia". *Schizophrenia Bulletin* 25:4, 67–692.
Hogarty, G.E.; Flesher, S. (1999b): "Practice Principles of Cognitive Enhancement Therapy for Schizophrenia". *Schizophrenia Bulletin* 25:4, 693–708.
Hogarty, G.E.; Greenwald, D.P.; Eack, S.M. (2006): "Cognitive Enhancement Therapy: Durability and Mechanism of Effects". *Psychiatric Services* 57:12, 1751–1757.
Ichinose, M.; Park, S. (2019): "Mechanisms Underlying Visuospatial Working Memory Impairments in Schizophrenia". *Current Topics in Behavioural Neurosciences* 41: 345–67.
Johnson, R. (1978): *The Anatomy of Hallucinations*. Chicago, IL: Nelson Hall.
Jones, P.; Murray, R.; Rodgers, B.; Marmot, M. (1994): "Child Developmental Risk Factors for Adult Schizophrenia in the British 1946 Birth Cohort". *The Lancet* 344:8934, 1398–1402.
Keshavan, M.S.; Mehta, U.M.; Padmanabhan, J.L.; Shah, J.L. (2015): "Dysplasticity, Metaplasticity and Schizophrenia: Implications for Risk, Illness and Novel Interventions". *Development and Psychopathology* 27:2, 615–635.
Kraepelin, E. (1919): *Dementia Praecox and Paraphrenia*. 8th ed. Edited by G.M. Robertson, trans. by R.M. Barclay. Edinburgh: HardPress Publishing.

Lundgrén-Laine, H.; Salanterä, S. (2010): "Think-Aloud Technique and Protocol Analysis in Clinical Decision-Making Research". *Qualitative Health Research* 20:4, 565–575.

McGurk, S.R.; Twamley, E.W.; Sitzer, D.I.; McHugo, G.J.; Mueser, K.T. (2007): "A Meta-Analysis of Cognitive Remediation in Schizophrenia". *American Journal of Psychiatry* 164:12, 1791–1802.

Medalia, A.; Aluma, M.; Tryon, W.; Merriam, A.E. (1998): "Effectiveness of Attention Training in Schizophrenia". *Schizophrenia Bulletin* 24:1, 147–152.

Medalia, A.; Revheim, N.; Casey, M. (2000): "Remediation of Memory Disorders in Schizophrenia". *Psychological Medicine* 30:6, 1451–1459.

Medalia, A.; Revheim, N.; Herlands, T. (2002): *Remediation of Cognitive Deficits in Psychiatric Patients: a Clinician's Manual*. New York: The Authors.

Medalia, A.; Thysen, J. A. (2010): "Comparison of Insight into Clinical Symptoms versus Insight into Neuro-Cognitive Symptoms in Schizophrenia". *Schizophrenia Research* 118:1–3, 134–139.

Meichenbaum, D.; Cameron, R. (1973): "Training Schizophrenics to Talk to Themselves: A Means of Developing Attentional Controls". *Behavior Therapy* 4:4, 515–534.

Meichenbaum, D. (1977): "Cognitive Behaviour Modification". *Cognitive Behaviour Therapy* 6:4, 185–192.

Mohn, C.; Torgalsbøen, A.K. (2018): "Details of Attention and Learning Change in First-Episode Schizophrenia". *Psychiatry Research* 260, 324–330.

Oken, B.S.; Salinsky, M.C.; Elsas, S. (2006): "Vigilance, Alertness, or Sustained Attention: Physiological Basis and Measurement". *Clinical Neurophysiology* 117:9, 1885–1901.

Rayner, L.H.; Lee, K.H.; Woodruff, P.W. (2015): "Reduced Attention-Driven Auditory Sensitivity in Hallucination-Prone Individuals". *The British Journal of Psychiatry* 207:5, 414–419.

Robbins, T.W. (20055): "Synthesizing Schizophrenia: A Bottom-up, Symptomatic Approach". *Schizophrenia Bulletin* 31:4, 854–864.

Sánchez, P.; Peña, J.; Bengoetxea, E.; Ojeda, N.; Elizagárate, E.; Ezcurra, J.; Gutiérrez, M. (2014): "Improvements in Negative Symptoms and Functional Outcome after a New Generation Cognitive Remediation Program: A Randomized Controlled Trial". *Schizophrenia Bulletin* 40:3, 707–715.

Sass, L. A., & Parnas, J. (2003). "Schizophrenia, Consciousness, and the Self". *Schizophrenia Bulletin*, 29(3), 427–444.

Savla, G.N.; Vella, L.; Armstrong, C.C.; Penn, D.L.; Twamley, E.W. (2013): "Deficits in Domains of Social Cognition in Schizophrenia: A Meta-Analysis of the Empirical Evidence". *Schizophrenia Bulletin* 39:5, 979–992.

Schaefer, J.; Giangrande, E.; Weinberger, D.R.; Dickinson, D. (2013): "The Global Cognitive Impairment in Schizophrenia: Consistent over Decades and Around the World". *Schizophrenia Research* 150:1, 42–50.

Stratta, P.; Mancini, F.; Mattei, P.; Casacchia, M.; Rossi, A. (1994): "Information Processing Strategy to Remediate Wisconsin Card Sorting Test Performance in Schizophrenia: A Pilot Study". *American Journal of Psychiatry* 151:6, 915–918.

Turkington, D.; Lebert, L.; Spencer, H. (2016): "Auditory Hallucinations in Schizophrenia: Helping Patients to Develop Effective Coping Strategies". *BJPsych Advances* 22:6, 391–396.

Wykes, T.; Huddy, V.; Cellard, C.; McGurk, S.R.; Czobor, P. (2011): "A Meta-Analysis of Cognitive Remediation for Schizophrenia: Methodology and Effect Sizes". *American Journal of Psychiatry* 168:5, 472–485.

Short Biography

Betzamel López is a medical student at Harvard Medical School in Boston, MA, USA. She obtained her undergraduate degree at Northeastern University, also in Boston, where she studied Economics along with the pre-medical curriculum, focusing on coursework such as biology and chemistry. While in college, she completed an internship at the Beth Israel Deaconess Medical Center and the Massachusetts General Hospital, where she studied schizophrenia and the effectiveness of different treatment modalities to treat this disease. She will be graduating from Harvard with a Doctor of Medicine (M.D.) degree in the spring of 2022, and her clinical interests include improving patient access to health care and patient outcomes as it pertains to the existing socioeconomic disparities in medicine.

Luis Sandoval, PhD, is a clinical researcher psychologist faculty at Harvard Medical School. He is an expert in assessing and treating neurocognition, mood, and psychosis spectrum disorders. Dr. Sandoval has demonstrated experience in designing, coordinating, and implementing clinical-research protocols in multidisciplinary settings and with multiculturally diverse teams. Dr. Sandoval is known for his keen ways to combine classical therapies in mental and human behavior with digital tools. Over the past 15 years years, he has designed, applied, and led innovative research in Digital Psychiatry and e-health. His background in clinical practice and academia has given him unique skills to understand how to integrate current technologies with mental health care, and most importantly, how to humanize digital tools so they can be better received and used by clients. Currently, Dr. Sandoval trains and supervises several hospitals across New England on how to improve neurocognition in clients with psychosis spectrum disorders.

Bas de Boer
Chapter 9
Attending to Your Lifestyle: Self-Tracking Technologies and Relevance

Abstract: This chapter intends to develop a phenomenological analysis of how self-tracking technologies structure attention. The goal of this analysis is to highlight how the use of such technologies – ones that are already used frequently, and will likely become part of the lives of many more individuals in the future – turn our body and habits into objects of relevance. The outline is as follows: First, I discuss some technological developments that intend to promote a healthy lifestyle (1). Second, I suggest, elaborating on insights from postphenomenology, that technologies mediate what appears as relevant by shaping the relation between human beings and the world (2). Third, I further unpack the implications of this idea by connecting it to Alfred Schütz's theory of relevance. I suggest how self-tracking technologies shape systems of topical, interpretational, and motivational relevances (3). Fourth, I argue that what stands out as relevant is constituted through sedimented habits, as well as the projects in which people engage (4). Fifth, I suggest that technologies that are designed to help people pursue a healthy lifestyle turn the attention of users towards their own body in ways that often remain unnoticed, such that it might become increasingly difficult to turn attention towards other objects and projects that might be equally relevant (5). In conclusion, I suggest that such technologies privilege a particular view of health that might become an unquestionable element of the lifeworld, and that this view might remain unnoticed when not being subject to careful analysis (6).

Introduction

What we attend to, how we attend to it, and when we do so often involves a relation with a technology of some kind. For example, the algorithms of streaming services such as Netflix make suggestions a which movies and TV shows to watch on the basis of data of what we have watched earlier, thereby hooking our attention to the specific platform. Furthermore, companies like Amazon send targeted advertisements on the basis of the products we have bought in the past, often at moments when we are concerned with something else. Currently, many technologies are

https://doi.org/10.1515/9783110647242-010

deliberately designed to make people pay attention to certain things rather than others, making increasingly many of our daily activities structured around the profit-making efforts of commercial enterprises (Zuboff 2019). This development has been critically analyzed using terms like 'the attention economy' or 'the commodification of attention,' which express that the particular structuring of our attention through technological devices must essentially be seen as a form of commercial exploitation (e.g., Stiegler 2010; Crogan and Kinsley 2012).

Not all technologies that intend to make it that something specific becomes the center of our attention are specifically designed for commercial profit or exploitation. Under banners like 'health empowerment' (e.g., Nelson et al. 2016) or 'behavior change for health' (e.g., Zhao et al. 2016), technologies (often in the form of mobile self-tracking applications) are designed to help people to live a more healthy lifestyle. This is often framed as a positive development. To quote Neelie Kroes, the former vice-president of the European Commission, such technologies "will reduce costly visits to hospitals, [and] help citizens take charge of their own health and wellbeing" (European Commission 2014, 1). Typical examples of such technologies are so-called self-tracking applications – e.g., mobile applications that allow for monitoring one's calorie intake or one's daily steps (cf. Ruckenstein and Schüll 2017). Broadly speaking, it can be said that such technologies are designed to draw unhealthy habits out of their transparency by making them visible to users in a particular way, such that they can become objects of concern (de Boer 2019). This development, then, shows that the structuring of our attention through technologies is not necessarily driven by commercial interests, and can even be beneficial to individuals (at least that is the suggestion). The point of this chapter is not to juxtapose technologies designed for commercial purposes and those designed for health purposes, but instead to focus on what such technologies have in common: both help imposing a certain attentional structure onto users.

This chapter intends to develop a phenomenological analysis of how self-tracking technologies structure attention. The goal of this analysis is to highlight how the use of such technologies – ones that are already used frequently and will likely become part of the lives of many more individuals in the future – turn our body and habits into objects of relevance. The outline is as follows: First, I discuss some technological developments that intend to promote a healthy lifestyle (1). Second, I suggest, elaborating on insights from postphenomenology, that technologies mediate what appears as relevant by shaping the relation between human beings and the world (2). Third, I further unpack the implications of this idea by connecting it to Alfred Schütz's theory of relevance. I suggest how technologies, including ones having the purpose of helping people to live a healthier lifestyle, shape systems of topical, interpretational, and motivational relevances (3).

Fourth, I argue that what stands out as relevant is constituted through sedimented habits, as well as the projects in which people engage (4). Fifth, I suggest that technologies that are designed to help people pursue a healthy lifestyle turn the attention of users towards their own body in ways that often remain unnoticed, such that it might become increasingly difficult to turn attention towards other objects and projects that might be equally relevant (5). In conclusion, I suggest that such technologies privilege a particular view of health that might become an unquestionable element of the lifeworld, and that this view might remain unnoticed when not being subject to careful analysis (6).

1 Technological Solutions for Living a Healthy Lifestyle

Globally, populations are both growing and are getting increasingly older, giving rise to a situation in which an increasingly large group of individuals will be in need of care in the (near) future. This development will likely put a burden on healthcare systems, and also likely gives rise to a severe increase in healthcare costs. On top of this, individuals obviously prefer to live a life with a minimum of disease-related complaints and hospital visits. To mitigate the burden of a growing elderly population on our healthcare systems, as well as to prolong the quality of life of individuals, healthcare professionals and policy makers maintain that medicine should be turned into preventive medicine: its primary goal should no longer be to *cure* disease or illness, but instead to *prevent* them from occurring in the first place (cf. European Commission 2018).

A central idea within the paradigm of preventive medicine is that individuals should pursue living a healthier lifestyle that makes them less vulnerable to future diseases. In order to live a healthy lifestyle, so it is argued, individuals should abstain from certain activities (e.g., smoking, alcohol intake, high calorie food intake), or engage in certain other activities more often (e.g., physical exercise). Technologies such as self-tracking applications (e.g., step counters, calorie tracking applications) are thought to be of crucial importance in helping the pursuit for a healthy lifestyle. Of course, such technologies alone are insufficient to realize this; their success is dependent on individuals that are willing to use them and have the ability to do so. And here, at least from the perspective of app developers and healthcare professionals, is where things tend to go awry: people tend to abandon tracking technologies after a couple of months, and quickly return to the eating or fitness patterns they had before purchasing the applications (Attig and Franke 2020). As a result, the project of using technologies to counter the presence of diseases and illness thus far seems unsuccessful.[1] Two prevalent solutions to this problem are (a) designing technologies in such a way that they help realizing the individual's own goals, instead of having pre-set goals embedded in them, or (b) offering personalized feedback on the individual's 'success' towards a certain goal, preferably at moments when individuals are most likely to engage with, or are most vulnerable to, such targeted feedback (Antezena et al. 2020).

Both of these solutions essentially attempt to make self-tracking technologies turn particular aspects of our lives into visible objects that must be continuously attended to. In doing so, they have the potential to shape which objects become the focus of our attention, as well as which projects we engage in when attending to those objects. In the remainder of this chapter, I attempt to analyze phenomenologically how self-tracking technologies, as well as the solutions proposed to promote their ongoing use, structure our attention. I do so by showing how our body is made visible as an object of attention in a particular way through the technologies involved.

2 Sedimentation and Technological Mediations of Relevance

It is a well-established idea in the philosophy of technology, especially in postphenomenology, that technologies (including the abovementioned) are no neutral intermediaries, but instead actively shape how people experience and understand themselves and the world around them by co-constituting particular forms of intentionality. Postphenomenology shares with phenomenology a focus on concreteness and experience, but dismisses the idea prevalent in phenomenology that technology and science alienate human beings from more authentic ways of being in the world;[2] it is rather interested in how scientific and

---

[1] In this chapter, I am not concerned with arguing for or against the idea that this attempt *should* be successful and/or whether self-tracking technologies are an adequate means towards realizing the goals of preventive medicine. Rather, I explore how these technologies structure attention by making one's body relevant in a particular way.

[2] It is beyond the scope of this chapter to discuss the adequacy of this diagnosis of how earlier phenomenologists approached science and technology. In the philosophy of technology, a distinction tends to be made between 'empirical' and 'classical' philosophies of technology. The latter are often accused of treating 'technology' and 'science' as monolithic structures, thereby neglecting how particular technological and scientific developments help constituting a particular way of revealing the world. For a critical analysis of this distinction, see Cressman (2020).

technological developments constitute particular human-world relationships (Rosenberger and Verbeek 2015, 10–13). A central concept in postphenomenology is that of *technological mediation*: technologies mediate how the world becomes present to people relating to them (e.g., Ihde 1990; Verbeek 2005; de Boer 2021, 27–31). The starting-point of this idea is that technologies have a form of technological intentionality: they shape how we perceive and understand the world by disclosing particular aspects of it. Particular mediations have consequences for what we attend to: "*[F]or every revealing transformation there is a simultaneously concealing transformation of the world, which is given through a technological mediation.* Technologies transform experience, however subtly, and that is one root of their *non-neutrality*" (Ihde 1990, 49; emphasis in original). According to Ihde (1990, 47), these transformations often remain unnoticed, because technologies often become transparent in use, making users unaware of how their relation with the world is shaped by the technologies related to. In other words, technologies co-constitute what stands out as relevant to us, and what will – through sedimentation and habituation – remain of ongoing relevance.

The insight that technologies often withdraw from view when used can be traced back to Heidegger's phenomenological analysis of tool use in *Being and Time*, where he famously distinguishes between tools appearing as "ready-to-hand" (*Zuhanden*) or as "present-at-hand" (*Vorhanden*), the former being the most primary way of relating to tools (Heidegger 1962, 95–115). In the former case, the tool (or technology) used is itself not an object of concern, but becomes transparent: it is something through which I engage with the world (e.g., a knife that can be used to slice bread easily), making certain other qualities recede from view (e.g., the warmth of the bread when touching it with my bare hands). In the latter case, the tool itself becomes an object of concern, drawing our attention to its particular qualities, an instance, which, according to Heidegger, primarily occurs when a tool ceases to work (e.g., when the knife no longer fits the purpose of slicing bread).

As Robert Rosenberger has recently argued, transparency is not a quality of technologies as such, but is something crucially dependent both on the particular design of them and the extent to which they are sedimented into our lives. Rosenberger defines sedimentation as "the degree to which it has become automatic for a user to approach a particular device through a particular organization of experience, with a particular field composition, and particular aspects of using taking on particular levels of transparency" (Rosenberg 2019, 307). For example, an experienced cyclist is not explicitly thematizing the bicycle (s)he is using, but instead has her experience organized in such a way that the road and traffic are what (s)he is paying attention to. Another example would be that of an experienced smartphone user who is not concerned with the design of the smartphone, but instead uses it as a window into something different, such as news channels, messenger programs, self-tracking applications and so on. These can be considered as successful forms of sedimentation in which a technological device itself disappears from view.

As stated, through sedimentation, the technology that is used becomes itself transparent. Here one of the ways in which the non-neutrality of technologies manifests is through the specific forms of sedimentation they give rise to and which draws our attention towards particular aspects of the world, thereby making others recede from view. As Rosenberger puts it: "[D]esigns can incline our perceptual focus on particular things, predisposing other things to withdraw and shaping our field of awareness" (Rosenberger 2019, 312). As I show in the next sections, Schütz's theory of relevance helps adding an extra layer to Rosenberger's analysis, because it helps showing that how technologies shape our field of awareness is dependent on the *projects* that they help instantiating.

3 Outlining Schütz's Theory of Relevance

In his *Reflections on the Problem of Relevance*, Alfred Schütz attempts to analyze why certain objects within our field of perception catch the eye, and why these objects are interpreted in a particular manner.[3] Schütz intends to do so through focusing on what makes our experience of certain objects thematic or nonthematic (horizontal). He strives to expand Husserl's analysis of attention in *Ideas I* that proposes to understand attention in terms of the centralization of a particular object through its being fixated through an attentional ray (Husserl 1983, 223; Schütz 2011 [1970], 95).[4] On Schütz's account, when analyzing how

---

[3] In this chapter, I am primarily concerned with how Schütz's theory of relevance can help showing how being attentive towards certain objects presupposes an orientation towards particular projects, and how thematizing certain objects can be understood as a way of further pursuing these projects. I thereby leave out many aspects of Schütz's work, most notably his analysis of how different provinces of meaning each contain different systems of relevance. For further analysis of Schütz's theory of relevance, see for example Campo (2015) and Kwang-Ki and Berard (2009).

[4] Wellner (2019) has argued that the idea that attention consists of delineating a figure from a background is itself informed by a particular technological development: photography. She argues that this idea of attention unjustifiably presupposes that people can only attend to one particular figure, and that new technologies such as computers and smartphones necessitate

something becomes the center of our attention, it needs first to be analyzed why a specific object starts standing out as relevant.[5] Understanding why an object becomes relevant requires, so Schütz holds, to address "(1) the question of the relationship between theme and horizon within the field of consciousness at any given moment of inner time, and (2) the motives by means of which this structurization has been initiated" (Schütz 2011 [1970], 95). In other words, an analysis of attention both concerns *how* objects appear to us as relevant in our field of consciousness, and *why* certain objects, rather than other ones, stand out as relevant in a given field of consciousness.

As a starting-point, Schütz (2011 [1970], 103–104) takes an example from the skeptic philosopher Carneades that describes the process of *periodeusis* (i.e., the hesitation between doubtful representations): a man enters a not well-illuminated room during the winter, and notices a pile of rope in the corner of the room. Yet, he does not see the object clearly, which makes him wonder if the object really is a pile of rope or a serpent instead. Since either interpretation is equally probable, the man starts distrusting his initial interpretation, making his thought oscillate between the two possible alternatives. When approaching the object he thinks of it as a rope, since it does not move. Yet, he also notes that the color of serpents is very similar to the perceived object and that snakes are made rigid by the cold. Still, none of the alternatives stands out as a more probable interpretation of the object. Eventually, the man may take a stick and hit the object. Now, when the object still does not move, an interpretation like: 'No, surely this is not a serpent' is warranted, making the man end his inquisition by concluding that it is indeed a pile of rope that initially drew his attention.

Schütz formulates several questions that need to be answered for clarifying why the object in the corner of the room becomes an object of relevance. These are: Why does this particular object become the central theme in our field of consciousness? Why is it that an interpretative choice needs to be made between two (or more) particular alternatives? Why is it that a certain course of investigation (e.g., investigating the object with a stick rather than with something else) is

---

reconsidering this view, since they point to the possibility of attending to multiple objects simultaneously, something she terms *digital multitasking* (Aagaard 2015). Whereas I agree with the idea that the capacity of digital multitasking is something frequently practiced nowadays, it seems that multitasking presupposes a rapid shifting of attention that can be understood as a shifting between different objects of relevance, thereby not being necessarily in contrast with the approach developed in this chapter.

5 For Schütz (and for Husserl), turning attention towards something, or the standing out of a particular object as relevant, is strictly speaking always a matter of *shifting* attention or relevance: "[T]hematic structure [. . .] is essential to consciousness; that is, *there is always a theme within the field of consciousness*." (Schütz 2011 [1970], 112)

entered into rather than another? Answering these questions requires, so Schütz holds, to investigate the structurization that makes possible a particular configuration of the field of consciousness.

3.1 Topical Relevances

Schütz distinguishes between three systems of relevance to answer the above-mentioned questions: topical relevances, interpretative relevances, and motivational relevances.

Topical relevances are those relevances through which something is constituted as problematic, thereby giving rise to a particular structurization of the field of consciousness. Becoming "problematic" here means that an object is segregated "from the background of unquestionable and unquestioned familiarity which is simply taken for granted" (Schütz 2011 [1970], 107). In our field of perception, a multiplicity of objects is always potentially present, either as the theme of our attention, or as the horizon of it. Objects that we are familiar with, often recede in the horizon, since these do not invite questioning, because their function, meaning and qualities are not relevant to the situation at hand. Put differently, familiarity helps forming "the demarcation line which the subject draws between that segment of the world which needs and that which does not need further investigation" (Schütz 2011 [1970], 108).

This demarcation line is drawn relative to the stock of knowledge available for the subject. The term "stock of knowledge" demands a short explanation. For Schütz, familiarity is directly constitutive of how we experience our surroundings, and it is through a stock of knowledge that familiarity is possible. Now, one's stock of knowledge is 'objectively' present as the social background in which one lives (e.g., as potentially transmittable through teachers, families, etc.), as well as 'subjective', or biographical, namely present as the actual experience of certain objects that have had certain effects on us in the past (Schütz 1945, 1953). For example, I might have learned that a glass is something that I can use to drink, that glasses come in a variety of shapes, and that I can recognize these varieties as belonging to a similar group (e.g., objects that can be used to drink); as a result, they do not have to be explicitly thematized anymore. Furthermore, I might have had the personal experience that a glass broke when falling onto the floor, opening the possibility to make this glass a theme of my attention when it is dangerously close to the edge of a surface. In the latter case, then, this object starts standing out as relevant in virtue of a particular topification, leading to a particular configuration of my field of consciousness.

A further distinction that is explanatory for why a particular object becomes of topical relevance is the distinction between *voluntary* and *imposed* relevance. In the former case, attention is directed towards something within a given theme (i.e., within a certain situational context that the subject engages in, such as the making of coffee, during which a sugar cube can become an object of relevance when the subject happens to drink its coffee with sugar). Here, the situation to which the subject orients remains relatively stable. In the latter case, on the contrary, the initial theme is thrown over by a particular event "that necessitates discontinuing the idealizations of 'and so on' and 'again and again'" (Schütz 2011 [1970], 109), such that a new thematization comes into being, which results in a new configuration of the field of consciousness.

3.2 Interpretational Relevances

As we saw, for Schütz, certain topical relevances are relative to situational themes, resulting in the transformation of the field of consciousness in which specific objects start to stand out as relevant. However, an explanation of how topical relevances make it that a particular object stands out as relevant in a given situation does not address the interpretational choices being made when an object is considered to be this or that object. This is where the concept of *interpretational relevances* comes in. In terms of Carneades' example this can be framed in terms of the question: why is it the case that the interpretation oscillates between a snake and a pile of rope?

A specific scheme of interpretation is required to make sense of the phenomenon that stands out as relevant. In doing so, the subject draws on its actual stock of knowledge, but additionally needs to demarcate which aspects of this stock of knowledge are connected with the particular object that is to be interpreted. Many aspects of my stock of knowledge are completely irrelevant when interpreting a particular object of attention. For example when deciding if a certain food item qualifies as healthy, my knowledge that the sun rises every day in the east, or that Zeus is an Olympian god married to Hera, is completely unconnected to the experience of the perceived object. Therefore, the particular way in which an object stands out as relevant, and the interpretational possibilities present in this situation arise in virtue of the subject *drawing selectively* from its actual stock of knowledge.

What this process of selective drawing results in not only depends on the actual stock of knowledge of the subject, but also on the particular situational context which it is in. In Carneades' example, the unusual object and the choice between interpretational alternatives are not only dependent on previous experiences of piles of ropes resembling the current object of relevance that may or may not have had a similar shape and color, or the previous experience with snakes from which it can be inferred that snakes move, but also by the specific place where the subject is located. For example, so Schütz (2011 [1970], 114–115) maintains, the system of interpretation one draws from differs significantly were the man that Carneades describes situated on board of a vessel, at a sailor's house, or at my own house (that happens to be a house in which neither a sailor nor a snake-lover lives). These different situations each shape the system of interpretation one draws from, as well as the likelihood of favoring a particular interpretation over another. After all, the plausibility of encountering a pile of rope on board of a vessel is far greater than it would be at my own house. Hence, practically, the rope-snake dilemma is never a dilemma between two a priori equally probable interpretations, because these probabilities are concretely shaped relative to specificity of where one is located.

3.3 Motivational Relevances

Now, in Carneades' example, the man is unable to come to an interpretative decision on the basis of glancing at the relevant object alone, and eventually hits the object with a stick, thereby engaging in one possible course of inquiry amongst many. It is for explaining this specific manoeuver that Schütz introduces the notion of *motivational relevances*. This notion attempts to lay bare why it is important to interpret the particular object correctly, and why a specific course of inquiry is started to determine what the most probable interpretation is.

The reason why the man in Carneades' example is interested in establishing with a high degree of certitude which object he is confronted with is that the outcome of his inquiry stands out as relevant in terms of his future conduct. For example, it would make a major difference if the interpretational dilemma was between a pile of rope and a pile of clothes than between a pile of rope or a snake. In the former case, acquiring certainty seems not really important as both interpretations do not potentially harm the man, such that it can be solved quite easily by walking towards the object of relevance and picking it up manually. In this case, certainty is obtained relatively quickly. Now, note how different the situation is in the rope-snake dilemma. Here, the subject might draw on its actual stock of knowledge to make certain inferences about the relevant object (e.g., snakes may bite, snakes are dangerous, I am afraid of snakes). These inferences, in turn, shape the subject's conduct: on the one hand, by engaging in a particular form of inquiry involving a stick. On the other hand, the object stands out as relevant in the first place in a particular way, because if it turns out to be a snake, the man might leave the room, or ask the snake to be removed from the room.

Here, it is (again) important to highlight the situated nature of relevance in Schütz's work. Whereas the notions of interpretational and topical relevances denote that relevance is relative to the spatiotemporal location in which the subject is situated, the notion of motivational relevance denotes that situations are never states of affairs independent of the subject, but are constituted relative to the subject's biography and the particular goal that it wants to obtain. Put differently, the subject has a certain motive to engage with the relevant object, which will result in a particular kind of action.

Schütz (2011 [1970], 120–121) distinguishes between two types of interrelated motives that structure how the relevant object is perceived: *in-order-to motives* and *genuine because motives*. The former refers to the imagined state of affairs to be achieved; we act in order to bring about a specific state of affairs. For example, the removal of the object present in the room while remaining out of danger; this explains both what grounds the inquiry, as well as why it takes a particular shape. The imagined situation here is to be able to move safely in the room in the future, without getting in direct contact with the potentially dangerous qualities of the relevant object. The latter type of motive is the *because of which* the investigation into the relevant object starts: for example, my desire for coming to a decision about the relevant object might be constituted by my fear of snakes that is the initiative through which I engage into investigating the object in the first place, and structures how this investigation proceeds.

3.4 Self-Tracking Technologies and Systems of Relevance

How, then, do technologies designed to help people living a more healthy lifestyle shape systems of relevance? As we saw before, topical relevances make a specific object stand out by making the subject draw a demarcation line between what is familiar and what demands further investigation. This is dependent on the actual stock of knowledge available to the subject in a given situation. Given that self-tracking technologies provide data about certain health parameters, they co-shape the subject's actual stock of knowledge, thereby attempting to familiarize subjects with the world around them in a particular way (e.g., the parameters used in a calorie tracking application invite observing the similarities and differences of food items in terms of differences in calories). Through the use of feedback mechanisms they transform habitual patterns into objects of relevance (e.g., through motivational messages stating that a user is in danger of having not made a 'sufficient' amount of daily steps), thereby imposing specific forms of problematization onto the subject. Taken together, this implies that these technologies constitute the horizon against which objects can be problematized, as well as the parameters implied when objects start standing out as relevant.

Qua interpretational relevance, it can be said that self-tracking technologies make objects stand out as relevant by soliciting specific ways of drawing selectively from one's actual stock of knowledge at hand by co-constituting a particular situation relative to which certain interpretations appear as probable. They help doing so by making several aspects irrelevant. For example, a calorie monitoring application makes it less relevant to interpret an object as suitable for a particular dish, but instead suggests a rather binary interpretational scheme, because the only relevant interpretational scheme one is solicited to oscillate between is 'healthy vs. not-healthy'. On top of this, the parameters used to make this distinction are narrowed down to those inscribed in the technology used. As a result, these technologies can be considered as mediating the process of selectively drawing from one's actual stock of knowledge.

Finally, self-tracking technologies shape both the in-order-to and the genuine because motives of the subject by suggesting a certain type of future conduct to be the ultimately desirable one, as well as structuring the course of inquiry when users are confronted with a particular relevant object by positing a certain states of affairs that is to be achieved. For example, in the case of using a step counter, the in-order-to motive for engaging in walking is the creation of a state of affairs in which one has reached the goal of having walked 10.000 steps. In doing so, they allude to what is taken to be a genuine because motive, namely the desire to live a long and healthy life.

4 Sedimentation, Projects, and Relevance

In the previous section, we saw that phenomena stand out as relevant relative to the extent to which we are familiar with objects in a given situation. In this section, I suggest that familiarity is best understood in terms of *sedimentation*. Doing so paves the way for a more intersubjective – or social – theory of relevance, which Schütz did not develop in the unfinished manuscript of *Reflections on the Problem of Relevance* in which many descriptions, as Schütz acknowledges, involve the unrealistic assumption that the subject experiences the world around it, as well as relevant objects in it, more or less privately (Schütz 2011 [1970], 132; Nasu 2014, 60). The notion of "familiarity" is key here: for Schütz, something is not encountered within a certain familiarity structure in isolation, but instead as a member of a relevant group. The world that is taken for granted by someone is always a world that it shares with other group members as well as with technologies, because of which every situation is in a sense a common situation emerging against a common horizon: individuals draw from this commonality to constitute their actual stock of knowledge that makes possible the standing out of a certain object as relevant (Campo 2015, 144–145).[6]

Relative to this commonality, the subject engages into certain projects that structure what stands out as relevant, as well as develops sedimented habits that make possible to leave the majority of objects unquestioned: we stipulate that, when pursuing a certain project, certain objects work 'as usual', thereby aiding us in realizing our projects, as well as helping our course of inquiry when a specific relevant theme is encountered. For example, when eating, I do not usually pay explicit attention to the cutlery, because what is relevant to me is the food that I consume, rather than the objects that make possible the engagement in this very project. I expect that these tools work as they do 'usually', based on both my skills in dealing with them and the meaning projected onto them. It is through my sedimented habits that I am capable of leaving these objects unquestioned, focusing my attention on other aspects of the world, which makes it possible to proceed eating my dinner.

What constitutes a project for Schütz demands a brief note. In the world of the everyday life, we are always

> 'wide-awake' [individuals] and immersed in our tasks and chores, working among fellowmen, performing actions which gear into the world, changing and modifying it in numerous ways. This [. . .] is the home base and starting point of our existence.
> (Schütz 2011 [1970], 152)

The notion of 'wide-awakeness' denotes that the subject is always already engaged in a certain project, such that the world is already constituted for it relative to a particular perceptual field. Furthermore, relating to the world in terms of a particular project implies that one's dealings with the world always imply the realization of a goal of some kind. The horizon and theme of our perceptual field, then, are particularly constituted in relation to this goal. And, how this goal can be realized in a given situation is dependent on other subjects and objects present in a given situation, as well as our sedimented habits and skills that enable working towards the goal to be realized.

---

6 That this might be a plausible assumption seems to a certain extent legitimated by the famous breaching experiments of the ethnomethodologist Harold Garfinkel (1967, 36–44) in which he asked participants to specifically problematize that which is occasionally treated as familiar, giving rise to problematic and highly uncomfortable situations.

Yet, this does not mean that our engagement with the world can be reduced to our pragmatic encounters with it. For Schütz, this would be to reduce human existence to motivational in-order-to motives, thus neglecting the other systems of relevance that are constitutive of one's actual stock of knowledge in a given situation. A phenomenological analysis of how objects appear as relevant remains incomplete when reduced to a pragmatic one, because the genesis of particular structures of familiarity that divide our field of perception into theme and horizon is not made subject of investigation. Doing so requires, according to Schütz, the capacity to make one's own stream of consciousness through which the world is encountered into a theme of investigation: to adopt a reflective attitude. Borrowing from Husserl, Schütz maintains that there are two ways of grasping the meaning of our previous experiences:[7] we can grasp them *monothetically* or *polythetically*. Either as a single ray of experience (monothetic), or as a process that is built up step by step in relation to my previous experiences (polythetic). Understanding the genesis of the projects that we engage in, as well as the relevant objects becoming thematized in those, requires to investigate the polythetic steps through which our habitual knowledge was acquired, and which gives rise to the very possibility of monothetic experience (Schütz 2011 [1970], 139).

It is through this distinction that Schütz starts explicitly questioning his initial 'putting between brackets' of the social world. When analyzing the polythetic steps leading to habitual knowledge, it becomes immediately clear that "only a very small part of my stock of knowledge at hand originates in my own personal experience of things" (Schütz 2011 [1970], 141). It is often through others that we trust that we attain knowledge of the objects around us, and in which the habits we developed are grounded. These others, then, communicated to us that their previous dealings with objects were of a specific nature, or draw on the tradition within which they are raised to articulate how meaning must be ascribed to the world around them. Our own experiences thus build upon the habitual knowledge of other inhabitants of our lifeworld, and can to a certain extent be articulated when adopting a reflective stance towards what we consider to be familiar.

However, the lifeworld always remains opaque to a certain extent, in the sense that many of its potentially knowable aspects remain unknown to the subject at any

---

[7] An analysis of Schütz's relation to Husserl's work is beyond the scope of this paper, as I am primarily concerned with Schütz's own analysis of how objects starts standing out as relevant, and how his approach can help articulating how technologies structure attention. For a discussion of the relationship between Schütz and Husserl, see for example Cox (1978) and Gros (2017).

given moment, as well as that they may disappear in the course of time. When engaging in the act of breaking down one's actual stock of knowledge into the polythetic steps leading to a particular meaning structure,

> it may frequently turn out that these traditional, habitual items of knowledge are such only as regards the monothetic meaning pertaining to the things supposedly known, whereas the tradition which contains the polythetic steps leading to this sedimentation (i.e., to the monothetic meaning) has been lost.　　　　　　　(Schütz 2011 [1970], 141)

One of the important elements of the lifeworld that constitutes how we make sense of the world around us (i.e., one of the things that is constitutive of monothetic meaning and sedimentation) are the technologies that we relate to. How these technologies do so often remains unnoticed, because an a priori rationality of efficiency is ascribed to them, such that their functioning appears as both necessary and void of pre-inscribed meaning structures (cf. Feenberg 2017). As I suggested earlier, by adopting specific parameters, self-tracking technologies exactly do inscribe meaning structure by pre-structuring what it means to be healthy, thereby privileging certain ways of turning our habits into projects.[8] As a result, they both structure which project should be carried out (caring for one's healthy body), as well as the conduct that people should engage in for realizing it. When considered as such, their constitutive role in making certain aspects of the world stand out as relevant at the expense of others threatens to remain unnoticed, precisely because their workings become sedimented and remain transparent accordingly.

5 Health as Theme and Project: Making One's Body Stand Out as Relevant

At the time of writing *Some Reflections on the Problem of Relevance* (between 1947 and 1951), Schütz was still able to write that most of our routines are not "within the thematic kernel at all. They remain, to the contrary, in the margin. I may think over my vital practical or theoretical problems while *walking, eating, shaving, smoking a cigarette*, and so on" (Schütz 2011 [1970], 173; my emphasis). It is noticeable that he relates this idea to how our body is not present in the kernel of the thematic field of consciousness (or attention) in the world of

---

[8] This might also explain why users of self-tracking technologies tend to abandon them after a few months: these apparently insufficiently align with their genuine because motives.

routine, but remains always present in its margin (Schütz 2011 [1970], 175). As we saw in this chapter, the current move towards preventive medicine and the technologies used to establish it, exactly attempts to turn this situation upside down by turning our body and our habits (e.g., walking, eating, smoking a cigarette) into explicit objects of relevance.[9]

In the phenomenology of health, it is often argued that the being transparent of one's body is a condition of possibility for health (Svenaeus 2001; Carel 2016; de Boer 2020). That is, the transparency of our body is something through which transcending into the world becomes possible, such that particular projects can be engaged in. In health, so it is maintained, we are unproblematically able to do so, because the body is itself not thematized in our experience. This is an idea that phenomenologists of health have in common with Schütz (2011 [1970], 175) who holds that although permanently present, both our body and our socially instituted existence are not present in the kernel of the thematic field of consciousness, but form the margins of it. Put differently, although our body is a central vehicle in making possible for objects to stand out as relevant, it is, in health, seldom turned into a relevant object itself.

However, what technologies designed to help people to pursue a healthy lifestyle intend to do, making them interesting from a phenomenological point of view, is precisely turning our body and our sedimented habits into an explicitly relevant theme (Van Den Eede 2015; de Boer 2020). These technologies have the explicit aim of turning one's body and the habits exercised through it into a project (or site of hermeneutic inquiry) through which it is constituted as something continuously warranting care and attention. As became clear earlier, when the body is turned into a project through self-tracking technologies, this still leaves many aspects of it horizontal, such that only some aspects relevant to both the constitution of one's body and the world experienced through it start standing out. One of the key goals of technologies designed to help people pursuing a healthy lifestyle is to make specific habits sedimented, such that users become capable of making certain choices (i.e., choices taken to be healthy) without them having to explicitly reflect on them (cf. Stawarz et al. 2015).

---

9 Although I have started with a slightly different theoretical framework, this idea comes close to Wehrle and Breyer's (2016) call for a dynamic approach to attention, in which they argue that what stands out as relevant is also constituted by a specific *noetic horizon* consisting of the motivations, attitudes, and activities of the subject. A postphenomenological approach can be of value to this dynamic approach by highlighting how technologies are co-constitutive of a specific noetic horizon.

In Schützian terms, one can say that these choices become unquestioned elements of the lifeworld, such that they cease to stand out as objects of relevance. When this happens, the particular way in which health is defined through these technologies might become unavailable to the wide-awake individual. This, then, might hamper the subject's possibility of reconstructing the polythetic steps leading to the constitution of the body as a project, as well as the actual stock of knowledge it draws on in carrying out this project. The analysis undertaken in this chapter can be considered an explication of how attention is mediated by self-tracking technologies by initially making certain objects (and qualities of them) stand out as relevant that might later become inaccessible when their use has become transparent.[10]

6 Conclusion

As I have argued in this chapter, the technological devices that preventive medicine allegedly relies upon mediate the relationship between human beings and the world in a specific way by explicitly turning one's body and habits exercised through it into an object or relevance. Drawing on postphenomenology and Schütz's theory of relevance, I have attempted to show that the question of how such technologies structure our attention can be reframed in terms of the question of how objects start standing out as relevant, and which actual stock of knowledge the subject draws from when relating to such relevant objects. Furthermore, I suggested that by being integrated into the stock of knowledge available to the subject (i.e., by becoming transparent), technologies designed to help people pursue a healthy lifestyle mediate the subject's relationship to the world. In doing so, technologies shape one's field of awareness due to their particular design, thereby drawing attention towards certain aspects of the world at the expense of others. This, in turn, has consequences for both the projects that the subject searches to engage in, as well as which objects are relevant in carrying out these projects. A Schützian analysis of these technologies showed how they constitute one's body as a particular kind of project in which

---

[10] This can be considered a move close to Schütz's in his essay *The Stranger*, where he states that if the newcomer's "process of inquiry succeeds, then [the group's] pattern and its elements will become to the newcomer a matter of course, an unquestionable way of life, a shelter, and a protection. But then the stranger is no stranger any more, and his specific problems have been solved" (Schütz 1944, 507). Similarly, when self-tracking technologies become unquestionable, their use, as well as the way they disclose the world, cease to be an object of inquiry.

certain health parameters are privileged. This particular privileging has the potential to become an unquestioned element of the lifeworld when self-tracking technologies become transparent through sedimentation, through which they constitute a contingent, yet reified understanding of health (Adams 2019). As I have suggested, when sedimented, the solicitations offered by such technologies refrain from being explicitly reflected upon, leading to an uncritical acceptance of a rather specific way of disclosing the world.

What is gained from the analysis undertaken in this chapter? First, that phenomenological analyses of attention can benefit from analyzing how technologies make specific aspects of the world stand out as relevant, and structure attention accordingly. Second, that a Schützian approach to attention (i.e., an approach that understands attention in terms of relevance) can be fruitfully applied the processes through which certain aspects of the world start to stand out as relevant. Third, that analyses of attention that conceptualize it as a commodity that is increasingly used for profit-making purposes can be augmented with an analysis of how technologies also shape the particular projects that people engage in, revealing potentially an even more profound influence of how technologies structure attention in our daily lives. Fourth, that phenomenological analyses of health should focus on how our bodies are increasingly becoming non-transparent in our everyday life due to the development of self-tracking technologies, as well as the paradigm shift in medicine towards preventive medicine.

There are several important limitations to the analysis undertaken in this chapter. I mention here what I take to be the two most important ones, each of which points to some questions for further investigation. First, when approaching attention from Schütz's theory of relevance, very little can be said about the embodied nature of attention (D'Angelo 2020), and neither about how technologies (self-tracking technologies amongst them) structure embodied attention through an increased focus on one's habits. Second, "technologies designed to help people pursuing a healthy lifestyle" is too broad a category to do justice to particular nuances in different technologies, especially in light of the aim of postphenomenology to be strongly grounded in concrete case-studies (Rosenberger/Verbeek 2015). How different health-related technologies constitute the body as a different kind of project, and give rise to particular forms of relevance sedimentation could be explicated in future investigations. As I have suggested in this chapter, such future investigations could benefit from focusing on which forms of transcending in the world are privileged in the design of technologies before they become reified as unquestionable norms.

References

Aagaard, J. (2015): "Media Multitasking, Attention, and Distraction: A Critical Discussion". *Phenomenology and the Cognitive Sciences* 14:4, 885–896.

Adams, M.L. (2019): "Step-Counting in the 'Health-Society': Phenomenological Reflections On Walking in the Era of the Fitbit". *Social Theory & Health* 17:1, 109–124.

Antezena, G.; Venning, A.; Blake, V.; Smith, D.; Winsall, M.; Orlowski, S.; Bidargaddi, N. (2020): "An Evaluation of Behavior Change Techniques in Health and Lifestyle Mobile Applications". *Health Informatics Journal* 26:1, 104–113.

Attig, C.; Franke, T. (2020): "Abandonment of Personal Quantification: A Review and Empirical Study Investigating Reasons for Wearable Activity Tracking Attrition". *Computers in Human Behavior* 102, 223–237.

Campo, E. (2015): "Relevance as Social Matrix of Attention in Alfred Schutz." *SocietàMutamentoPolitica* 6:12, 117–148.

Carel, H. (2016): *Phenomenology of Illness*. Oxford: Oxford University Press.

Cox, R.R. (1978): *Schutz's Theory of Relevance: A Phenomenological Critique*. The Hague: Martinus Nijhoff.

Cressman, D. (2020). "Contingency and Potential: Reconsidering a Dialectical Philosophy of Technology". *Techné: Research in Philosophy and Technology* 24:1/2, 138–157.

Crogan, P.; Kinsley, S. (2012): "Paying Attention: Towards a Critique of the Attention Economy". *Culture Machine* 13, 1–29.

D'Angelo, D. (2020): "The Phenomenology of Embodied Attention". *Phenomenology and the Cognitive Sciences* 19:5, 961–978.

de Boer, B. (2019): "Health Monitoring Applications and the Transparency of Health". *Delphi – Interdisciplinary Review of Emerging Technologies* 2:3, 129–134.

de Boer, B. (2020): "Experiencing Objectified Health: Turning the Body Into an Object of Attention". *Medicine, Health Care and Philosophy* 23:3, 401–411.

de Boer, B. (2021). *How Scientific Instruments Speak: Postphenomenology and Technological Mediations in Neuroscientific Practice*. Lanham: Lexington Books.

European Commission. (2014): *Healthcare in Your Pocket: Unlocking the Potential of mHealth*. https://ec.europa.eu/commission/presscorner/detail/en/IP_14_394 (last accessed 10 November 2021).

European Commission. (2018): *On Enabling the Digital Transformation of Health and Care in the Digital Single Market; Empowering Citizens and Building a Healthier Society*. https://ec.europa.eu/digital-single-market/en/news/communication-enabling-digital-transformation-health-and-care-digital-single-market-empowering

Feenberg, A. (2017): *Technosystem: The Social Life of Reason*. Cambridge: Harvard University Press.

Garfinkel, H. (1967): *Studies in Ethnomethodology*. Cambridge: Polity Press.

Gros, A.E. (2017): "Alfred Schutz on Phenomenological Psychology and Transcendental Phenomenology". *Journal of Phenomenological Psychology* 48:2, 214–239.

Heidegger, M. (1962): *Being and Time*, trans. by J. Macquarrie; E. Robinson. Oxford: Blackwell.

Husserl, E. (1983): *Ideas Pertaining to a Pure Phenomenology and to a Phenomenological Philosophy: First Book: General Introduction to a Pure Phenomenology*. Collected Works II, trans. by F. Kersten. The Hague: Martinus Nijhoff.

Ihde, D. (1990): *Technology and the Lifeworld: From Garden to Earth*. Indianapolis: Indiana University Press.
Kim, K.; Berard, T. (2009): "Typification in Society and Social Science: The Continuing Relevance of Schutz's Social Phenomenology". *Human Studies* 32:3, 263–289.
Nasu, H. (2014): "Alfred Schutz and a Hermeneutical Sociology of Knowledge". In: M. Staudigl; G. Berguno (eds.): *Schutzian Phenomenology and Hermeneutic Traditions*. Dordrecht: Springer, 55–68.
Nelson, E.C.; Verhagen, T.; Noordzij, M.L. (2016): "Health Empowerment Through Activity Trackers: An Empirical Smart Wristband Study". *Computers in Behavior* 62, 364–374.
Rosenberger, R. (2019): "The Experiential Niche: Or, On the Difference Between Smartphone and Passenger Driver Distraction". *Philosophy & Technology* 32:2, 303–320.
Rosenberger, R.; Verbeek, P.-P. (2015): "A Field Guide to Postphenomenology". In: R. Rosenberger; P.-P. Verbeek (eds.): *Postphenomenological Investigations: Essays on Human-Technology Relations*. Lanham: Lexington Books, 9–41.
Ruckenstein, M.; Schüll, N.D. (2017): "The Datafication of Health". *Annual Review of Anthropology* 46, 261–278.
Schütz, A. (1944): "The Stranger: An Essay in Social Psychology". *American Journal of Sociology* 49:4, 499–507.
Schütz, A. (1945): "On Multiple Realities". Philosophy and Phenomenological Research 5:4, 533–576.
Schütz, A. (1953): "Common-Sense and Scientific Interpretation of Human Action". *Philosophy and Phenomenological Research* 14:1, 1–38.
Schütz, A. (2011 [1970]): "Reflections on the Problem of Relevance". In: L. Embree (ed.): *Alfred Schutz: Collected Papers V. Phenomenology and the Social Sciences*. Dordrecht: Springer, 93–197.
Stawarz, K.; Cox, A.L.; Blanford, A. (2015): "Beyond Self-Tracking and Reminders: Designing Smartphone Apps That Support Habit Formation". In: *Proceedings of the 33rd Annual ACM Conference on Human Factors in Computing Systems*. New York: ACM, 2653–2662.
Stiegler, B. (2010): *Taking Care of the Youth and the Generations*, trans. by S. Barker. Stanford: Stanford University Press.
Svenaeus, F. (2001): *The Hermeneutics of Medicine and the Phenomenology of Health: Steps Towards a Philosophy of Medical Practice*. Dordrecht: Springer.
Van Den Eede, Y. (2015): "Tracing the Tracker: A Postphenomenological Inquiry Into Self-Tracking Technologies". In: R. Rosenberger; P.-P. Verbeek (eds): *Postphenomenological Investigations: Essays On Human-Technology Relations*. Lanham: Lexington Books, 143–158.
Verbeek, P.-P. (2005): *What Things Do: Philosophical Reflections On Technology, Agency, and Design*, trans. by. R.P. Crease. Pennsylvania: The Pennsylvania State University Press.
Wehrle, M.; Breyer, T. (2016): "Horizonal Extensions of Attention: A Phenomenological Study of the Contextuality and Habituality of Experience". *Journal of Phenomenological Psychology* 47:1, 41–61.
Wellner, G. (2019): "Onlife Attention: Attention in the Digital Age". In: K. Otrel-Cass (ed.): *Hyperconnectivity and Digital Reality: Towards the Eutopia of Being Human*. Cham: Springer, 47–66.
Zhao, J.; Freeman, B.; Li, M. (2016): "Can Mobile Phone Apps Influence People's Behavior Change? An Evidence Review". *Journal of Medical Internet Research* 18:11, e287.
Zuboff, S. (2019): *The Age of Surveillance Capitalism: The Fight for a Human Future at the New Frontier of Power*. London: Profile Books.

Short Biography

Bas de Boer, PhD, is a philosopher of technoscience who works at the University of Twente, The Netherlands. He is interested in how technoscientific developments shape the ways in which people understand themselves and the world around them. His current research focuses on how technologies shape how people experience and understand their health and well-being. He is the author of *How Scientific Instruments Speak: Postphenomenology and Technological Mediations in Neuroscientific Practice* that was published in 2021.

Galit Wellner
Chapter 10
Attention and Technology: From Focusing to Multiple Attentions

Abstract: The common discourse on attention tends to be bound to a dichotomist approach that oscillates between focusing and distraction. In this context, distraction is frequently understood as the destruction of an innate attentional capability to concentrate for long periods of time over a single object. Moreover, the common discourse usually places technology on the side of distraction and hardly refers to the fact that we seldom find ourselves with no technologies around us. In the age of the Internet and smartphones, it is time for a new approach that would reflect a more balanced attitude to technologies. Such an approach can be based on the principles of postphenomenology, a relational theory according to which technologies and humans are in a constant process of co-shaping. Such a basis would allow us to replace the monolithic critical duo of attention-distraction by a complex understanding of attention.

This chapter outlines a genealogy of modern attention problematizing two variables – a mode-of-attention and a prevalent media technology. The genealogy's 'moment zero' is epitomized in the bound book and the practice of silent reading as described by Augustine. In modernity, 'moment one' is situated in the late nineteenth century with the emergence of the cinema. This media technology transpires when attention is understood as the opposite of distraction, and synonym of focusing. Next, in the mid-twentieth century, radio and television are accompanied by a new concept of attention which moves fast between several objects. It is served by the metaphor of the searchlight and attracts criticism on the practices of browsing the Internet. Today, we are in 'moment three', when we experience the rise of smartphones on the technology side, and 'multi-attentions' on the mode-of-attention side. This mode-of-attention indicates our capacity to pay attention to more than a single object. Here attentions come in the plural, similarly to Donna Haraway's 'knowledges' (1998). Realizing that attention can come in several forms enables us to avoid a dystopian analysis of media technology and appreciate the various ways in which attention is practiced.

https://doi.org/10.1515/9783110647242-011

Introduction

While searching for materials on attention, I came across a blog that referred to some interesting articles, running by the name 'Nir and Far'.[1] The name is not a mistake. It is a playful variation on the blogger's name, Nir Eyal. He describes himself as someone who practices a new profession that he terms 'behavioral designer.' And the behavior that interests him the most is that which relates to attention. He has collected his insights in two books: The first is called *Hooked*, explaining how to design technologies that can capture maximum attention from their users; the second, *Indistractable*, addresses the users, aiming to assist them in controlling their attention. Each book is directed to a distinct audience. The books also differ on a deeper level which relates to the approach to attention. From regarding attention as a currency within the framework of attention economy, Eyal shifts in his second book to assigning to the users a more active role. Yet, he is still bound to a dichotomist approach that oscillates between attention and distraction. In both books, the underlying assumption is that technology generates distraction (directly or indirectly), thereby 'ruining' the users' 'natural' attention.

In his blog and books, Eyal reflects the common attitude to attention and technology that subconsciously regards technology as an alien entity against which humans should struggle. But can we separate humans from their technologies? Philosophy of technology goes in the opposite direction and conceives technology as an essential part of everyday life, as that which makes us human (Ihde 1990; Stiegler 1998). Hence, technologies do not necessarily damage our attention. Technologies are neither good nor bad, according to Melvin Kranzberg's (1986) first law of technology. Technologies mediate the world for us (Ihde 1990; Verbeek 2005), change how we move in the world, how we view it, and how we pay attention. These insights of philosophy of technology seem to be unknown to most philosophers of attention – they tend to ignore the role of technology. Those who do refer to technology usually hold a critical position close to that of Eyal, according to which technology causes distraction thereby presumably ruining an innate attentional capability (e.g., Aagaard 2019).

My interest in attention stems from the complex relations between attention and technology, especially media technologies. These technologies not only mediate content but also direct their users towards a certain mode-of-being-attentive to such contents. This directedness leads to a co-shaping process, that is – to a mutual shifting of technologies and modes-of-attention. Thus, I find

---

[1] https://www.nirandfar.com/.

correlations between the prevalent mode-of-attention as it may be from time to time on one hand, and media technologies that enable such an attention – books, cinema, later radio and television, and today the internet and cellphones – on the other. These correlations mean that our understanding of attention changes over time and cannot be considered a-historical. Attention as a concept has a genealogy.

In this chapter, I present a genealogical analysis that overviews the experience of attention and how it has been transformed. My aim is to analyze how a certain mode-of-attention has become dominant through a co-shaping process with media technologies. The analysis is performed by probing several points of intersection between a mode-of-attention and a specific media technology typical to a certain era. Each intersection is a 'moment' in the genealogy of attention. Each moment is comprised of a technology and the corresponding philosophical understanding of attention.

The genealogy offered here runs along three major moments of modern media technologies: the cinema; radio and television; internet and cellphones. To better understand them, a 'moment zero' serves as an opening, discussing attention at the time of the emergence of the bound codex. The end point is 'moment three' that involves contemporary media technologies and their related mode-of-attention which I term here 'multi-attentions'. The analysis is intended to show that we should redefine our approach to attention and conceive attention in the plural as multi-attentions. In order to better understand how the contemporary mode-of-attention has been formed, the shift from 'moment two' to 'moment three' is described through two additional sub-moments: 'moment 2 ½' probing the rise of the internet and a mode-of-attention known as hyper-attention; and 'moment 2 ¾' depicting smartphones and multi-tasking, thereby becoming very close to 'moment three'. These sub-moments set a more detailed background to multi-attentions and assist in viewing this new mode-of-attention more favorably.

1 A Genealogy of Attention and Technology

What is genealogy? Rooted in historiography, this methodology facilitates the linking of sets of definitions to a given era, thereby enabling the arrangement of diverse definitions and approaches along a time axis (Koopman 2013). Genealogy assists in showing that attention is a dynamic concept that changes over time. It reveals that what was true and relevant in the nineteenth century might be obsolete in the twenty-first century.

When associating the definitions of attention with certain eras, it becomes clear that each era is 'stamped' by some media technologies that were invented and developed more or less at the same time, yielding a certain mode-of-attention. Hence, each genealogical 'moment' exemplifies an entanglement of two variables: a prevalent mode-of-attention on one hand and a certain type of media technology on the other.[2] To compare, when Michel Foucault employs genealogy, he examines the variables of power and knowledge.

1.1 Moment Zero: Augustine and the Book

When should a genealogy of attention and technology commence? Where can the 'point zero' of the timeline be set? A possible beginning of this genealogical investigation can be located in early Christianity: The variable of mode-of-attention is that which is described by Saint Augustine, and the variable of technology is exemplified by the book which in that period was transformed from scrolls into bound codices.

Augustine understood attention as an active state of consciousness that relates to the senses and the subjective time (Crary 1999; Silva 2014). It is an active state, because it can be activated at will. This aspect of activeness is exposed in Augustine's writings by phrases like "I directed my attention to" (Augustine 1955, Bk 7, Cha. III, #5). Compare the selection of the verb 'direct' to attention economy's choice of verbs, like 'pay', in which the action is receptive and less active. For Augustine, attention is an active performance of directedness, and not a response to some external call to look, hear or buy. No less important is the link between attention and the subjective time. Attention, according to Augustine, has a duration in the present positioned between the future and the past. He specifically states: "Our attention has a continuity" (Augustine 1955, Bk 11, Cha. XXVIII, #37), which occurs 'now'.

Attention as a state of mind is exerted in reading. And reading became an interesting topic in the days of Augustine as it dramatically transformed from oral to silent reading. This changeover is reflected in Augustine's reports on how the socially acceptable form of reading was shifting from a practice performed in public and aloud, to an act executed in private and in silence.[3]

---

[2] In the theory of genealogy, this is the problematization (Koopman 2013).
[3] Augustine notes in the *Confessions* that Ambrose read silently (1955, Bk 6, Cha. 3, #3). The way he describes the situation may lead contemporary readers to the conclusion that this practice of reading was considered back then very unusual (Manguel 1997, 43–44).

In parallel, the book technology itself transformed from roll to codex (Petroski 1999; Stroumsa 2012). This technological innovation enabled readers to 'jump' directly to the desired page, bypassing the need to scroll and read linearly. Guy Stroumsa defines this change as "nothing less than the most dramatic event in the history of the book in the western world until Johann Gutenberg's commercialization of printing [. . .] in the mid-fifteenth Century" (2012, 153). The codex technology paved the way for smaller books that could be easily carried around. These books "were not meant for cultic display" (Stroumsa 2012, 153). They were intended for private use.

The book technology coincided with the practice of silent reading. The bound codex enabled a meditative form of reading that was based on repeating a given passage while freeing the hands from holding the scroll open. The attention could now be focused on the reading of the text itself, as the single activity to be performed. It is similar to The activity of mediation that was analyzed by Michel Foucault (2001) as part of his study of early Christianity's ascesis. He showed that silence was considered a mark of meditation and thinking.[4] In this framework, reading was meant to be repetitive, like a "constant prayer" (Stroumsa 2012, 156), far from today's concepts of reading and attention that are meant to produce knowledge. The reading of early Christianity was an act of self-management, intended to ensure that the reader is absorbed in a form of meditation, and through the meditation – in thinking of God. Accordingly, in the early Christian tradition of the first centuries, the bound book was meant to be read in private and in silence as part of religious practices.

This kind of reading was developed into a certain mode-of-attention that filters out any unnecessary sensorial stimulus, thereby directing the attention of the reader exclusively to the content of the book. Attention was conceived as a dichotomy: either one is attentive, or not.

1.2 Moment One: Dewey and the Cinema

The end of the nineteenth century brought two developments that are important for the genealogy introduced here: normative positions that regard attention as "a major problem in accounts of subjectivity" (Crary 1999, 19); and novel media technologies, i.e., the cinema. This link between the two developments is at the

---

[4] Foucault (2001) ties between three elements of ascesis: listening, reading (and writing), and speech. All three are related to meditation, truth and a specific mode of subjectivation. They are all part of the exercise of the self. Attention, however, is not specifically analyzed in this work of Foucault, and can be only deduced.

heart of Jonathan Crary's book *Suspensions of Perception: Attention, Spectacle and Modern Culture* (1999) where he details how the notion of attention was metamorphosed in the last quarter of the nineteenth century. He focuses on that period because then attention became a central area of academic research and emerged as a complex concept. The attentional state that was practiced by monks in early Christianity ('moment zero') now evolved into an ideal to be practiced by everyone and everywhere. It extended beyond religious practices and beyond mere reading into a general question of subjectivity.

Crary points to the emergence of large-scale industrial factories including their associated policies and work procedures. Back then, the factory workers faced difficulties in remaining attentive to the monotonous work near the manufacturing machines for long hours. When their minds wandered elsewhere, this reaction was deemed to be a human deficit rather than a sign for a bad interface design. It was defined as inattention, or distraction, and was considered negative.[5] Consequently, inattention "began to be treated as a danger and a serious problem, even though it was often the very modernized arrangements of labor that produced inattention" (Crary 1999, 13). The negative tone drifted from the employment arena to other fields such as education. Accordingly, pupils were expected to sit still for long hours in school, even when the teaching materials and methodologies were simply boring. If they expressed boredom, it was interpreted as distraction, i.e., lack of attention, i.e., a problem. Crary notes that even the definitions of attention were phrased in negative terms like 'exclusion' and 'unperceived' (1999, 24–25).

Focusing was heralded as the optimal mode-of-attention in the late nineteenth century. In the early writings dated from 1882–1898, John Dewey (1967) echoed this paradigm when he coined the lens metaphor of attention, for this technology could concentrate light and heat in a single point (cf. Crary 1999; Jennings 2012). Consistent with Augustine's understanding of attention, Dewey viewed attention as a function that could handle only one object at a time. There was a clear distinction between that object and the rest of the world. One could either be attentive or distracted. This mode-of-attention became known as the foreground-background structure (Kelly 2005 and the references there).

It is not a coincidence that the cinema was invented in the late nineteenth century. It has functioned from its very beginning as an 'attention machine',

---

[5] Only recently was daydreaming acknowledged as a process separated from the attention-distraction duo (Dario and Tateo 2020). The researchers define daydreaming as 'mind-wandering' and stress that it is a psychic process rather than an attention deficit or a compensation for boredom. It is a complex mental experience in which several aspects of body-mind and social interaction converge into elaborated meaning-making processes.

optimized to capture the viewers' maximal attention. The mode-of-attention that excludes everything but one object, resides in the core mechanism of the cinema. The cinematic experience reinforced the desired 'focusing' by placing the viewers in a sedentary position in a dark hall. Their gaze was attracted to the large illuminated screen in front of them. All this was designed to ensure that the viewers concentrate on the movie and only on the movie. The cinema 'consumed' its audiences' attention more extensively than the book in the days of Augustine. This mode remained ideal throughout the twentieth century and has persisted until today, as exemplified in the works of Eyal who is committed to the dyad of attention-distraction.

Consequently, in modernity, attention is constructed as a binary concept according to which one can be either attentive or distracted, either focused on an object or wander outside the scope of the lens. That which attracts the attention is positioned in the foreground and the rest of the world needs to withdraw to the background.

1.3 Moment Two: Merleau-Ponty, Radio and Television

Crary's analysis ends at the beginning of the twentieth century as it aims to mark the starting point of modern attention. My genealogy continues from that point and locates the second moment in the mid-twentieth century, when radio was present in many homes and multi-channel television entered the living rooms. Here the prevalent mode-of-attention mutates into a superficial scanning and fast switching of foci, from one broadcasting station to another. The early models of radios and televisions had a dial with which one could smoothly move between the stations which were organized according to their broadcast frequency. The whole process, from scanning to listening (in the case of the radio) or watching (in the case of the television), required the attention of the listeners/viewers. Moreover, the content was consumed while sitting and preferably concentrating on the broadcast content. This posture is very similar to that of the cinema, but allowed more freedom (or distraction, depending on the terminology used), as it was performed at the comfort of the home.

The philosophers were interested in the scan and search phase. In their critical analyses, they focused on the fast pace at which attention switches between objects. The leading metaphor for attention became a searchlight. In 1945, in the midst of the age of the radio and at the beginning of the age of television, Maurice Merleau-Ponty grappled with this mode-of-attention. He described it as "the function which reveals, as a searchlight shows up objects pre-existing in the darkness" (Merleau-Ponty 1962, 26). Merleau-Ponty belittled this mode-of-attention. He claimed that it assumed an objective world that already existed. Moreover, not only the existing world was taken as unchangeable, but also the searchlight effect was constant, no matter which surface was illuminated. Merleau-Ponty maintained that this formulation of attention had a uniform revealing force that scanned only the surface of the world.

His concept of the ideal attention was different. Whereas according to the searchlight metaphor, a second 'visit' of attention should have yielded the same impression, in practice a second visit does provide a different impression. Therefore, he claimed, attention "first of all presupposes a transformation of the mental field, a new way for consciousness to be present to its objects" (Merleau-Ponty 1962, 29). As a creative force, attention "creates for itself a *field*, either perceptual or mental, which can be 'surveyed' in which movements of the exploratory organ or elaborations of thought are possible" (Merleau-Ponty 1962, 29). With such an understanding of attention, the target objects are no longer pre-given. His approach fits the meditative silent reading of moment zero above. In both cases, attention functions as an internal force of revealing, may it be god or the outside world.

Merleau-Ponty's contribution to the genealogy of attention and media technology became applicable to the internet which arrived some decades later, as will be described in the next section. His critique of the searchlight metaphor is duplicated in analyses of Internet browsing practices. Instead of correlating the act of the searchlight to the fast switching between broadcast stations, now this act is paralleled to clicking hyperlinks in order to hop from one website to another. From radio to television to the internet, this mode-of-attention became paradigmatic for the Western culture in the second half of the twentieth century. As Crary notes, "Part of the cultural logic of capitalism demands that we accept as *natural* switching our attention rapidly from one thing to another" (1999, 29–30). The searchlight metaphor thus replaced the attention-inattention dichotomy of the late nineteenth century.

1.4 Moment 2 ½: Hayles and the Internet

The third genealogical moment begins to emerge at the end of the twentieth century with the rise of the internet, especially with its practice of browsing through hyperlinks. But before reaching the third moment, the searchlight approach gains power, though via different metaphors, most of them related to the cinema cameras. In this renewed set of metaphors, the camera does more than Dewey's lens that concentrates light into a single point. And it does more than merely lightening up the object as the searchlight did. The camera *dynamically* makes the object look more vivid and sharp while moving from one object to another. We find such camera metaphors in Friedrich Kittler's (1999) work which uses the close-up paradigm, and in Hubert Dreyfus' concept of 'shifting attractors' (2007, 25).

Sometimes the searchlight dominates the analysis of internet-related practices even without mentioning any metaphor. For example, in 2007, N. Kathrine Hayles formulated the ideal mode-of-attention as 'deep attention' contrasted against 'hyper attention', implicitly echoing Merleau-Ponty's analysis of the searchlight described above as 'moment two'. Hayles formed these terms to describe reading habits of students: the former is the prevalent mode of the past; the latter is performed by today's students. Although both modes exist in most civilizations, deep attention is associated with 'developed societies' and accompanied by "assumptions about its inherent superiority" (Hayles 2007, 188), thereby preserving the logic of 'moment zero'. The other mode of hyper attention is frequently understood as seeking stimulation beyond what is currently considered as normal and hence tends to be classified as a mental disorder.[6]

Five years later, Hayles presented a softer approach that can be understood as an acceptance of hyper attention as socially desirable. In her book *How We Think* (2012), she still considered deep attention as 'essential' for specific tasks "such as mathematical theorems, challenging literary works and complex musical compositions" (Hayles 2012, 69). The shift was with regard to hyper attention that became regarded as "useful for its flexibility in switching between different information streams, its quick grasp of the gist of material, and its ability to move rapidly among and between different kinds of text" (Hayles 2012, 69). Eventually, these two modes-of-attention are necessary as each is optimized to perform different types of activities. She stressed that "the problem [. . .] lies not in hyper attention [. . .] as such but rather in the challenges the situation presents to parents and educators to ensure that deep attention [. . .] continue[s] to be [a] vibrant component of our reading cultures" (Hayles 2012, 69). Her conclusion was that children, students and adults should practice both modes-of-attention, and one mode does not preclude the other. Whereas Hayles studied educational strategies and her conclusions referred to reading practices, this direction can be applicable to attention as practiced with other technologies like cellphones.

---

[6] Mainstream scientific research still approaches attention in the context of malfunctioning, as in the case of Attention Deficit Disorder (ADD) that is considered a mental problem. Whereas Crary (1999, 35) is critical with regard to the blindness to the social construction of certain behaviors, my focus in this chapter is on the role of media technologies in such a construction.

Hyper attention in its positive form can and should be extended beyond reading in order to understand how we live with contemporary digital technologies. Take for example the discourse on cellphone use which is still bounded within the searchlight metaphor and the negative tone of the original hyper attention concept of 2007. Such a discourse assumes that attention operates by being directed to one object and then turning quickly to another. It cannot accommodate paying attention to two objects at the same time (Aagaard 2019). It is based on the assumption that one can talk on the cellphone only when sitting and fully concentrating on the conversation (Rosenberger 2014). To think about two objects of attention, we need to turn to the concept of multi-tasking. We need to break free from the understanding of attention as necessarily equal to concentration and focusing.

1.5 Moment 2 ¾: From Multi-tasking to Multi-attentions

As the objects of attention multiply, the debate is whether attention(s) are practiced in parallel, i.e., via multi-tasking, or – like in the previous modes-of-attention – serially. According to *Online Etymology Dictionary*, the term of 'multi-tasking' originated in computer science, denoting the ability of a computer to process more than one task at a time. In computers, this feature is implemented as several processing units working side by side. But the early computers implemented multi-tasking in a single processor by switching very fast from one task to another.

The scholars who criticize multi-tasking compare multi-attention to the early computers consisting of one processing unit. When the mind is paralleled to a single processor computer, attention is conceived as something that cannot be 'divided' or 'split'. The argument continues and regards the 'pseudo split' as leading to a reduced performance thereby weakening the whole mental system. Following this logic, multi-tasking is no more than fast switching between tasks and hence 'true' multi-tasking is regarded as 'mission impossible'.

But the brain is not a computer (Cisek 1999) and the fact that we have one brain does not mean that it functions like a single processor of a computer. Our metaphors should not limit our understanding, and they can be updated every now and then. Accordingly, when discussing attention in terms of computer multi-tasking, it can mean that attention is paid to two or more objects at the same time.

Today, the debate on multi-tasking in humans oscillates between two poles: either attention is quickly diverted as the searchlight conceptualizes; or actually paid to more than one object. Critical thinkers remain with the single processor computer scheme and view attentional multi-tasking as unworkable. They maintain a dichotomist attitude according to which one can be either on-task or off-task as per a single task. For them, attention cannot be divided, only diverted (cf. Carr 2010; Stiegler 2010; Citton 2017; Aagaard 2019). They assert that when the switching is fast, one gets the (false) impression of simultaneousness. Usually the switching is considered as too fast, leading to the likelihood of performance decreasing in at least one of the tasks. Apparently, nineteenth century's negative tone still dominates the philosophical discourse on attention. Technologies remain on the 'problematic' side, so that media multi-tasking and digital distraction become equivalents (cf. Aagaard 2015, 2019).

The common philosophical discourse regards the effects of multi-tasking as negative and even dangerous (cf. Rosenberger 2012, 2014). This is what I termed elsewhere 'the horcrux logic' (Wellner 2014) borrowing from the Harry Potter series[7] where an object becomes magical if a fragment of the magician's soul is 'invested' in it. The more objects (and fragments) utilized by the magician, the more he or she loses their human traits. Correspondingly, once Voldemort (the dark magician in the series) splits his soul into seven objects, he becomes less human and starts resembling a snake. This logic can be found in the critique against multi-tasking: the more the attention splits between objects, the weaker it becomes (cf. Aagaard 2019). For the opponents of multi-tasking, it is a zero-sum game.

Dario Salvucci (2013), a computer scientist who studies computational models of human cognition and behavior, is one of those who claim that multi-tasking is impossible. For him it is no more than fast switching. Salvucci bypasses the (false) equation between the brain and a computer processor by referring to the body. He explains that in many cases a task simply requires the whole body. The body functions as a 'motor interface' that makes multi-tasking impossible. His examples include: a chef who needs both hands to prepare a dish and thus cannot work on several dishes at the very same time; and the computer's graphical interface where the user can type only in one window at a given point in time. This argument represents a very narrow understanding of technology that disregards the human entanglement with technologies (Ihde 1990; Verbeek 2005; Ihde and Malafouris 2019). The chef can prepare several dishes at the same time, as she stirs while the food processor mixes; she can also talk on the cellphone with her kids while cooking. And some windows

---

[7] Indebted to the Harry Potter universe, this section is named 'moment 2 ¾' after train platform 9 ¾ from which students boarded to the secret train that brought them to their school Hogwarts.

work in the background, like those managing the backup of my computer. Eventually, Salvucci admits that "while the basic psychological literature has made clear that cognitive processing can be a central bottleneck in multitasking performance, the translation of these simple laboratory paradigms to real-world domains has proven difficult" (2013, 60). Regrettably, such self-reflection and self-questioning are not very common among those who preach against multi-tasking. The relational theory of postphenomenology according to which technologies and humans are in a constant process of co-shaping (Ihde 1990; Verbeek 2005) would allow us to replace the monolithic critical stance by a complex understanding of attention.

1.6 Moment Three: Multi-Attentions and the Smartphone

At the beginning of the twenty-first century the cellphone turned into a smartphone and started serving as a dominant media technology. As such, it has transformed the way content is accessed and consumed. Unlike the early books that were read as part of one single activity of a meditative prayer, a smartphone is often used while doing something else. A much discussed example is driving a car while talking on the cellphone (also known as 'celling') or navigating with the help of a GPS-based app (Besmer 2014; Irwin 2014; Michelfelder 2014; Wellner 2014). These usage modes can be easily classified as multi-tasking. Once we think of the concept of multi-tasking in the context of new technologies, we realize that multi-tasking as such is not a new phenomenon. It has been present before the internet and the cellphone in mundane acts like talking to someone while reading the newspaper or washing the dishes (Tun and Wingfield 1995); playing football, which requires paying attention to the ball, the player's group members, the other group members and the referee (Tripathy and Howard 2012); or driving while talking to the other passengers (Irwin 2014). In these everyday situations, attention must be paid to more than a single object simultaneously.

Multi-tasking could not have been conceived as a legitimate mode-of-attention under the regimes of foreground-background or the searchlight: in the foreground there could be only one object; the searchlight beam could illuminate only one activity. The option of multiple attentions (hereinafter 'multi-attentions') departs from a dramatically different assumption according to which attention can come in the plural. This option follows the footsteps of Donna Haraway's (1998) notion of knowledges. In both cases the addition of the 's' might be read by the spelling checker as an error, but it aims at pointing at a fundamental shift in the common view.

Embracing the possibility of paying attention to two objects simultaneously, a new question arises: Is it a matter of a split attention or multiple attentions?[8] The question is deeper than mere usage of words. Discussing the situation in terms of splitting means there is one attention, and it can be split to cover two, three or more objects (cf. Cave et al. 2010; Tripathy and Howard 2012). This option was described above as 'the horcrux logic' resulting in a certain reduction of the attention as a whole. Inversely, the multi-attentions option accommodates the ability to pay attention to two or more objects without suffering from disastrous degradation.

Multi-attentions turned into a viable mode-of-attention when the cellphone, and especially the smartphone, became part of our everyday life. One account of multi-attentions is provided by Sebastian Watzl (2011a, 2011b, 2011c, 2017). In his terminology, attention "consists in the activity of regulating priority structures" (Watzl 2017, 3). His model calls for degrees of attention, exemplified by the listening to jazz music, where "you might focus your attention on either piano or saxophone, but remain conscious of both in either case" (Watzl 2011a, 723). The concept of multi-attentions does not mean that attention is necessarily spread equally. In Watzl's example, "While arguably both piano and saxophone receive some degree of attention, it makes a phenomenal difference which one you pay *more* attention to" (Watzl 2011a, 723). Likewise, while commuting, reading the headlines on the news app (or my friends' statuses on a social networking app, or any other activity involving the cellphone's screen) attracts most of my attention, I am still attentive to the happenings around me and the stations. I can notice who among the other passengers stood up, and can realize (almost always) when the train or the bus is nearing the station where I should disembark. Attention, according to Watzl, is a set of attentions given to several objects to various degrees and extents.

These degrees and extents can be organized in various ways. Here the analysis of Maren Wehrle and Thiemo Breyer (2016) is useful as they map the plurality of attention into two dimensions: one is at the personal experience level, where attention can be directed to more than a single object; the other is at the conceptual level, where the notion of attention diffracts into several sub-phenomena. Roughly and very schematically, they refer to the objects of attention as 'noematic horizons' on the temporal and spatial axes on one hand, and the motives, dispositions, habits and interests as 'noetic horizons' on the other. The two types of horizons enable them to explain the gaps between testing multi-tasking in a psychological lab vs. real life scenarios. Hence, they suggest that experiments should examine not only the number of objects of attention but also the sub-personal

---

8 I'd like to thank Oren Bader for pointing to this difference which is reflected in his work (Bader 2016).

factors. They explain: "Attention is not primarily directed to empty spaces, distinct objects, or specific features in the life-world of everyday action. Rather, it functions in a situational and bodily manner" (Wehrle and Breyer 2016, 45). Therefore, walking down the street while talking on the cellphone can hardly be compared to psychological laboratory's experiments involving watching geometrical shapes moving on a screen.

Wehrle and Breyer's 'noematic horizons' may call for a new framing of experiments' results. Once the researchers embrace the possibility of multi-attentions, new findings can arise. Take for example neurological experiments of vision. Under the strict spotlight assumption, subjects are expected to see only one object at a time. Brain scientists Stephanie McMains and David Somers (2004) demonstrate via fMRI recordings from the human visual cortex that multiple spatial attention can be spotted in separate zones of the visual cortex. It means that the brain can attend to two separate locations concurrently. They determine: "The spotlight can be split" (McMains and Somers 2004, 678). Limited to the metaphor of the twentieth century, they suggest that the 'spotlight' of spatial attention can be in fact two spotlights. They specifically refer to the possibility to attend to multiple spatially distinct regions of space at once.

In a comment on these findings, Frank Tong (2004) brings a real-life example of a duplication of the visual attention. It is an urban legend according to which Elvis Presley had a TV room with three television sets lined up along a wall, each showing a different program simultaneously. Tong explains that for those who support the spotlight model, this TV room is useless because they would assume that attention cannot be divided across multiple locations. They would further conclude that people can attend to only one region in space – i.e., TV set – at a time. The experiment results, however, prove that subjects can attend to two separate regions and selectively modulate visual activity in a top-down fashion. The fast switching explanation used by the opponents to multi-tasking could not effectively describe the experiment conducted by McMains and Somers because the subjects were able to attend to the two locations simultaneously with minimal cost, even within the same hemisphere.

What are the practical implications of multi-attentions? In my work on attention, I examined the interaction between attention and two technologies: the cellphone and online advertising. In relation to cellphones, one of the heated debates is on the safety of driving while talking on the cellphone (celling). In philosophy of technology, this debate was presented in a special issue of *Techné Research in Philosophy and Technology* (Volume 18 issue 1/2) through nine articles that examined the multiple aspects of driving while celling and related topics. On the single-attention side, Robert Rosenberger (2014) claimed that such a behavior is dangerous based on empirical cognitive research data and a postphenomenological justification he developed for these findings. On the multi-attentions side, I presented (Wellner 2014) a contestant view that favors driving while celling by using genealogy similar to that presented here and my own interpretation of postphenomenology. The genealogy highlighted how attention had attracted negative vocabulary in the nineteenth century (Crary 1999) and showed how this negativity is duplicated into the discussion of driving while celling. Postphenomenology enabled me to examine attention as a multi-stable phenomenon and argue that it can be split, so that a driver can drive while celling, while listening to the radio or while talking to the other passengers. Now I differentiate between split attention and multi-attentions as discussed in this chapter, and find justifications to driving while celling in the work of Wehrle and Breyer that offers a critical perspective on the cognitive research brought by Rosenberger.

Next I became interested in the more political aspects of attention. This is a direction I pursued in a later work (Wellner 2019) kicked-off as a response to the Onlife Manifesto. The manifesto describes the new forms of subjectivity in the digital age and how information and communication technologies call for re-distribution of tasks and responsibilities between humans and their technologies. However, attention is still conceived in the Manifesto in modernist terms of the nineteenth century, as a problem of distraction. This is the underlying assumption of the Manifesto and the departure point to criticize the attention economy and its abuse of traditional forms of attention. The criticism is not followed by a suggestion for the next step or an alternative. My aim was to show how multi-attentions can serve as a subversive response to the Internet's attention economy. I argued that the logic of online advertisements is based on the foreground-background model (i.e., attention-distraction), assuming full attention is given to an advertisement. Instead of resisting the attention economy regime by returning to the old mode-of-attention of focusing, I suggested to develop the concept of multi-attentions and encourage multitasking in which only a small fraction of the attention is paid to the advertisement. My analysis extended the distributedness aspect into the domain of attention and positioned such a behavior politically.

2 Concluding Remarks

In this chapter I analyzed attention as a dynamic notion that changes over time. I mapped the various references to attention using genealogy. My variables were the prevalent mode-of-attention and media technologies. This analysis exposed the complex co-shaping processes between the established mode-of-attention (as it may be from time to time) and the dominant media technologies. The different moments demonstrated how a new mode-of-attention emerges. Table 10.1 summarizes the major moments.

Table 10.1: Major moments in the genealogy of attention and technology.

Moment	Media technology	Thinker	Mode-of-attention
0	Bound book	Augustine	Meditation
1	Cinema	Dewey	Focus, foreground-background
2	Radio	Merleau-Ponty	Spotlight
3	Smartphone	Watzl, Wehrle and Breyer	Multi-attentions

After describing Moments 0, 1, and 2, I traced the birth of a new mode-of-attention typical to contemporary media technologies like smartphones. These are moments '2 ½' (Internet and hyper attention) and '2 ¾' (multi-tasking). My goal was to carefully construct the notion of multi-attentions and to show that attention can come in the plural.

Describing this mode-of-attention with the vocabulary of the nineteenth century leads to a duplication of that period's unfavorable tone. For example, the negative link between distraction and technology such as cellphones can be traced back to the nineteenth century and the demand for pupils and factory workers to remain attentive. Such an account of attention places technologies on the side of distraction and refers to them as a problem. At this point, Watzl's construct of levels of attention enables us to exceed the dichotomy by assessing *how much* attention is paid to each object in a given scene.

It is important to keep in mind that the nineteenth century's focusing-as-attention is not the only legitimate mode-of-attention. Nor is the fast switching paradigm of the searchlight. At this point I agree with Wehrle and Breyer who identify multiple forms of attention as well as multiple objects for attention. Conceiving attention in the plural reflects a human capability to deal with several objects or processes that are dynamically kept at various degrees of clarity and focus. Moreover, attention can be multiple with the *help* of technologies, so that we can pay attention to several objects simultaneously.

As Hayles noted, we need more than one type of attention as different scenarios require different attentional responses. Heralding one mode-of-attention as the best or the only legitimate mode does not assist in better living with technologies. The solution offered here is a 'bouquet' of modes-of-attention, each optimized to some technologies and scenarios: reading an academic article is probably best done with focusing; finding a piece of information on the Internet is best done with the searchlight paradigm; and managing an online class via video conferencing tool is probably best done with multi-attentions.

References

Aagaard, J. (2015): "Media Multitasking, Attention, and Distraction: A Critical Discussion". *Phenomenology and the Cognitive Sciences* 14:4, 885–896.
Aagaard, J. (2019): "Multitasking as Distraction: A Conceptual Analysis of Media Multitasking Research". *Theory & Psychology* 29:1, 87–99.
Augustine (1955): *Confessions*, trans. by Albert C. Outler. Philadelphia: Westminster Press.
Bader, O. (2016): "Attending to Emotions is Sharing of Emotions – A Multidisciplinary Perspective to Social Attention and Emotional Sharing. Comment on Zahavi and Rochat (2015)". *Consciousness and Cognition* 42, 382–395.
Besmer, Kirk (2014): "Dis-Placed Travel: On the Use of GPS in Automobiles". *Techné: Research in Philosophy and Technology* 18:1/2, 133–146.
Carr, N. (2010): *The Shallows: What the Internet Is Doing to Our Brains*. New York, London: W. W. Norton & Company.
Cave, K.R.; Bush, W.S.; Taylor, T.G.G. (2010): "Split Attention as Part of a Flexible Attentional System for Complex Scenes: Comment on Jans, Peters, and De Weerd (2010)". *Psychological Review* 117:2, 685–696.
Cisek, P. (1999): "Beyond the Computer Metaphor: Behaviour as Interaction". *Journal of Consciousness Studies* 6:11/12, 125–142.
Citton, Y. (2017): *The Ecology of Attention*. Cambridge, UK / Malden, MA: Polity Press.
Crary, J. (1999): *Suspensions of Perception: Attention, Spectacle and Modern Culture*. Cambridge, MA: MIT Press.
Dario, N.; Tateo L. (2020): "A New Methodology for the Study of Mind-Wandering Process". *Human Arenas* 3:2, 172–189.
Dewey, J. (1967): *The Early Works of John Dewey, 1882–1898. Volume 2: Psychology*. Southern Illinois University Press.
Dreyfus, H.L. (2007): "Why Heideggerian AI Failed and How Fixing it Would Require Making it more Heideggerian". http://cid.nada.kth.se/en/HeideggerianAI.pdf (last accessed 1 May 2020).
Foucault, M. (2001): *The Hermeneutics of the Subject: Lectures at the Collège de France 1981–82*. New York: Picador.
Haraway, D. (1998): "Situated Knowledges: The Science Question in Feminism and the Privilege of Partial Perspective". *Feminist Studies* 14:3, 575–599.
Hayles, N. K. (2007): "Hyper and Deep Attention: The Generational Divide in Cognitive Modes". *Profession*, 187–199.
Hayles, N. K. (2012): *How We Think: Digital Media and Contemporary Technogenesis*. Chicago, London: The University of Chicago Press.
Ihde, D. (1990): *Technology and the Lifeworld: from Garden to Earth*. Bloomington, Indianapolis: Indiana University Press.

Ihde, D.; Malafouris, L. (2019): "Homo Faber Revisited: Postphenomenology and Material Engagement Theory". *Philosophy & Technology* 32:2, 195–214.
Irwin, S.O. (2014): "Technological Reciprocity with a Cell Phone". *Techné: Research in Philosophy and Technology* 18:1/2, 10–19.
Jennings, C.D. (2012): "The Subject of Attention". *Synthese* 189, 535–554.
Kelly, S.D. (2005): "Seeing Things in Merleau-Ponty". In T. Carman (ed.): *Cambridge Companion to Merleau-Ponty*. Cambridge: Cambridge University Press, 74–110.
Kittler, F. (1999): *Gramophone, Film, Typewriter*, trans. by Geoffrey Winthrop-Young & Michael Wutz. Stanford: Stanford University Press.
Koopman, C. (2013): *Genealogy As Critique: Foucault and the Problems of Modernity*. Bloomington: Indiana University Press.
Kranzberg, M. (1986): "Technology and History: Kranzberg's Laws". *Technology and Culture* 27:3, 544–560.
Manguel, A. (1997): *History of Reading*. New York: Penguin Books.
McMains, S.A.; Somers, D.C. (2004): "Multiple Spotlights of Attentional Selection in Human Visual Cortex". *Neuron* 42, 677–686.
Merleau-Ponty, M. (1962): *Phenomenology of Perception*. 1978, trans. by Colin Smith. New Jersey: Routledge & Kegan Paul Ltd.
Michelfelder, D. (2014): "Driving While Beagleated". *Techné: Research in Philosophy and Technology* 18:1/2, 117–132.
Online Etymology Dictionary, https://www.etymonline.com/word/multitasking (last accessed 12 April 2020).
Petroski, H. (1999): *The Book on the Bookshelf*. New York: A. A. Knopf.
Rosenberger, R. (2012): "Embodied Technology and the Dangers of Using the Phone While Driving". *Phenomenology and the Cognitive Sciences* 11:1, 79–94.
Rosenberger, R. (2014): "The Phenomenological Case for Stricter Regulation of Cell Phones and Driving". *Techné: Research in Philosophy and Technology* 18:1/2, 20–47.
Salvucci, D.D. (2013): "Multitasking". In: J. D. Lee; A. Kirlik (eds.): *The Oxford Handbook of Cognitive Engineering*. Oxford: Oxford University Press.
Silva, J.F. (2014): "Augustine on Active Perception". In: J.F. Silva; M. Yrjönsuuri (eds.): *Active Perception in the History of Philosophy. From Plato to Modern Philosophy*. Cham: Springer, 79–98.
Stiegler, B. (1998): *Technics and Time, 1: The Fault of Epimetheus*, trans. by Richard Beardsworth & George Collins. Stanford: Stanford University Press.
Stiegler, B. (2010): *Taking Care of Youth and Generations*. Stanford: Stanford University Press.
Stroumsa, G.G. (2012): "Augustine and Books:". In: M. Vessey (ed.): *A Companion to Augustine*. Oxford, UK: Wiley-Blackwell, 151–157.
Tong, F. (2004): "Splitting the Spotlight of Visual Attention". *Neuron* 42, 524–526.
Tripathy, S.; Howard, Ch.J. (2012): "Multiple Trajectory Tracking". *Scholarpedia* 7:4, 11287.
Tun, P.A.; Wingfield, A. (1995): "Does Dividing Attention Become Harder with Age? Findings from the Divided Attention Questionnaire". *Aging, Neuropsychology, and Cognition* 2:1, 39–66.
Verbeek, P.-P. (2005): *What Things Do: Philosophical Reflections on Technology, Agency and Design*. University Park, PA: The Pennsylvania State University Press.
Watzl, S. (2011a): "The Philosophical Significance of Attention". *Philosophy Compass* 6:10, 722–733.

Watzl, S. (2011b): "Attention as Structuring of the Stream of Consciousness". In: Ch. Mole; D. Smithies; W. Wu (eds.): *Attention: Philosophical and Psychological Essays*. Oxford: Oxford University Press, 145–173.
Watzl, S. (2011c): "The Nature of Attention". *Philosophy Compass* 6:11, 842–853.
Watzl, S. (2017): *Structuring Mind: The Nature of Attention and How It Shapes Consciousness*. Oxford: Oxford University Press.
Wehrle, M.; Breyer, T. (2016): "Horizontal Extensions of Attention: A Phenomenological Study of the Contextuality and Habituality of Experience." *Journal of Phenomenological Psychology* 47:1, 41–61.
Wellner, G. (2014): "Multi-Attention and the Horcrux Logic: Justifications for Talking on the Cell Phone While Driving". *Techné Research in Philosophy and Technology* 18:1/2, 48–73.
Wellner, G. (2019): "Onlife Attention: Attention in the Digital Age". In: K. Otrel-Cass (ed.): *Hyperconnectivity and Digital Reality: Towards the Eutopia of Being Human*. Cham: Springer, 47–65.

Short Biography

Galit Wellner, PhD, is an adjunct professor at Tel Aviv University. Galit studies digital technologies and their inter-relations with humans. She is an active member of the Postphenomenology Community that studies philosophy of technology. She published several peer-reviewed articles and book chapters. Her book *A Postphenomenological Inquiry of Cellphones: Genealogies, Meanings and Becoming* was published in 2015 by Lexington Books. She translated to Hebrew Don Ihde's book *Postphenomenology and Technoscience* (Resling 2016). She also co-edited *Postphenomenology and Media: Essays on Human–Media–World Relations* (Lexington Books 2017).

Cor van der Weele
Chapter 11
How Can Attention Seeking Be Good? From Strategic Ignorance to Self-Experiments

Abstract: Receiving attention is widely recognized as a vital human need, closely connected to recognition as a requirement for developing a sense of safety, value and self-esteem. But what to think of instagrammers and others who are actively – and seemingly insatiably – seeking it? Why do we tend to condemn it? This chapter attempts to take a closer look at how we approach and evaluate attention seeking. I will discuss its bad moral reputation and argue that this encourages avoidance rather than the interest and curiosity that it needs, particularly in the light of problematic attention inequalities. I will argue for a transition from widespread strategic ignorance of attention seeking to more sympathetic curiosity. In this plea, Adam Phillips' recent book on attention seeking is an important ally. Realizing how important it is for all of us to receive attention, and how ashamed we tend to be of our attempts to get it, Phillips explores a more welcoming attitude. While his approach has a psychoanalytical background combined with literary sources, my own argument builds on conceptual considerations as well as empirical observations from various disciplines and historical periods. It will proceed in four steps.

The first step deals with some consequences of the fact that it is impossible to give attention to everything; attention is inevitably selective. In step two, I will turn to attention in social contexts, where we not only do or do not 'give' or 'pay' attention but also hope to receive it, which leads to mechanisms of reciprocity or exchange as well as social inequalities. In step three, I introduce Adam Phillips' (2019) approach to seeking. Starting from the widely accepted insight that receiving attention is vital for us, he thinks we often seek attention without knowing precisely what we are looking for, and what kind of attention will help us. I will connect these considerations in step four, arguing that strategic ignorance of attention seeking makes us miss a lot that may help us deal more openly and responsibly with attention seeking. With the help of my students' experiences I will suggest that a more positive approach to attention seeking may lead to surprising new learning experiences concerning the quality of mutual social attention.

Acknowledgements: I thank Steven Kraaijeveld, Beatrijs Haverkamp, Henk van den Belt and the editors for helpful questions and suggestions.

https://doi.org/10.1515/9783110647242-012

Introduction

A doctor receives a young girl with her mother. The mother tells the doctor that the girl has a pain in her ears, that she does not sleep and that she will need some cure. The doctor can find nothing wrong with the ears and notes that the girl starts to chat happily when he asks her about school. The doctor to the mother: "Forget about a cure, your daughter is overreacting, what she really needs is attention." The mother: "Attention? What for?"

(Koelewijn 2020)

In an essay about 'selfie-tourism', Christiaan Weijts (2020), together with his young son, visits a Dutch tulip field that is maintained and opened to the public in order to allow selfies in the midst of colourful tulips. He observes his son and the other visitors and reflects on what he sees: people not coming for the tulips but for the selfies-with-tulips. The instagrammer edits and exhibits reality in order to create images that produce attention; "Your own eyes and those near to you won't suffice." Yet, the author wonders, isn't that what he himself is also doing in the essay: make things a little more colourful, a little funnier and more meaningful in order to produce external validation?

"I think calling someone an 'attention seeker' is often a bullshit way to gaslight or shame extroverts, depressed people and those that choose to be different."

(Title of blogpost by justnotcool, 2020)

Receiving attention is widely recognized as a vital human need, closely connected to recognition as a requirement for developing a sense of safety, value and self-esteem. But what to think of instagrammers and others who are actively – and seemingly insatiably – seeking it? Why do we tend to condemn it? This chapter attempts to take a closer look at how we approach and evaluate attention seeking. I will discuss its bad moral reputation and argue that this encourages avoidance rather than the interest and curiosity that it needs, particularly in the light of problematic attention inequalities. Attention has become a scarce resource in many ways (Goldhaber 1997; Davenport and Beck 2001; Schroer 2019). Much competition can be observed to get it, which often results in unequal distributions.[1] It has been found – to mention an arbitrary example – that 20 % of twitterers receive well over 90% of followers, retweets and mentions (Zhu and Lerman 2016). It is also suggested that inequalities of wealth and attention are problematic in intertwined ways (e.g., Yun Family Foundation 2019).

---

[1] Not only among individuals, but also among companies, cities, universities, etcetera, for reasons that differ in overlapping ways.

Thinking about attention traditionally focuses on how we (selectively) attend to the world, and for me, too, this will be the starting point. I am specifically interested in a form of non-attending, 'strategic ignorance', a term that designates selective avoidance of aspects of the world that are embarrassing, painful or otherwise unwelcome. The avoided aspect this chapter focuses on is attention seeking, again a phenomenon of attention. It takes place in social contexts, where we are receivers as well as givers of attention – a bidirectionality that is a key aspect of social attention. The associated moral ideal calls for mutual attention, to which attention seeking is a threat.

I will argue for a transition from widespread strategic ignorance of attention seeking to more sympathetic curiosity. In this plea, Adam Phillips' recent book on attention seeking is an important ally. Realizing how important it is for all of us to receive attention, and how ashamed we tend to be of our attempts to get it, Phillips explores a more welcoming attitude. While his approach has a psycho-analytical background combined with literary sources, my own argument builds on conceptual considerations as well as empirical observations from various disciplines and historical periods. It will proceed in four steps.

The first step deals with some consequences of the fact that it is impossible to give attention to everything; attention is inevitably selective – in fact, patterns of attention are mixtures of detailed interest, shallow interest, indifference, ignorance, distractedness and many other varieties of attention and inattention. Just as inevitably, such patterns have moral significance, even though moral motives can be cloudy, as in the case of strategic ignorance; the 'strategic' character of such non-attention can appear in different grades of conscious awareness.

In step two, I will turn to attention in social contexts, where we not only do or do not 'give' or 'pay' attention but also hope to receive it, which leads to mechanisms of reciprocity or exchange. In this exchange, attention giving holds the moral high ground. Receiving attention is beneficial for the receiver, but actively seeking it or becoming too dependent on it are socially and morally problematic. It is often argued, for example by Charles Derber (1979/2000), that in present-day society the pursuit of attention is getting out of hand. The moral ideal on the background, a balance between giving and seeking, has deep sources and has been thoroughly investigated by Adam Smith. In his book *The Theory of moral sentiments* (1976 [1759]) he had much to say about attention exchange, intimately connected with sympathy-exchange, and the search for balance in this exchange. He struggled with the extent of the human need to be seen and approved and was deeply disturbed by the big societal inequalities of attention he observed.

According to this analysis, attention seeking cannot in itself be good. Yet we do it. As a result, we develop complicated relations with it, and given the human tendency to avoid our own weak spots, widespread strategic ignorance about attention seeking is only to be expected.

In step three, I introduce Adam Phillips' (2019) proposal to approach attention seeking in more interested and sympathetic ways: "Attention seeking is the best thing we do, even if we have the worst ways of doing it" (Phillips 2019, 7). Starting from the widely accepted insight that receiving attention is vital for us, he thinks we often seek attention without knowing precisely what we are looking for, and what kind of attention will help us. He also explores the shameful relation we tend to have with attention seeking. Shame comes with avoidance, while at the same time it signals importance.

I will connect these considerations in step four, arguing that strategic ignorance of attention seeking makes us miss a lot that may help us deal more openly and responsibly with attention seeking. With the help of my students' experiences I will suggest that a more positive approach to attention seeking may not lead – as we may fear – to an aggravation of competitive attention trends and increasing inequalities. It may, on the contrary, lead to surprising new learning experiences concerning the quality of mutual social attention.[2]

1 Selective Attention and Strategic Ignorance

The attention we give to the world is necessarily selective. William James robustly established this insight when he noted that selective interest shapes our experience and that attention, like a spotlight, involves withdrawal from some things in order to deal effectively with others (James 1890, 404). Attention is not an all-or nothing matter; the spotlight metaphor suggested by James pictures an intuitive kind of attention (focus) but it is not the only one – Freud (1900) introduced the importance of 'free-floating' attention for psycho-analysis, and there are more forms, such as fleeting or indifferent attention. An inventory is not my goal; instead, I will focus on an implication of selectivity: that there are also very many things we do *not* attend to. This results in combinations of

---

[2] An underlying theme of the chapter is plurality; there are many interesting phenomena and varieties of attention, historical continuities and discontinuities, cultural similarities and differences, and abundant relations with approval, esteem, belonging, acceptance, and so on. Besides there are many distinctions, sources and disciplines that can be helpful for understanding. The chapter will inevitably have many loose ends.

attending and ignoring that build up to patterns and routines. "My experience is what I agree to attend to," James also famously wrote (James 1890, 402). While 'agree to attend' misleadingly suggests that this choice is always conscious, the idea that our attention shapes our experience does alert us to the basic importance in our lives of what we do and do not attend to. It also immediately suggests attention's moral significance. What does not appear to us will not affect us.

When it comes to morality, an intuitive idea is that our attention needs to be cast on the world broadly and impartially – a clear challenge, in the light of unavoidable selectivity. Whether such a broad view should include ourselves is a different matter. A large tradition is concerned with the importance of knowing and caring for the self, associated also with the art of living. Yet there are also strong associations between morality and outward attention. In this latter line, Iris Murdoch, partly building on Simone Weil, argues that "the direction of attention should properly be outward, away from the self" (Murdoch 1991 [1970], 59). This kind of attention is open, also to unpleasant realities, while it should lead us away from selfish fantasies – it is a form of 'un-selfing' (Murdoch 1991 [1970], 84), cf. also Freeman 2015).

A rich palette of outward as well as inward attention is required by Martha Nussbaum's plea based on Henry James's novels (Nussbaum 1985). Obtuseness and 'refusal of vision' are the vices to be overcome, while fine awareness as a basis for responsible lucidity is the ideal. In order to be 'richly responsible' in any situation, one needs to be a person 'on whom nothing is lost', as James put it. Just as in a work of art, he thought, we must render reality precisely, and be thoroughly committed to the real.[3]

This ideal of complete and realistic awareness typically focuses on specific persons and situations. Even though James had to admit that people who are so richly aware are extremely hard to find, he said that the point was that we should use our imagination to create them, which according to Martha Nussbaum is a moral achievement on behalf of our community as it teaches us what a responsible moral life is ideally like (Nussbaum 1985, 529).

These calls for rich attention giving are thus primarily meant to inspire, rather than describe. But even if we manage to live up to them, we still ignore most of the rest of the world. Attention, however fine-grained in what it focuses

---

[3] Nussbaum compares James with Aristotle on artistic realism: "The realism in question is internal and human; its raw material is human social experience, which is already an interpretation and a measure. But it is realism all the same – and this is what makes the person who does the artistic task well so important for others. In the war against moral obtuseness, the artist is our fellow fighter, frequently our guide." (Nussbaum 1985, 528).

on, simply cannot exist without non-attendance to other things. Many further questions are therefore left open in this line of thinking, about what we do and do not attend to and how selectivity is compatible with responsibly broad, balanced or impartial attention – or whatever moral criteria we decide on.

It is also relevant that we are not completely in control of our attention. Tech companies for example have been found to manipulate it on a large scale; they earn money by attracting, holding and selling their client's attention, which is now often referred to as attention theft. Tim Wu (2016), who showed that this business model is not nearly so new as we might think, refers to William James's view that what we pay attention to determines our life experience. Although we may not fully realize it, we are at risk of "living lives that are less our own than we imagine", he says. Taking back control is his aim: "We must act, individually and collectively, to make our attention our own again, and so reclaim ownership of the very experience of life" (Wu 2016, 353).

But has our attention ever been our own?[4] Humans have always had to deal with many stimuli, sources of information, temptations, dangers and other attentional forces that we did not make or choose ourselves. Even if we need to deal with more information than earlier generations, it is not very plausible that former humans were in control of their attention all the time or even most of the time. Consciously or unconsciously we manage, by ignoring most of what we might have attended to. This ignoring, too, escapes full control. We ignore things that are not urgent or important, but we also ignore things that are important but unwelcome – which takes us to strategic ignorance.

Strategic ignorance (also referred to by other names, such as avoidance) makes a connection between attention and knowledge; while ignoring something means not paying attention, ignorance refers to not knowing. Strategic ignorance is a way of avoiding information that is unwelcome, uneasy, confusing or confronting, for example because it leads to uncomfortable ambivalence or unwelcome responsibility. Strategic ignorance is a complicated and paradoxical phenomenon, since in order to avoid information you have to know enough to 'decide' that you do not want to know more – it typically takes place around the threshold of consciousness. One example is that people who love meat while they are also concerned about animal welfare, often avoid information about animal farming; their ambivalence is uncomfortable as their love of meat leads to potentially unwelcome moral choices (Onwezen and Van der Weele 2016). Strategic ignorance can be understood as a way to evade responsibility

---

4 A more general question is how well we know ourselves. There is much reason to say 'not too well'; see e.g., Timothy Wilson's *Stranger's to Ourselves* (2002).

for such choices, a form of self-protection.⁵ This is not just an individual phenomenon. In her book on climate denial, Kari Norgaard (2011) describes avoidance as organised by cultural traditions and routines. If knowledge about climate change is threatening, for example because people do not know or do not like what they can do to deal with that information in meaningful ways, collective strategic ignorance affords shelter.

When we are confronted with more than we can deal with, avoidance can clearly be a very useful mechanism. Yet it can also be morally problematic. It has been amply documented how governments and companies – or people within them – deny or avoid certain kinds of information, for example about pollution, neglect or fraud, in order to avoid important responsibilities (Heffernan 2011; McGoey 2019). Note that in such cases, ignorance can be very deliberate; *Willful Blindness* is not accidentally the title of Heffernan's book.⁶

The above examples referred to information from outside – about meat, climate, pollution – but we can also strategically ignore confusing or otherwise unwelcome questions and information about ourselves. In a book on distraction, Damon Young (2008) argues that while we blame technology or tech companies for our distraction, we may partly use their products for distracting ourselves, notably from difficult issues we struggle with – or do not know how to struggle with – for example about what we really value in life. Avoiding existential questions through distraction, he argues, has always existed; it's just that nowadays we turn to social media for that goal.⁷

In a paper about 'affected ignorance' (another term for strategic ignorance), Michelle Moody-Adams (1994) wrote about how we tend to avoid, or even refuse, unwelcome perspectives on ourselves and how we thus tend to resort to wishful thinking about ourselves. Avoiding our human fallibility is perhaps "the most common form" of affected ignorance, she thinks, and she reminds us

---

5 Such evasion protects against discomfort of ambivalence, which is felt especially when there are difficult choices to be made. Ambivalence is related to cognitive dissonance; strategic ignorance serves similar functions as other mechanisms that reduce conflicting states of mind. In a paper on meat and cultured meat, we discuss the ambivalence we encountered and reflect on its important role in processes of moral change (Van der Weele and Driessen 2019).
6 I am presenting the phenomenon in generalized ways. Strategic ignorance appears in many contexts, in many grades of consciousness and it can take different forms: avoiding clear language, avoiding certain questions, certain sources of information, certain insights. and more. There is also much more to say about differences between individual and collective forms of avoidance, about affinities with other mechanisms of 'coping' such as dissonance reduction (see previous note) and about relations with other terminology, for example 'moral disengagement'.
7 For avoidance of existential issues, more specifically death, see also Ernest Becker (1973) and the ensuing 'terror management theory'.

how this tendency, together with the desire to suppress the convictions of others, was at the basis of Mill's plea for freedom of thought and expression (Moody-Adams 1994, 301–302).

In the next step I will turn to attention seeking. Charles Derber wrote about it as a problem getting out of hand. His moral ideal of a balanced mutuality then takes us to Adam Smith, whose thorough analysis of attentional inequalities illustrates that problematic imbalance is not a new phenomenon.

2 Reciprocity

When we think about attention in social life, a basic starting point is that in this context, rather than just selectively allocating attention to 'the world', we exchange attention with others; we are givers/allocators as well as receivers/seekers of attention. This leads to new complexities in the perspective on attending and ignoring.[8]

The importance of receiving attention can hardly be overstated; it is widely thought that we need to receive (sympathetic) attention to develop a sense of self-worth. Receiving attention is closely connected to and associated with such things as self-esteem, social approval, and recognition, as well as with status and reputation. All these things clearly not only affect our sense of self, but also, interconnectedly, our place in society.

The examples at the beginning of the chapter illustrate some extremes. The first reminds us that some people, even if they are parents, have not learned to acknowledge how vital it is for children to receive attention. In the second example, people are insatiable in their search for social attention and approval. This commonly meets with disapproval; while receiving attention may be vital, actively seeking it and/or depending on it is problematic, at least potentially, or a even a psychiatric condition. Jealousy, lack of self-esteem, loneliness and personality disorders are seen as potential causes.[9] Attention seeking always seems to be in danger of crossing borders;[10] moral condemnation is never far

---

[8] This was also already relevant – but not yet thematized – when I discussed Nussbaum.
[9] See for example the issue about attention seeking on healthline.com; https://www.healthline.com/health/mental-health/attention-seeking-behavior. Histrionic personlaity disorder has attention seeking as a central characteristic (French and Shresta 2020).
[10] Where precisely to locate these borders is a question with different answers for different people and in different cultures and areas. In science, for example (but not only there), it is relevant if we, through our work, deserve attention. In the footsteps of Robert Merton (1973), much has also been written about the Matthew effect, and attention issues more generally, in

away. Such condemnation is not only directed at individuals. Societies may also go astray. Attention seeking is sometimes called "the defining need of our times" as people are more lonely and ignored, which is also a basis for crimes that can be understood as cries for attention (Benedictus 2018). Yet the disapproval is also contested, as the third example at the beginning of the chapter shows. It will return in the last section.

In the 1970s, Charles Derber began to worry about the consequences of the rising need for attention in American society. In *The Pursuit of Attention* (1979/ 2000), he wrote about his micro-sociological studies on informal dinner conversations. His central observation was that a fair and mutualistic exchange of attention, which he considered as a basic element of healthy social life, was giving way to a more self-absorbed and competitive atmosphere in which attention-giving initiatives, such as asking supportive questions, no longer seemed to be required, except when someone expressed great pain or distress. In the second part of the book, he took a more timeless and 'formal' approach to attention, focusing on structural inequalities. His central point here was that dominant (more powerful/wealthy) groups are also dominant receivers of attention, while subordinate groups find themselves primarily in attention-paying roles. One of his examples concerned attentional inequality between men and women. Traditionally, women not only learn to give attention to others, but they often also experience "some degree of doubt, fear or guilt when taking it or accepting it for themselves" (Derber 2000, 45).

When the book was re-issued in 2000, Derber wrote a new foreword, focussing primarily on the competitive attention trends of the first part of the book: his original (1979) worry had developed into a cry of alarm. Self-absorbed attention seeking behaviour in American society had broadened and deepened, with Donald Trump as the model of the self-promoting CEO (Derber 2000, xx),[11] while for many ordinary

---

science. But this chapter is not about science, nor will I pursue moral borders in different cultures. Some of these questions, loose ends of this story, might well put attention seeking in a different light.

11 Note that Trump at that time (2000) was 'just' a CEO. The presidential attention strategies that Derber commented upon were those of Reagan and Clinton. He saw Clinton as one of those people who lack an inner core (with reference to Rieman et al's *The Lonely Crowd*) and are completely other directed, that is to say dependent on "being seen and embraced" (Derber 2000, xxiii), which turned the Presidency into "an eternal campaign for public attention and acclaim. [. . .] When politics becomes theater and politicians celebrities, it signals a degradation of public discourse and a threat to meaningful democracy." A later observer wrote about Trump as president, in corona-time: "At a time when the country craves someone of balance and intellect, someone to reassure and calm, someone who can admit mistakes and rectify

Americans the pursuit had taken the form of conspicuous consumption, webcams in dorms, etcetera. The US had become a celebrity society in which the pursuit of attention was now all pervasive. He explained this by pointing to overly individualistic tendencies in an increasingly competitive capitalist society. Insecure individuals must increasingly cope on their own and are desperately and competitively in pursuit of attention for gaining a sense of worth. As a consequence, many people must now cope with intolerable levels of being invisible and ignored.

Is attention seeking getting ever worse, with Donald Trump and selfie-making tourists as just two symbols? Is increasing competition leading to increasing inequality, with a few winners and large numbers of losers? The Yun Family Foundation certainly thinks so. In 2019 it called attention to increasing attention inequality, interwoven with inequality in wealth. They, too, make a connection with loneliness, as well as with technology: "The foundation aims to address the epidemic of loneliness and alienation that is spreading around the world in the technology age." Others, too, mention loneliness as a prominent societal evil – to be remedied by more equal patterns of attention and by new senses of community and connection (see e.g., Murthy 2020).

Yet, again, some aspects of current developments, especially structural inequalities associated with wealth and power, are not completely new. It can be helpful to turn to what Adam Smith wrote about inequalities of attention in the 18th century. He struggled extensively with human motivation, the roles of wealth and power, and inequalities of attention.

In *The Theory of Moral Sentiments* (1759 and later editions) Adam Smith posits sympathy as the basis of morality. He sees sympathy not – as Hume did – as an automatic contagion caused by the feelings of others, but as an act of the imagination: sympathy is actively accomplished by imagining ourselves in other people's situation. He immediately adds that we likewise need others to sympathize with us and that nothing mortifies us more than being deprived of such sympathy – for example when we tell a joke and nobody laughs (Smith 1976 [1759], 14). He discusses how we are constantly looking for a balance of mutual sympathy.[12] This is not a trivial search, as it does not simply have a meeting point somewhere in the middle. Some situations are far easier to

---

them, someone who can lead, we're sadly left with a man who craves attention and affirmation at any cost." (Dunn 2020)

12 They search is for so called 'points of propriety'. The propriety of the expression of passions and the amounts of sympathy expected and given, depends on the views of the impartial spectator (the imaginary 'man within our breast', see next paragraph), and differs for different passions. The 'point of propriety', where the exchange is in balance, is determined by what the impartial spectator "can go along with" (Smith 1976 [1759/1790], 27).

sympathize with than others, and outcomes can be very unsatisfactory. The biggest problem gets to the core of our theme; Smith observes that 1. We are so extremely interested in receiving sympathy that it is often a dominant motive and 2. Success (wealth, power, joy) attracts more sympathetic attention than misery or poverty. Therefore: "It is because mankind are disposed to sympathize more entirely with our joy than with our sorrow, that we make parade of our riches, and conceal our poverty." (Smith 1976 [1759/1790], 50) He explains that it would be misguided to think that we are ambitious because we need great wealth (a normal job will supply what we need) or because it will make us sleep better (the opposite is evidently the case). No, the purpose of "all the toil and bustle" of this world, all the "avarice and ambition, the pursuit of wealth, of power, and preheminence" is "to be observed, to be attended to, to be taken notice of with sympathy, complacency and approbation [. . .] It is the vanity, not the ease, or the pleasure, that interests us" (Smith 1976 [1759/1790], 50).

As a result, the rich and powerful receive an overdose of approving attention, which makes them develop a graceful and easy deportment, as well as an undeserved sense of superiority (Smith 1976 [1759], 54).[13] As they also tend to avert their attention from 'human wretchedness',[14] the poor and unfortunate are left feeling ignored and ashamed. Smith again uses the word 'mortifying' here, for "to feel that we are taken no notice of, necessarily damps the most agreeable hope" (Smith 1976 [1759], 51). He sees this whole disposition – worshipping the rich, despising and neglecting the poor – as "the great and most universal corruption of our moral sentiments" (Smith 1976 [1759],61).[15]

In these and similar passages, Adam Smith seems to agree with Bernard Mandeville, who in *The Fable of the Bees* (1714), had suggested that people are mainly motivated by reputation, called 'vanity' by Mandeville, and that this is what keeps society going. Smith struggles with Mandeville's view that this is our main motive, as he does, less explicitly, with Rousseau's view that society has corrupted our original good nature by triggering the mutual comparison that leads to the total

---

13 "As he [the young nobleman] is conscious how much he is observed, and how much mankind are disposed to favour all his inclinations [. . .] his air, his manner, his deportment, all mark that elegant and graceful sense of his own superiority, which those who are born to inferior stations can hardly ever arrive at." (Smith 1976 [1759], 53–54)
14 "The fortunate and the proud wonder at the insolence of human wretchedness, that it should dare to present itself before them, and with the loathsome aspect of its misery presume to disturb the serenity of their happiness. The man of rank and distinction, on the contrary, is observed by all the world." (Smith 1976 [1759], 51)
15 Although it is "necessary both to establish and to maintain the distinction of ranks and the order of society" (Smith 1976 [1759], 61). I leave aside the complications of Smith's view that Nature has designed everything for the best.

dependence on the opinion of others and the insatiable struggle to surpass them.[16] He agrees with both, as we just saw, that reputation is a main human motive and even remarks that Mandeville's views "border upon the truth" (Smith 1976 [1759], 313). Yet he refuses moral cynicism or despair, by rejecting that this is all. The basic difference with Rousseau is that Smith thinks we are social beings to the core, and that living with others is a need as well as a fundamental source of pleasure. The basic objection to Mandeville is that Mandeville overlooks that we also have other motives, most importantly to really sympathize with others, and that we can practice the virtue of a more impartial motivation. It is again the imagination that helps us: our ability to imagine ourselves in the shoes of others likewise enables us to look through the eyes of an 'impartial spectator' (an imaginary 'man within our breast') who teaches us to attend not just to our own interests but also to those of others. In addition, the impartial spectator assists in becoming more independent of the opinion of others, by teaching us the difference between 'the love of praise' and 'the love of praiseworthiness' (Smith 1976 [1759]; cf. also Van der Weele 2011).

The Theory of Moral Sentiments is a mixture of prescription and description. Throughout the book, Smith made it clear that he saw peace of mind and agreeable company as the biggest goods in life. He therefore sings the praise of a modest life, in which these goods are easily within reach, and he is worried by the unfortunate tendencies that make the views of Mandeville border upon the truth. Although he is convinced that with the help of the impartial spectator we are able to do better in principle, in many places he does not look very optimistic in practice. Our need of seeing ourselves in a positive light is so big that we are not especially inclined to honestly acknowledge our own weaknesses and follies. His description *avant la lettre* of this form of strategic ignorance is that "it is so disagreeable to think ill of ourselves, that we often purposely turn away our view from those circumstances which might render that judgment unfavourable" (Smith 1976 [1759], 158).

Only a few people may be fearless enough to develop independence from these mechanisms and from the opinion of others. For them there is one way ("and perhaps but one"), for being and remaining free and independent: "Never enter the place from whence so few have been able to return; never come within the circle of ambition; nor ever bring yourself into comparison with those masters

---

16 At the end of his discourse on inequality, Rousseau summarises this by saying that "the savage lives within himself, while social man lives constantly outside himself, and only knows how to live in the opinion of others, so that he seems to receive the consciousness of his own existence merely from the judgment of others concerning him" (Rousseau 1755). Compare Derber's view of president Bill Clinton, see note 11.

of the earth who have already engrossed the attention of half mankind before you." (Smith 1976 [1759], 57)

The 'timeless' anthropological mechanisms that Smith writes about, which lead to morally problematic inequalities of status, are being affirmed in contemporary research. In a review article, Koski et al (2015) present evidence that both monkeys and humans prefer to look at higher compared to lower ranked group members and that higher ranked group members tend to be liked better, even though they behave worse: "Several studies show that high status or upper class individuals are more likely to make unethical decision, break the law while driving, and even lie or cheat to get their way." (Koski et al, 2015, 12)

Attention inequalities thus emerge as deeply rooted and hard to fight. It does not follow that they are unchangeable. In the eyes of Derber and others, they are getting ever worse in our technological age and/or our individualizing economic system, where we need to compete for attention in order to secure our sense of self-worth. But more optimistic voices can also be heard. In *The Struggle for Recognition*, Axel Honneth (1995) argues for recognitional justice; societies should create conditions for the mutual recognition of all groups and individuals. According to him, struggles for recognition from the side of marginalized groups have already "greatly expanded the possibilities for self-realization" in developed societies (Honneth 1995, 179). This, too, is a relevant approach to societal attention trends. I will not pursue an analysis and evaluation of societal trends, leaving this issue here as one the loose ends of the story.

The writings of both Derber and Smith contain an implicit answer to the question whether and how attention seeking can be good. It looks as if it cannot be good on its own, but only when it is sufficiently balanced by attention giving. Mutuality, reciprocity and balance are the moral key words. The picture is implicitly quantitative and economic, a suggestion that is strengthened by terms such as 'paying' and 'stealing' attention.

The bad reputation of attention seeking means that it is something we tend not to like about ourselves, something we are ashamed of and are not eager to attend to. In order not to 'think ill of ourselves', in the words of Adam Smith, we may rather strategically avoid thinking about it too much. The next section is dedicated to a call for more curiosity.

3 Attention Seeking as Experiment in Living

> Attention is a form of recognition. We seek attention because we want recognition for something, but often we don't know for what precisely. We grow through recognition, through

exchange. So we need people who pay attention, and we need to give attention to others. It is our medium of exchange and desire, the place where we can find real pleasure.

These words by Adam Phillips in a Dutch newspaper interview (Eaton 2020, my translation) were occasioned by his book *Attention Seeking* (Phillips 2019). They are in line with the idea of reciprocity of the previous paragraph; both Charles Derber and Adam Smith would agree wholeheartedly that attention exchange gives pleasure. The selectivity of attention is another accepted idea that Phillips takes for granted. But the small book, written in his characteristic wide-probing style that combines psycho-analysis with literary analysis, surprises not only by taking mutuality simply for granted but also by embracing attention seeking: "Attention seeking is one of the best things we do, even if we have the worst ways of doing it" (Phillips 2019, 7). In a conversation about the book (London Review Bookshop 2019), the interviewer expresses her great happiness about this sentence, saying that it felt as if "the whole weight of my existence was lifted off my shoulders", and wondering also why nobody had ever mentioned this to her before; "my desire for attention has always been a profound source of shame".

The book arose from second thoughts about attention seeking's bad moral reputation and the complicated, indeed shameful ways we therefore deal with it. It is, Phillips says in the London interview, "something we should learn not to do, and yet we need attention, so we are in a bind and we have to find acceptable ways of getting the attention we want or need". As he explains in the book: "It has to be disguised as something else – as art, say, or manners, or prayer, or success – so a lot of our so-called creativity involves us in finding acceptable ways of finding the attention we desire." (Phillips 2019, 8)

Our shameful relations with attention seeking are prominent in his thinking. Shame, he writes, is often so unbearable that it is "unusually difficult to think about" (Phillips 2019, 39); it tends to stop thinking, attention and conversation, and thus creates blind spots, which may be part of its function; it may serve as a kind of vanishing act for the self. At the same time, shame points to the very importance of the things we feel ashamed about, and though we rarely talk about enjoying feeling ashamed, he says, we should: "it is our most visceral protests against our most demeaning confinements" (Phillips 2019, 45–46).

Another prominent theme in the book is that we often do not know what kind of attention we want, and for what. We send out all sorts of signals that we are not clearly aware of, and we are dependent on other people to see what gets picked up and comes back – always hoping for audiences that will respond to our signals in ways that keep us going. Seeking attention is therefore – with a reference to John Stuart Mill – an 'experiment in living': through it, we try to find out more about ourselves and what we find. During this process, the focus may start to shift from ourselves to our attention and perhaps even further away from our selves, towards explorations of ways of attending; Phillips is very interested in 'wide-angled attention', attention without focus, desire or purpose. He refers to writer and psycho-analyst Marion Milner here. The idea also contains similarities with Iris Murdoch's ideal of 'un-selfing' or ideals of selflessness in Eastern philosophies.

Attention seeking thus emerges as a double investigation: we seek attention for ourselves, but we also try to find out what deserves our attention and how. Phillips builds on Freud's idea that his patients were in search of a new kind of attention, for symptoms that had previously "misfired" because nobody picked them up as valuable signals (Phillips 2019, 75). His psycho-analytical background makes him attend to signals such as dreams, shame, distractions, ambivalences and spontaneous thoughts, but Phillips also wants to open up, for example by shedding off Freud's 'big idea' of the sexual significance of all these symptoms. His search breathes an appetite for exploring new areas of attention and reality.

The book points in new directions for answering the question how attention seeking can be good. Attention seeking is valuable as an experimental way of looking for new and liberating kinds of attention to ourselves and the world and each other and also to attention itself. Meanwhile, reciprocity remains basic, though it is mentioned only in passing, as a self-evident condition, so that attention inequality is not a theme. I will return to it in the next and final section.

4 Quality of Attention

We seek attention in order to receive something precious, something we need, but from prevailing psychological as well as moral perspectives, it is better to receive it unsolicited; active seeking is always in danger of crossing borders and is then associated with self-centredness in combination with a lack of autonomy and too much dependence on others. Attention giving is the good thing, and ideally people live in a balance of mutual giving. Yet in practice, the rich and powerful receive far more attention from the poor and vulnerable than the other way around. Thus, though subordinate groups may find themselves on the morally good side, the price is that they do not get the attention and recognition they need.

It is an interesting question how societies differ and how changes in societies influence general patterns. Yet rather than evaluating societal trends, I will stay with the evaluation of attention seeking and suggest how more curiosity will not make us embrace attention seeking in all its forms but will rather affirm the promises and attractions of a balanced mutuality.

Attention seeking's bad reputation makes it an ideal subject for strategic ignorance about ourselves. It is a strategy that backfires: if we avoid thinking about our attention seeking behaviour, we are in a bad position to find better ways of dealing with it, and with the needs that motivate it. We may remain locked in the worst ways of doing it, as Phillips expresses it – think of the ruthlessly competitive attention seeking described by Derber, or the shameless ways in which many rich and powerful people accept and enjoy attention inequalities. Their existence suggests that the shameful relation described by Phillips is not all-pervasive and that shameless attention seeking deserves just as much scrutiny as shameful forms.

Might a revaluation of attention seeking encourage more of the worst? Phillips' search for better ways may be in the air. Let us look at some recent cries of protest against the view that attention seeking is necessarily something bad, wondering what they are calling for. Here is again the blogpost title from the start of the chapter: "I think calling someone an 'attention seeker' is often a bullshit way to gaslight or shame extroverts, depressed people and those that choose to be different." (justnotcool 2019) Or see this text on a therapeutic website:

> You know what makes me crazy? The phrase 'attention-seeking behaviour'. We all know what people are talking about when they use this phrase. It's the child throwing a fit, the woman who creates drama and starts gossip, it's the teen who knows what to say in order to get people to swarm. But I reject the notion that it's 'attention-seeking'. No. It's 'attention-needing'. (Wilson 2019)

In the same spirit, Alain de Botton's *School of life* offers a video on "how to get attention without attention-seeking" (School of life 2019). It hopes to teach us more acceptable and effective ways of getting the attention we need, by encouraging us to become aware of our needs for love and reassurance and expressing those needs in vulnerable rather than angry ways.

These calls focus on our underlying needs, which indeed looks promising if we want to turn attention seeking into 'one of the best things we do' – Phillips' words again. But there are many additional issues. To what extent do we know what we need? What about the needs of others? What do we deserve? How much is enough? What determines quality of attention? What are we aiming for and what is needed to get there? . . . This could clearly become a long list.

A sympathetic curiosity will inevitably reaffirm the moral touchstones of mutuality and reciprocity as key guides concerning social attention. But further, such curiosity immediately enables learning. Let me conclude by illustrating this through my students' experiences.

In a course I teach on selective attention, students are expected to do two one-week 'self-experiments'. In each of them they change one thing in their attention habits. They keep a logbook, reflect on the results, preferably also in the light of some literature, and hand in a report. The first experiment is about something they would like to change in their habits of attention allocation (popular themes are e.g., spending less time on social media and Netflix, go for daily walks or try meditation, more attention for family). The second experiment is about changing something in their habits as attention seekers, in conversations or other situations. In this second experiment, students could roughly be divided into those who felt unseen or unheard and decided to seek more attention, and those who thought they could do with a bit less and therefore experimented with listening more, asking more questions, or let silence be silence for a while. The basic approach was thus quantitative; they went for more attention, or for less.

Interestingly, almost all of them reported changes in the quality of social exchanges too, invariably in positive ways. Not surprisingly, the students who wanted more attention got a real boost from making themselves more visible and better heard. Their experiments often took quite some courage – some of them had paired up to create safe rehearsal spaces with just the two of them. It was encouraging to read in their reports that already after a few days, asking for attention or for help, or thinking of things to say about their daily lives, became easier and began to feel more natural.

Most students who decided to go for less attention were taken by surprise by the gains in quality of social interactions; most of them reported unexpected feelings of social 'room' and ease, noting that conversations as well as they themselves became more relaxed, or they noted how friends explicitly appreciated their new behaviour,[17] or how they learned more from and about others. Someone discovered that he 'loved to listen', another student discovered how some of his friends needed a little more time to express their thoughts.

---

[17] I had urged them a. to stop and/or ask for help if the experiment became too stressful and b. to explain the character of the experiment to anyone who commented on changes in their behavior.

These results suggest that once people are encouraged to pay attention – in friendly ways – to their own attention seeking behaviour, and think about it and experiment with it, they will not stop at quantitative observations but quickly develop a keen sense for qualitative aspects of social exchange. Their experiences richly suggest directions for responsible ways of attention seeking, including a search for a balance of seeing and being seen, giving and asking room, stepping forward and stepping back as well as a (re-) discovery of the pleasure of such mutuality. A greater interest in attention seeking opens up this type of learning and discovery, and building on such beginnings, there is evidently much more to discover. Experimenting with attention seeking may make it good.

References

Becker, E. (1973): *The Denial of Death*. New York: Free Press.
Benedictus, L. (2018): "Look at Me: Why Attention-Seeking is the Defining Need of Our Times." The Guardian, February 5. https://www.theguardian.com/society/2018/feb/05/crimes-of-attention-stalkers-killers-jihadists-longing Last assessed on January 12, 2021
Davenport, T.H.; Beck, J.C. (2001): *The Attention Economy. Understanding the New Currency of Business*. Cambridge, MA: Harvard Business Review Press.
Derber, C. (1979/2000): *The Pursuit of Attention. Power and Ego in Everyday Life*. Oxford: Oxford University Press.
Dunn, M. (2020): "As a Tabloid Editor, I Covered Trump – and His Ego. He Hasn't Changed a Bit". The Guardian, April 8. https://www.theguardian.com/commentisfree/2020/apr/08/trump-new-york-coronavirus-crisis-newspapers Last assessed on January 12, 2021
Eaton, A. (2020): "Internet is als een extreem veeleisende moeder". Interview with Adam Phillips. *NRC*, April 10.
Freeman, M. (2015): "Beholding and Being Beheld: Simone Weil, Iris Murdoch, and the Ethics of Attention". *The Humanistic Psychologist* 43: 160–172.
French, J.H.; Shrestha, S. (2020): "Histrionic Personality Disorder." *National Center for Biotechnology Information* (NCBI) https://www.ncbi.nlm.nih.gov/books/NBK542325/ Last assessed on January 12, 2021.
Goldhaber, M.H. (1997): "The Attention Economy and the Net". *First Monday* 2/4. https://firstmonday.org/ojs/index.php/fm/article/view/519 last assessed at January 12, 2021
Heffernan, M. (2011): *Willful Blindness. Why We Ignore the Obvious at Our Peril*. London: Simon & Schuster.
Honneth, A. (1995): *The Struggle for Recognition. The Moral Grammar of Social Conflicts*, trans. by J. Anderson. Cambridge UK: Polity Press.
James, W. (1890): *The Principles of Psychology*. New York: Henry Holt & Co.
Justnotcool (2019): "I Think Calling Someone an 'Attention Seeker' is Often a Bullshit Way to Gaslight or Shame Extroverts, Depressed People, and Those that Choose to Be Different". https://www.reddit.com/r/TrueOffMyChest/comments/egsjhj/i_think_calling_someone_an_attention_seeker_is/ Last assessed on January 12, 2021.

Koelewijn, J (2020): "Ze luisteren naar mij, maar ze geloven jou". *NRC*, March 21.
Koski, J.E.; Xie, J.; Olson, I.R. (2015): "Understanding Social Hierarchies; the Neural and Psychological Foundations of Status Perception". *Social Neuroscience* 10:5, 527–550.
London Review Bookshop (2019): "Adam Phillips on 'Attention Seeking'." https://www.london reviewbookshop.co.uk/events/past/2019/7/attention-seeking-adam-phillips-and-devorah-baum Last assessed on January 12, 2021.
Mandeville, B. (1714): *The Fable of The Bees, or Private Vices, Public Benefits*. Available at https://www.earlymoderntexts.com/assets/pdfs/mandeville1732_1.pdf Last assessed on January 12, 2021.
McGoey, L. (2019): *The Unknowers. How Strategic Ignorance Rules the World*. London: ZED Books.
Merton, R.K. (1973): "The Matthew Effect in Science." In: R.K. Merton (ed.): *The Sociology of Science*. Chicago: The University of Chicago Press, 439–459.
Moody-Adams, M. (1994): "Culture, Responsibility and Affected Ignorance." *Ethics* 104: 291–309.
Murdoch, I. (1989 [1970]): *The Sovereignty of Good*. London and New York: Routledge.
Murthy, V.H. (2020): Together: Loneliness, Health and What Happens When We Find Connection. New York: HarperCollins Publishers.
Norgaard, K.M. (2011): : *Living in Denial. Climate Change, Emotions and Everyday Life*. Cambridge, Mass: MIT Press.
Nussbaum, M. (1985): "'Finely Aware and Richly Responsible': Moral Attention and the Moral Task of Literature". *Journal of Philosophy* 82:10, 516–529.
Onwezen, M.C.; Van der Weele, C.N. (2016): "When Indifference is Ambivalence: Strategic Ignorance about Meat Consumption." *Food Quality and Preference* 52: 96–105.
Phillips, A. (2019): *Attention Seeking*. UK: Penguin.
Rousseau, J-J. [1755: Discours sur l'origine et les fondements de l'Iinegalité parmi les hommes.] *Discourse on Inequality*, translated by G.D.H Cole, availbale at https://www.aub.edu.lb/fas/cvsp/Documents/DiscourseonInequality.pdf879500092.pdf Last assessed on January 12, 2021.
Schroer, M. (2019): "Sociology of Attention". In: W.H. Brekhus; G.I gnatow (eds): *The Oxford Handbook of Cognitive Sociology*. Oxford: Oxford University Press, 425–448
Van der Weele, C. (1999): *Images of Development. Environmental Causes in Ontogeny*. Albany: SUNY Press.
Van der Weele, C. (2011): "Empathy's Purity, Sympathy's Complexities; De Waal, Darwin and Adam Smith". *Biology and Philosophy* 26, 583–593.
Van der Weele, C,; Driessen, C. (2019): "How Normal Meat Becomes Stranger as Cultured Meat Becomes More Normal; Ambivalence and Ambiguity Below the Surface of Behavior". *Frontiers in Sustainable Food Systems*, 28 August, https://www.frontiersin.org/articles/10.3389/fsufs.2019.00069/full. Last assessed on January 12, 2021.
School of Life (2019): "How to Get Attention Without Attention Seeking". https://www.youtube.com/watch?v=771jjzt1vl0 Last assessed on January 12, 2021.
Smith, A. (1976 [1759]): *The Theory of Moral Sentiments*. Oxford: Oxford University Press.
Weijts, C (2020): "Alleen wat gezien wordt bestaat". *NRC*, March 21.
Wilson, B. (2019): "Attention Seeking Behavior: There's no such Thing". https://restorationho pecounseling.com/2019/02/21/attention-seeking-behavior-theres-no-such-thing/ Last assessed on January 12, 2021.

Wilson, T.D. (2002): *Strangers to Ourselves; Discovering the Adaptive Unconscious.* Cambridge, MA: Harvard UP.
Wu, T. (2017 [2016]): *The Attention Merchants. The Epic Struggle to Get Inside Our Heads.* London: Atlantic Books.
Young, D. (2014 [2008]): *Distraction.* London and New York: Routledge.
Yun Family Foundation (2019): "The Yun Family Foundation Introduces 'Attention Inequality Coefficient' as a Measure of Attention Inequality in the Attention Economy". (Press release). https://markets.businessinsider.com/news/stocks/the-yun-family-foundation-introduces-attention-inequality-coefficient-as-a-measure-of-attention-inequality-in-the-attention-economy-1028783738 Last assessed January 12, 2021.
Zhu, L.; Lerman, K. (2016): "Attention Inequality in Social Media". *ArXiv* abs / 1601.0720 https://arxiv.org/pdf/1601.07200.pdf Last assessed on January 12, 2021

Short Biography

Cor van der Weele, PhD, is professor of humanistic philosophy at Wageningen University, The Netherlands. She also has a background in biology. In her PhD-thesis in the philosophy of biology, she explored selective attention in explanations of embryological development (Van der Weele 1999). In recent years, her interest in processes of moral change – specifically concerning meat, and triggered by the idea of cultured meat – led to publications on the role of ambivalence and strategic ignorance in such processes. These themes are relevant more generally for e.g., humanistic perspectives on human nature and for ethics.

Lauren Hayes and Juan M. Loaiza
Chapter 12
Exploring Attention Through Technologically-Mediated Musical Improvisation: An Enactive-Ecological Perspective

Abstract: In this chapter we consider attention from an enactive-ecological perspective in which the organism-with-environment interdependencies that emerge in the process of living are fundamental and necessary for understanding cognition (Thompson 2010). While technological advances have often provided an impetus for empirical studies of attention, we propose, moreover, that such developments have enabled the facilitation of new highly participatory forms of musicking. In these types of technologically-mediated musical interactions, there is fertile ground for exploring attention within the types of organizational dynamics that emerge over time between group members within musical situations, especially those related to improvisation.

We propose three enactive-ecological themes that we develop in terms of three ranges of timescales of attention modulation. The purpose of the three themes and associated timescales is to provide a way of making distinctions and disentangling processes while doing justice to the complex interdependencies of organizational dynamics (Gahrn-Andersen et al. 2019). Firstly, we consider the notion of habit which affirms the interdependencies between organism and environment as a fundamentally embodied process of identity generation (James and Loaiza 2020). Secondly, we consider the role of attention within social interactivity, whereby attention can be said to emerge within the processes of participatory sense-making (De Jaegher and Di Paolo 2007). Finally, we propose that while attention has often been articulated through ocularcentric metaphors of focus or illumination, a more fruitful approach might involve articulating the speed, strength, or amplification of such dynamics as a processual, non-static paradigm. Technologically-mediated musical practices, particularly ones which involve improvisational modes of playing, offer not only an environment in which these ideas can be studied, but also provide participatory and experiential platforms for interdisciplinary research (Hayes 2019).

Introduction

Over the last couple of decades, there has been a significant growth in the study of musical performance and improvisation within cognitive science research. Yet, when Pressing offered one of the first formal attempts to theorize improvisation in 1988, the research was scant (Pressing 1988). While his formulation was indeed generalizing in its attempt to blend studies from psychology that were concerned with attention and skill within performance practice with literature that he identified as relevant from musicological research, he had pointedly noted four years earlier that it would be within the very practices of improvising artists that psychologists would unquestionably find "the single largest source of information" (Pressing 1984, 345) on which to base their studies. Pressing suggested that not only could analyses be drawn from the large corpuses of recorded media, comprising video and audio recordings, but moreover, that the phenomenological and experiential knowledge of the improvisers themselves could provide information on "proprioception and self-observation [. . .] issues as learning, training, the usefulness of imagery, muscular coordination, and cognitive processing" (Pressing 1984, 345). At around the same time, the proposal of the enactive framework was put forth as an alternative way of understanding cognition, compared to traditionally cognitivist or 'computational' approaches (Varela et al. 1991). Specifically, the enactive view proposes that cognition arises out of active participation and engagement between organisms and their environments, stressing the embodied nature of cognition, and the mutually affecting nature of such relationships. While not explicitly linked at the time, Pressing highlights the importance of the relationships between cognitive processes and both auditory and proprioceptive feedback loops within the ongoing activity of improvisation. Furthermore, in Pressing's writing we see some of the first discussion that connects musical improvisation to studies of attention in the field of psychology (Pressing 1988). Crucially, he suggests that it is within the "long-standing question of multiple attention" (Pressing 1984, 356) – namely whether and how it is possible to attend to two or more things simultaneously – that key work for understanding improvisation will need to be undertaken.

What can be understood about the processes of attention through an enactive-ecological account of musical improvisation? Research within the cognitive sciences has started to acknowledge and reflect on the importance of action within a model of musical improvisation (Linson and Clarke 2017). In addition to studies focusing on motor control, perception and cognition, some empirical work has begun to explore themes such as, for example, the coordination of joint action (Walton et al. 2015) and even the processes of creativity itself (Loui 2018).

Concurrently, neuroscientific research that is now several decades old has hinted at the importance of attention in connecting these various areas (Posner and Petersen 1990) and its ability to demonstrate that creativity and cognition are by no means skull-bound activities. In this chapter, we consider attention from an enactive-ecological perspective in which the interdependencies of the dynamical processes that emerge between organism and environment in the process of living are fundamental and necessary for understanding cognition (Chemero 2009; Thompson 2010).

We refer to the idea of attention modulation not from the cognitivist standpoint of 'neural activation' in relation to localized or isolated functional modules within the brain, but rather as part of the sensorimotor action-perception loops that cognition comprises. In what follows, we propose the notion of three ranges of timescales – or temporal ranges – of attention modulation which draw on three distinct yet related areas of discourse from within enactivist research. At the longest timescale, we consider the habitual and embodied processes of how identities are formed; at the mid timescale, we discuss issues of joint attention via collective participation in, for example, communities of practice; and at the fastest timescales, we consider how attention changes on the fly. While technological advances have often provided an impetus for empirical studies of attention at various key points historically, we propose that moreover, such developments have concurrently enabled the facilitation of new highly social forms of musicking. We argue that it is within these types of technologically-mediated musical interactions that fruitful opportunities for exploring attention amidst the organizational dynamics that emerge over time between group members will be found.

1 Technology and Improvisation

While all musical instruments can be thought of as 'technologies', we wish to focus this discussion around the changes that occurred with the development of electro-mechanical, electro-acoustic, electronic, and latterly digital technologies and how the affordances of these technical musical objects have impacted the practices of musical improvisation. Firstly, the process of transduction of kinesthetic information or gestural energy into electrical signals which appeared with some of the earliest electrical instruments around the late nineteenth century, meant that distinct acoustic parameters were no longer bound up with the sound-producing mechanisms of the instrument. For example, with the Theremin – which was patented in 1928 – moving one's hand within the vertical electro-magnetic field would produce changes in pitch, while moving the other hand near the horizontal electrode would change the amplitude. While acoustic instruments certainly afford changes in, for example, dynamics to the performer, these are generally not specifically related to a particular section or physical part of the instrument. Secondly – and in particular concerning digital musical instruments (DMIs) – the mechanisms that produce sound are no longer necessarily coupled to the mechanisms which transduce gesture. The action of striking a key on a piano results in a hammer being deployed onto a string, which produces an acoustic sound. However, there now exists a large variety of 'controllers' which can be mapped onto digital signal processing (DSP) within computer software in order to affect various parameters within a signal chain. For example, the x-y joystick of a game controller could be mapped both to a filter frequency on the x-axis, and a low-frequency oscillator on the y-axis. This would allow for extensive real-time timbral transformations to be produced and explored. Perhaps the most crucial paradigm shift – which of course had a profound impact on how music is shared and consumed – is the ability to record and playback sound. In terms of improvisation,[1] recording technologies have been utilized in order to not only remix external material from other performers, but also to capture, transform, and reproduce material within the course of the improvisation itself. In this way, the performer may have to attend to material which has not been created 'in real time', or indeed be aware of ongoing processes – such as recording or analysis – which are not yet audible.

In the domain of 'freely' improvised music involving technology, instruments are often highly personalized, may exist as singularities, and may comprise assemblages of acoustic, analogue, and digital elements. Moreover, a performer may have to navigate several different interfaces within a single instrument. In the case of digitally augmented acoustic instruments, for example, in addition to navigating the existing acoustic form, a performer may have to work with foot pedals, tabletop MIDI controllers, and a variety of other devices and interfaces. As previously noted, "this may expend the amount of time available to watch out for cues from the other players" (Hayes and Michalakos 2012, 39).

---

[1] Much has been written about the musical activity of improvisation, including discussions that aim to situate improvising on various – often unhelpful – axes including improvisation-composition, or freedom-restraint (cf., for example, Andean 2014). While it is beyond the scope of this chapter to discuss such trends, we will use the term "improvisation" simply to refer to musical activity that involves some form of real-time decision making where the timeliness of such choice making and action is crucial. Writing on improvisation from both embodied and ecological perspectives, Iyer summarizes that "we might take improvisation to denote that semi-transparent, multi-stage process through which we sense, perceive, think, decide, and act in real time" (Iyer 2016, 74).

Technologically-mediated musical practices, particularly ones which involve improvisational modes of playing, can also be multimodal in that they combine sonic, visual, and movement elements. For example, using sensors attached to the hands of a dancer, both musical and projected visual changes can be produced as derived from the analysis data of the sensors (cf. Hayes 2019, for further examples). Furthermore, sensor technologies can be employed to derive control data from biosignals using devices such as portable electroencephalogram (EEG) headsets or pulse sensors (Ortiz et al. 2011). The feedback loop can subsequently be closed as the near real-time analyses of such processes can be sent back to the performer as sonic output. Another corollary of human-machine improvisation is the ability for agency to be imbued within computer systems, to the extent that an improviser may feel as if they are improvising with another performer (Lewis 2000).

It is important to highlight the ways in which these technical developments have allowed for new forms of musical activity to take place given the increasing attention to musical improvisation within cognitive science research. Not only do such practices offer an environment in which issues of action, motor control, perception, and cognition can be studied, but they also provide participatory and experiential platforms for interdisciplinary research into the role of attention within these areas. Attention can be distinguished as perceptually-guided actions that alter the agent's openness to a multitude of further possibilities for action in relation to a task in a situation (van den Herik 2018). In the case of technologically-mediated musical improvisation, the performer's situation and environment comprises not only what may be hybrid or augmented acoustic-electronic-digital instruments and other musicians, but also the space in which the music is being created. This is important not only in terms of its acoustic properties but also the cultural history, norms, and possibly behaviors that have suffused within the site. An example of this might be the ways in which a particular venue will dictate whether the audience must sit in silence, or is 'permitted' to move around or make sound during musical play.

2 Attention from an Enactive-Ecological Perspective

We adopt an approach to understanding the modulation of attention which brings together insights from enactivism and contemporary versions of ecological psychology (Chemero 2009; Rietveld and Kiverstein 2014). This mixed enactive-ecological approach – or E approach – presents the idea that cognition is constituted by the skillful bodily engagement of an agent with a rich socio-material ecology (environment). Accordingly, attention is seen not so much as a particular cognitive faculty but as an aspect of the embodied engagement of the agent; it is an aspect of enaction or the bringing forth of a sensorimotor loop of perceptually-guided action (Varela et al. 1991). In this view the agent's skillful engagement manifests as the responsiveness, sensitivity, and selective openness to the opportunities for action afforded by the environment and the unfolding coupling between the agent and their niche (Bruineberg and Rietveld 2014). It is the skillfulness of such interactions that requires selectiveness. For an E approach, the point is not to show how attention works as a distinctive intracranial process of selection of focus or awareness. Rather, it is about embedding attention within the ongoing skillful activity and showing how ecologically-spread processes shape, control, amplify, and select the unfolding of the engagement as a whole.

The E approach, in a very general sense, consists of two complementary philosophical insights. On the one hand, it offers the idea that the agent's history of interactions – with materials, tools, particular settings in a situation, sounds, words, other agents' bodies, and so on – determines the agent's particular sensitivity to the possibilities for action in a given situation (Bruineberg et al. 2018). The agent's particular coupling of body and environment becomes a main source of significance and experience. This idea is a consequence of the emphasis on autonomy of the enactive approach (Di Paolo and Thompson 2014). On the other, the E approach proposes the idea of the direct perception of the kind of information in the environment that stands out in relation to the agent's bodily capacities. In other words, this entails the coming together of physical regularities or law-like properties of the environment and the features of the body (Chemero 2009). This idea – which is essentially relational – is captured by the term "affordance" in ecological psychology (Gibson 1977). Building on this, Rietveld and Kiverstein propose to see affordances not as individual aspects of the environment to which the agent attends one by one, but as constituting what the authors call "landscapes of affordances" (Rietveld and Kiverstein 2014, 326). With this move, they aim at the problem of so-called 'higher-cognition', addressing it in terms of "skillful activities in sociocultural practices and the material resources exploited in those practices. Skilled 'higher' cognition can be understood in terms of selective engagement – in concrete situations – with the rich landscape of affordances" (Rietveld and Kiverstein 2014, 326). Ultimately what Rietveld and Kiverstein achieve is the elimination of a hard distinction between 'lower-cognition' (e.g., motor control in a task) and 'higher-cognition' (e.g., planning). What changes along the continuum between 'lower' and 'higher' cognition is the observational focus on larger, more spatially and temporally distal relationships – that is, a social landscape of affordances – that enable the regularity of behavior.

The two ideas come together to account for a particular form of experience and intentionality. From an experiential perspective, the sensitivity to relations between body and environment is manifested as a felt concern or tension that solicits attention, recalibration of perception, and tension-reducing actions. In the phenomenological literature the experience of tensions – as the drive to adjust or act – is usually phrased in terms of what Merleau-Ponty called a process of finding an "optimal grip" (Merleau-Ponty 1962 [1945], 51). The classical example is the visitor in a gallery who adjusts their position in front of a painting as a means to maximize their aesthetic experience. Confronted with the painting as it is immersed in the light conditions of the art gallery, the visitor feels a need to reduce the tension with respect to the situation and is thus prompted to find the right angle of vision and distance of their body with respect to the painting. In the E approach vocabulary, the visitor enacts an optimal grip in the situation (Bruineberg et al. 2018). The key point of the E approach is that this particular sensitivity is not the result of a general inbuilt cognitive capacity for picking up and filtering stimuli but the effect of the active exercise of bodily habits that are acquired and stabilized through recurring interactions throughout the agent's life.

This view highlights bodily aspects in interaction, habitual ways of doing things, and the immersion of the agent in rich socio-material contexts. Agents develop forms of selective openness – attentional actions – in a way that is always engaged with a multiplicity of ongoing activities. This amounts to one of the principles of the E approach: perception and action are not separate processes linearly connected by cognitive modules. Rather, perception and action constitute a continuous dynamical loop (Chemero 2009). Thus, to perceive possibilities for action and to act is to generate the flows of information that can be sensed by means of, for example, moving in a gallery room. As agents are continuously immersed in ongoing activities the dynamic nature of their sensitivity and selectivity is constantly modulated by the parameters of interaction and engagement in those activities. Turning back to the example of the gallery visitor: each of their movements is equivalent to changes in the relation between body and environment – changes in the flow of energy array available in the gallery ecology – that alter simultaneously the perception of possibilities for further actions. Moreover, the history of interactions of the gallery visitor determines how certain features of the environment become salient as the visitor moves in the room. Such sedimented processes and background histories reveal themselves only when disrupted by, for example, artworks which aim to frame such interactions, or expose institutional confinements such as is the case in Daniel Buren's works which exceed "the physical boundaries of the gallery by having the art work literally go out the window" (Kwon 2004, 18). Here we notice that what might be described as 'cultural factors' are indeed part of Rietveld and Kiverstein's notion of affordance topographies which encompass not only affordances in the Gibsonian sense – opportunities for action – but also affordances with more distal relationships, such as those that may be encountered within, for example, cultural institutions, social conventions, and so on.

Spanning a diverse set of fields from human-computer interaction, systematic musicology, and music psychology, these key ideas from enactivist theory and ecological psychology that comprise the E approach have been hugely valuable in spearheading musical research that moves away from conceiving music as a purely cerebral, or skull bound activity. These themes have been discussed with respect to a variety of 'non-representational' musical practices; that is, musical activities that are not concerned with representing aspects of an 'original' reality – standing in for things – and that, moreover, favor and foster participation rather than detached observation (cf. Hogg 2013 for further discussion). Some pertinent examples include the design of new DMIs which focus specifically on embodied and tacit knowledge of everyday materials (O'Modhrain and Essl 2004); interdisciplinary improvisation using novel technologies involving musicians, visual artists, and dancers (Hayes 2019); research into musical emotion (van der Schyff and Schiavio 2017); music pedagogy focusing on embodied approaches (van der Schyff et al. 2016; Hayes 2017); and group improvisation through the lens of distributed cognition (Linson and Clarke 2017). This shift in focus not only sidesteps issues of cultural dominance – where Euro-American musicological research has privileged, for example, twelve-tone equal tempered scales – but also opens up interdisciplinary realms that challenge which forms of musical activity might be useful to study. Thus, music can be expanded to include all forms of organized sound, moving beyond focusing only on what can be represented within the Western notation system, which quickly demonstrates its limitations when timbral or spectral content is considered, for example.

3 Three Ranges of Timescales of Attention Modulation

4E approaches, in brief, radically reframe the notion of attention as selective openness in the continuous coordination between agents and a richly structured environment. They highlight how attentional actions are interwoven within the mutually shaping dynamics of patterns and regularities of the landscape of affordances (i.e., physical, but also socio-cultural regularities) and bodily features of the agent (including the self-organization of the central nervous system). They also bring to the fore the history of environments (particular socio-material relations that bear on the presence of affordances), history of bodies (idiosyncratic tendencies and habitual actions), and the emergent histories of interaction between agents and the world.

To understand modulation of attention in E approach terms thus requires a different strategy compared to conventional cognitivist approaches. Rather than privileging a description of modulation of attention based on a single process, often occurring on a single timescale, we propose to characterize at least three ranges of timescales (cf. Figure 12.1), each of which pick up a particular angle on the rich coordination between agents and environments. Following multiscalar frameworks of human interactivity and coordination (Gahrn-Andersen et al. 2019; Loaiza et al. 2020) we propose three temporal ranges associated with slow, mid, and fast modulation.[2] We outline these temporal ranges in relation to three enactive-ecological flavored themes as follows:

1. Firstly, we consider the notion of habit which affirms the interdependencies between organism and environment as a fundamentally embodied process of identity generation: this refers to the slowest or longest timescales of development and enculturation. The person, as a matter of development of personal and sensorimotor identity, becomes sensitive, or attentive to certain forms of interaction, expression, use of tools, and so on.
2. Secondly, we consider the role of attention within social interactivity, whereby attention can be said to emerge within the processes of participatory sensemaking (De Jaegher and Di Paolo 2007). This refers to mid timescales of participation in communities of practice and sociality. Here selectivity is modulated through joint attention.[3] In particular, this occurs through joint attention to and with mediating technologies, including not only musical interfaces but also environmental aspects such as venues.

---

[2] Loaiza, Trasmudi, and Steffensen (2020) define a temporal range as an observer-dependent set of timescales that can sufficiently capture the interdependencies and feedback loops of processes that take place across multiple scales. This can be understood in the usual terms of longer / slower timescales determining the boundary conditions of faster / shorter processes, but also in terms of reciprocating constraints on slower timescales reproduced on faster timescales.

[3] Here joint attention refers to the kind of social triangulation most commonly present in dyadic interactions with a shared task. For example, the dyad of child and caregiver in which interpersonal coordination comprises the engagement with a 'third' party: an object of the child and caregiver's joint attention. Joint attention is an important developmental step in what ecological psychologists have described as "education of attention", in which adults guide the child's discovery of affordances in the environment (van den Herik 2018).

3. Finally, we propose that while attention has often been articulated through ocularcentric metaphors of focus or illumination, a more fruitful approach might involve articulating the speed, strength, or amplification of such dynamics as a processual, non-static paradigm. This refers to the fast timescales of changing attention on the fly.

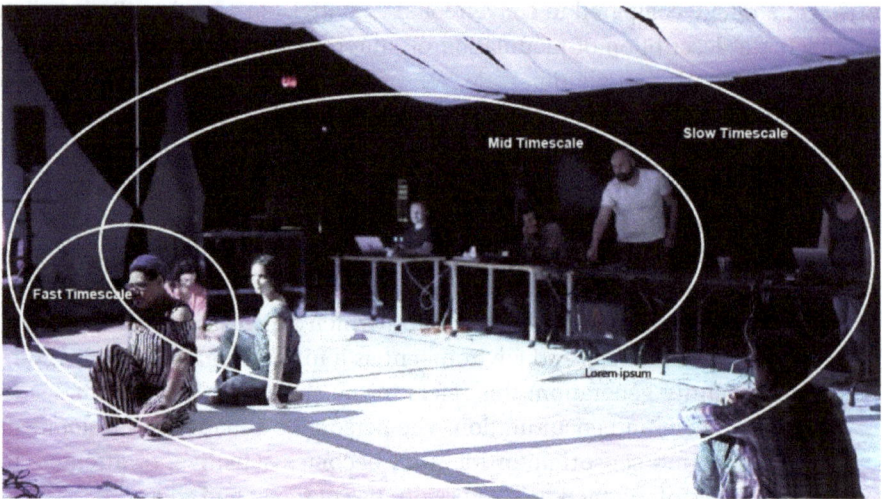

Figure 12.1: Example of multiscalar frameworks of human-environment / human-human coordination within interdisciplinary improvisation involving sound, movement, and visuals during LLEAPP 2018, Tempe, AZ. The slowest timescale includes the regularities and conventions of the performance space, the audience and performer etiquette, the material factors that enable this performance; the mid timescale concerns the community of practice (in this case, the various performers); the fastest timescale indicates how the individual performer changes attention on the fly, perhaps weakening the focus on physical movement while strengthening awareness of co-performers or visual changes that appear within the space.

The long timescales of modulation of attention along a person's development can be explored with the notion of habit. This term has been recently reincorporated into the literature on cognition from an enactivist perspective (Barandiaran and Di Paolo 2014). After the eclipse of behaviorism on account of the early forms of cognitivism in the 1950s and 60s, the notion of habit became associated only with the kinds of automatic and repetitive behavior or putative forms of 'lower' cognition. However, outside of the cognitive science discipline, the term remained in wider use. For example, echoing Bourdieu's use of *habitus*, ethnomusicologist Thomas Turino refers to habits when accounting for the acquisition of musical tastes and embodied styles of performative behavior (Turino 2008; Bourdieu 1977). For Turino, dispositions and habits are products of relations to the conditions around [the person] and her concrete experiences in and of the environment. Habits and dispositions guide what we think, do, and make (practices). Our practices and the things we produce affect, to greater or lesser degrees, our environment, which in turn affects our dispositions, which in turn affect our practices, which in turn affect external conditions, and so on. (Turino 2008, 120).

Turino uses this circularity in order to formulate both a definition of the "personal self", and of "cultural formations" and "cultural cohorts" (Turino 2008, 120). Ethnomusicologist Judith Becker similarly points to a "habitus of listening" (Becker 2004, 69) that accounts also for a totality of predispositions and tendencies of "communities of listeners" whereby "every hearer occupies a position in a cultural field not of his or her own making: every hearing is situated" (Becker 2004, 69). In short, for Becker musical perceptual attitudes are socially constituted.

Recent developments on habit have incorporated some of the core concepts of the enactive approach (Barandiaran 2008; Barandiaran and Di Paolo 2014; Ramírez-Vizcaya and Froese 2019; James and Loaiza 2020), in particular the organizational notion of autonomy that characterizes structures "between the biological and the psychological" (Egbert and Barandiaran 2014, 2). In the enactive view, habits are "self-sustaining patterns of sensorimotor coordination formed when the stability of a particular mode of sensorimotor engagement is dynamically coupled with the stability of the mechanisms that generate it" (Barandiaran 2008, 281). In this way, habit is not simply a kind of automatic or solidified behavioral program but a process that acquires a "life of its own" and thus is the cause and effect of itself (Barandiaran 2017). This circularity of self-production is precisely what characterizes the notion of autonomy in the enactive literature, the outcome of which is the complementary idea of self-distinction or identity.[4] Habits, thus, are self-producing and self-distinguishing patterns with lasting structuring effects on behavior.

We argue that in terms of long/slow timescales, attention is modulated by habits of listening, playing, and gesturing, amongst others. Following the enactive

---

[4] The notion of identity that we use here is a technical term of enactivism. It corresponds to the persistent patterns of organization of behavior that emerge from the operational closure occurring at distinct levels (Thompson 2010, 60). In this way, there's an identity for each cell at the level of closure of cell metabolism; there's an identity of the central nervous system at the level of closure of self-organized neurodynamics; and similarly, it is suggested that there are other forms of identities at the levels of the whole organism and beyond the individual: sensorimotor coordination, habitual patterns, personal linguistic use, participatory coordination (Di Paolo et al. 2018).

notion of emergent self-production, these habits are not simply copied from the sociocultural milieu as memes, nor do they otherwise express the implementation of a pre-established program. Instead, habits – on an individual level – emerge gradually and in an unprescribed way along the myriad of concrete encounters between persons in interaction and engaging with materials. Habits, once they have life of their own, strongly regularize behavior. But the way, quality, and intensity in which habits emerge in the first place are thoroughly contingent and idiosyncratic. Beyond single individual patterns, habits can form networks of mutually reinforcing activity (James 2020). Once they act as a network taking a significant part of the individual person's behavior, habits generate what James and Loaiza call a form of "enhabiting" (2020, 6).[5] Persons "enhabit" as they acquire habits and also are themselves the effect of enhabiting. In this view, the individuality of our personalities and unique musical experiences, tastes, and imagination, as well as the personal autonomy we can self-designate and claim credit for, all rest on a network of mutually self-sustaining habits in a form of an enhabiting process. Crucially, enhabiting – that is, the activity of a network of habits – has an identity, yet there is no central 'self' behind the network to which we can point to as the source or ultimate bearer of such identity. In other words, identity is precisely the overall effect of the self-sustaining and self-constraining activity of the whole network (Loaiza 2016).

What does this idea of enhabiting have to do with modulation of attention? Each individual habit implies a way of modulating attention as part of the circular reciprocity between action and perception. As a network of habits matures, it is thus also implied that such particular forms of modulating attention will mutually reinforce, with the effect of the emergence of a more general identity of the attentional activity. In this way, a person may have a network of associated habits of listening to particular melodic structures, habits of handling instruments and controllers, habits of use of vocalizations in performance. The network, given the particulars of an environment and the tasks at hand, may yield a more general pattern of attentional actions. For example, this may include patterns of marking points of interest in the production of sonic material

---

[5] The enactive approach to habit is a development of ideas found in – amongst others – Husserl, Piaget, and Merleau-Ponty. Merleau-Ponty sees habits "as incorporated styles of being-in-the-world [. . .] [which reveal] corporeal intentionality in contrast to notions of habits as blind automatisms" (Barandiaran and Di Paolo 2014, 6). What the enactive approach adds to the phenomenology of habit is a dynamical and operative account of the type of operational closure of processes that needs to be in place for habits to self-maintain and simultaneously be adaptive (James and Loaiza 2020). See previous footnote on the relation of operational closure and identity in enactivist terms.

by timing and integrating a certain detail in the way of handling the instrument, a particular idiosyncratic vocalization, or a certain way of anticipating a melodic structure. In our view, attention – as a global organization of awareness and selectiveness in a task – emerges in part as a self-organized network of mutually self-sustaining habits of listening, moving, seeing, touching, vocalizing, and so on. In this way, following Ganeri (2017), there is no need to think of a centralized entity (a self) or process that is the bearer of the attentional 'spotlight' or the fundamental source of attention.[6] Attention, at slower timescales, is modulated by the history and entrenchment of constellations of habits. What follows is to show how, although generally stable in slow timescales, habits, and thus attention, can also manifest plasticity in faster timescales. They come to be modulated by constraints brought forth in social participation. In this way, habits can also become part of skillful engagements with other persons within communities of (skillful) practice.

In their efforts to provide an enactive account of social interaction, Hanne De Jaegher and Ezequiel Di Paolo develop the notion of "participatory sense-making" (De Jaegher and Di Paolo 2007, 485). Their approach focuses on a move away from individualistic interactions between agents and environments, and instead deals with the crucial micro-interactions that comprise social interaction. Moreover, this is not limited to linguistic or even anthropocentric patterns. Specifically, they define social interaction as "the regulated coupling between at least two autonomous agents, where the regulation is aimed at aspects of the coupling itself so that it constitutes an emergent autonomous organization in the domain of relational dynamics" (De Jaegher and Di Paolo 2007, 493). Importantly, the autonomy of the agents involved – while it may grow or diminish – remains intact. Here, attention emerges within the mutually-affecting relationship between coordination and interaction. Echoing their example of

---

6 Ganeri (2017) presents a non-orthodox view of the relation between attention and self that in part reverses and overcomes the more classical European view in which attention implies a minimal form of authorship and self (cf. Zahavi 2014). Ganeri's position explores the Buddhist doctrine of "no-self" (Anātman) in light of contemporary philosophy of mind to argue against a conception of authorship in attention. In parallel to Ganeri, the enactive approach is rooted in the Buddhist Madhyamika tradition in which "emptiness" (Sunyata) and "(co-)dependent arising" (Pratītyasamutpāda) play an important part in the rejection of self/nonself essentialist distinctions. Enactivism reworks "codependent arising" in its notion of emergent identity of self-organized systems. To our knowledge, the connection between Ganeri's work on attention and the enactive approach to emergence of identity and self has not been worked out more fully in the literature (Ganeri 2017, 306, 309). The discussion reaches considerably beyond our present proposal. It is however necessary to mention the resonances and possible cross-pollination between the enactive approach and Ganeri's views. Thanks to one of the reviewers for pressing the question of the relation between attention and self (minimal or otherwise).

dancers taking on the role of leader or follower, so too can we illustrate this idea within musical practices. In freely improvised music, for example, there is a common approach where performers may take on these roles, either through direct assignment prior to starting to play, through a number of 'conducted' approaches, or through other ad hoc means. Yet despite the typical etiquettes determined by such roles, it becomes evident that each musician turns their attention to the other which affects how the other plays, how they themselves play, and how both, together, create, sustain, and modulate the ongoing stream of musical activity. This participatory framing can be contrasted with what has been described as a "spectatorial stance" (Schiavio and De Jaegher 2017, 32).

We define this mid-scale temporality of attention as referring to an agent's participation in communities of practice – that is, their participation with their various milieux. Crucially here, attention is modulated not only through the specific histories and experiences of the individual, but rather through collective processes of joint attention that occur among and between agents. Moreover, this occurs through jointly being attentive to various technologies of mediation, which may include not only musical instruments or sounding objects, but also the spaces in which the activity is taking place. Here, "musicians (and audience), as a coupled system, participate in, *and thus can form and transform* each other's sense-making, enacting unique shared worlds of meaning" (Schiavio and De Jaegher 2017, 33). Through engaging with mediating technologies, groups of improvisers develop habitual patterns of behavior[7] through informal or formalized attentive strategies. Group musical vocabularies are established and developed through playing together and listening practices, as well as through reflection and discussion. Individual musical practices and techniques are shared and subsumed within the group's evolution. For example, as part of the creative music practice research group Laboratory for Live Electronic Audio/Art Performance Practice (LLEAPP), musicians and movement practitioners engaged in various collectively defined attentive strategies (cf. Figure 12.1) using techniques drawn from both the performing and the visual arts (Hayes 2019). Here, a sound-reactive projected visual – an instrument developed by one of the musicians in the group – was repurposed in order to create rule-based movement strategies for the entire group (cf. Hayes 2019 for more details). Working together over several days led to the emergence of a group identity which was formed out of not only the unique histories and skillsets of the

---

7 See our earlier explanation (Footnote 4) of identity as a technical term within enactivism. This allows for a definition that encompasses, on one hand, the cellular level to issues of group identity via participatory coordination, on the other.

individual members – slow timescales – but also through mutually affecting relationships of participatory sense-making. The sociality of such "habitual identity" (Wehrle 2020, 1) is evident within such practices. As Wehrle notes, "As human bodily subjects we are situated, which is to say that we are embedded in and shaped by already existing ecological, historical, cultural and socio-economic environments, reflecting in turn specific social norms and power relations" (Wehrle 2020, 14).

Perhaps most prevalent within discussions of attention is what we define as the third time-scale: that of the fast, often near-instantaneous on the fly changes of attention. Typically, the sorts of metaphors that are commonly used to describe attentive processes are almost exclusively derived from ocularcentric notions of 'illumination', even within studies of attention in musical contexts. These can be found in studies that use the typical 'spotlight' metaphor, where participants are asked to shift their focus – in the same way that a spotlight can jump between different targets – from, for example, the movement of their body to the sounds that their movements produce (Duke et al. 2011); or where attention might be diverted by a secondary task (Norgaard et al. 2016). Another example is a comprehensive study dealing with auditory attention from both spatial and feature processing perspectives which similarly uses an "auditory searchlight" (Fritz et al. 2007, 441) metaphor. While such studies do important work from an interdisciplinary perspective, we suggest that this framing is limiting in its ability to describe more nuanced dynamics of the processes of attention. With this strategy, the focus – or location – of attention can be described but not how intensive or prominent this may be, nor how it may vary over time. Rather, we invoke metaphors of speed, strength, and amplification of such dynamics in their ability to help convey the richer possibilities that can be exemplified in attentional mediation within various forms of musical activity. Furthermore, this approach avoids the prevalent issues of how to account for divided attention or task switching (cf. Monsell 2003 for a neuroscientific discussion of such issues).

Numerous empirical studies point to the importance of multimodal feedback in the joint coordination of musicians (Schroeder et al. 2007; Eerola et al. 2018). Rietveld and Kiverstein's (2014) extended notion of landscapes of affordances – as more than opportunities for motor action alone – is particularly helpful in understanding the richly dynamic ways in which attention is mediated in the often highly complex situations of musical activity. As mentioned above, when performing with technologically-augmented instruments, musicians have the ability to work with non-real time aspects of what has transpired, or even with audio samples recorded in advance. Here, attention to memory extends beyond the current performance and involves not only the emergence of new musical material based on joint articulation, but also the ability to re-trigger musical events that have been heard in the past. Moreover, computer technologies mean that sounds can be performed beyond the physicality that is humanly possible: algorithmic automation can produce repeated percussion 'hits' every millisecond. Musicians must assess whether to respond to gestures that may not even produce sound, such as triggering a DSP analysis process on a laptop. Furthermore, ensembles can often be extremely large, and may not even be in the same location, such as in the field of telematic musical performance. While De Jaegher and Di Paolo's thesis is based on the joint articulation between two people – although they state that it can be extended to three or more – a musician may have to attend to what is transpiring between dozens of other improvisers, or more. When the world of music making opens itself to unconventional instruments, novel technologies, spontaneous configurations of performers, unusual spatial – often dislocated – arrangements, and so on, it becomes clear that 'openness to affordances consists of a readiness to act in ways appropriate to a particular concrete situation' (Rietveld and Kiverstein 2014, 347).

Rather than, then, the skillfulness of a musical improviser being determined by their ability to focus on a locus of activity, and shift the 'searchlight' of their attention to something else – or indeed divide this into several searchlights – we suggest that a dynamic network of attentive bonds that become strengthened, amplified, or diminished better describes this skillfulness. Moreover, these bonds may form and dissipate at varying speeds: a sharp interjection from another musician may command rapid attention, but more subtle development within textural sonic material, for example, may become noticed and responded to more gradually. Attention here is multimodal and embodied, as the musician attunes to a variety of physical gestures and sounds, or perhaps a more general rhythm, pulse, or vibe, and so on. The musician's skillful engagement is found within their ability to be sensitive and responsive to these changes in their environment. That is, it resides within their selective openness to the opportunities for action that constantly appear and disappear within the ephemerality of improvisation. Amplification of a particular musical motif – which literally gets louder – may trigger attentive amplification, but the skilled improviser may also strengthen their attentive bond to a quieter, subtler, perhaps almost inaudible sound that transpires simultaneously, or entrain their bodily movements to an irregular but palpable pulse. Similarly, when improvising with DMIs, rather than focusing on a particular parameter of an instrument or system, the skillful musician remains open to the affordances offered by what may be a highly complex musical software agent or instrument. This is done not by forming mental representations of, for example, interface to DSP mappings on the fly, but rather, by being able to respond by strengthening and weakening particular attentive bonds within this constantly evolving dynamic activity.

4 Conclusion

In this paper we have argued that technologically-mediated musical practices, particularly ones which involve improvisational modes of playing, offer a site in which attention can be explored as an embodied and dynamic process, rather than conceiving of it as a specific cognitive faculty. Through the enactive-ecological framework we have defined three ranges of scales which demonstrate the different ways in which the selective openness or responsiveness of a musician within their niche can be said to modulate attention. Within her writing concerning the modes of attention, composer and founder of Deep Listening, Pauline Oliveros (2010) has pointed to two contrasting types of attention: "focal" and "global". While her notion of focal attention aligns very much with the ocularcentric metaphorical paradigm – "attention to one point and nothing else" (Oliveros 2010, 29) – her concept of global attention very much aligns with our notion of attention as selective openness, as an "open receptive state [. . .] attention [is] expanded to a field" (Oliveros 2010, 29). We hope that this framework not only provides new avenues for the study of attention across all three ranges of timescales – as habitual maturation, joint sociality, and dynamic engagement – but also that it opens up collaborative possibilities that build upon the expertise found within the skillful and deeply social practices of techno-fluent improvising musicians. In this, we hope to see the emergence of interdisciplinary research that – by moving away from taking "spectatorial stances" – is itself simultaneously enriched.

References

Andean, J. (2014): "Research Group in Interdisciplinary Improvisation: Goals, Perspectives, and Practice". In: A. Arlander (ed.): *This and That – Essays on Live Art and Performance Studies*. Helsinki: Theatre Academy, 174–191.

Barandiaran X. E. (2008): *Mental life: A Naturalized Approach To The Autonomy Of Cognitive Agents*. Unpublished Ph. D. thesis, University of the Basque Country, Leioa.

Barandiaran X. E. (2017): "Autonomy and Enactivism: Towards a Theory of Sensorimotor Autonomous Agency". *Topoi* 36, 409–430.

Barandiaran, X. E.; Di Paolo, E. (2014): "A Genealogical Map of the Concept of Habit." *Frontiers in Human Neuroscience* 8.

Becker, J. (2004): *Deep Listeners: Music, Emotion, and Trancing* (Vol. 1). Bloomington, Indiana: Indiana University Press.
Bourdieu, P. (1977): *Outline of a Theory of Practice*. Cambridge: Cambridge University Press.
Bruineberg, J.; Rietveld, E. (2014): "Self-organization, Free Energy Minimization, and Optimal Grip on a Field of Affordances". *Frontiers in Human Neuroscience* 8, 599.
Bruineberg, J.; Chemero, T.; Rietveld, E. (2018): "General Ecological Information Supports Engagement with Affordances for 'Higher' Cognition". *Synthese* 196:12, 5231–5251.
Chemero, A. (2009): *Radical Embodied Cognitive Science*. Cambridge, MA; London: MIT Press.
De Jaegher, H.; Di Paolo, E. (2007): "Participatory Sense-making". *Phenomenology and the Cognitive Sciences* 6:4, 485–507.
Di Paolo, E. A.; Cuffari, E. C.; De Jaegher, H. (2018): *Linguistic Bodies: The Continuity between Life and Language*. London: MIT Press.
Di Paolo, E.; Thompson, E. (2014): "The Enactive Approach". *The Routledge Handbook of Embodied Cognition*. New York and London: Routledge, 86–96.
Duke, R, A.; Cash, C. D.; Allen, S. E. (2011). "Focus of Attention Affects Performance of Motor Skills in Music". *Journal of Research in Music Education* 59:1, 44–55.
Eerola, T.; Jakubowski, K.; Moran, N.; Keller, P. E.; Clayton, M.(2018): "Shared Periodic Performer Movements Coordinate Interactions in Duo Improvisations". *Royal Society Open Science* 5:2, 171520.
Egbert, M. D.; Barandiaran, X. E. (2014): "Modeling Habits as Self-Sustaining Patterns of Sensorimotor Behavior". *Frontiers in Human Neuroscience* 8, 590.
Fritz, J. B.; Elhilali, M.; David, S. V.; Shamma, S. A. (2007): "Auditory Attention – Focusing the Searchlight on Sound". *Current Opinion in Neurobiology* 17:4, 437–455.
Gahrn-Andersen, R.; Johannessen, C. M.; Harvey, M.; Marchetti, E.; Simonsen, L.M.; Trasmundi, S. B. et al. (2019): "Interactivity: Why, What and How?" *RASK–Int* 50, 113–136.
Ganeri, J. (2017): *Attention, Not Self*. New York, NY: Oxford University Press.
Gibson, J. J. (1977): "The Theory of Affordances". In R. Shaw and J. Bransford (eds.): *Perceiving, Acting, and Knowing*. New York, NY: Lawrence Erlbaum, 67–82.
Hayes, L.; Michalakos, C. (2012): "Imposing a Networked Vibrotactile Communication System for Improvisational Suggestion". *Organised Sound* 17(1): 36–44.
Hayes, L. (2017): "Sound, Electronics, and Music: A Radical and Hopeful Experiment in Early Music Education". *Computer Music Journal* 41(3), 36–49.
Hayes, L. (2019): "Beyond Skill Acquisition: Improvisation, Interdisciplinarity, and Enactive Music Cognition". *Contemporary Music Review* 38/5, 446–462.
Hogg, B. (2013): "The Violin, The River, and Me: Artistic Research and Environmental Epistemology in Balancing String and Devil's Water 1, Two Recent Environmental Sound Art Projects". *HZ Journal* 18, http://www.hz-journal.org/n18/hogg.html.
Iyer, V. (2016): "Improvisation, Action Understanding, and Music Cognition with and without Bodies". In: Lewis, George E.; B. Piekut (eds.): *The Oxford Handbook of Critical Improvisation Studies, Volume 1*. New York: Oxford University Press, 74–90.
James, M. M. (2020): "Bringing forth within: Enhabiting at the Intersection between Enaction and Ecological Psychology". *Frontiers in Psychology* 11.
James, M. M.; Loaiza, J. M. (2020): "Coenhabiting Interpersonal Inter-Identities in Recurrent Social Interaction". *Frontiers in Psychology* 11.
Kwon, M. (2004): *One Place after Another: Site-Specific Art and Locational Identity*. Cambridge, MA: MIT Press.

Lewis, G. E. (2000): "Too Many Notes: Computers, Complexity and Culture in Voyager". *Leonardo Music Journal* 10, 33–39.
Linson, A., Clarke, E. F. (2017): "Distributed Cognition, Ecological Theory, and Group Improvisation". In: E. Clarke; M. Doffman (eds.): *Distributed Creativity: Collaboration and Improvisation in Contemporary Music*. New York, NY: Oxford University Press.
Loaiza, J.M. (2016): "Musicking, Embodiment and Participatory Enaction of Music: Outline and Key Points". *Connection Science* 28, 410–422.
Loaiza, J. M.; Steffensen, S. V.; Trasmundi, S. B. (2020): "Multiscalar Temporality in Human Behaviour: A Case Study of Constraint Interdependence in Psychotherapy". *Frontiers in Psychology* 11, 1685.
Loui, P. (2018): "Rapid and Flexible Creativity in Musical Improvisation: Review and a Model". *Annals of the New York Academy of Sciences* 1423(1), 138–145.
Merleau-Ponty, M. (1962): *Phenomenology of Perception*. London: Routledge & Kegen Paul.
Monsell, S. (2003): "Task Switching". *Trends in Cognitive Sciences* 7:3, 134–140. https://doi.org/10.1016/S1364-6613(03)00028-7
Norgaard, M.; Emerson, S. N.; Dawn, K.; Fidlon, J. D. (2016): "Creating Under Pressure:Effects of Divided Attention on the Improvised Output of Skilled Jazz Pianists". *Music Perception* 33:5, 561–570.
O'Modhrain, S.; Essl, G. (2004): "PebbleBox and CrumbleBag: Tactile Interfaces for Granular Synthesis". In: *Proceedings of the 2004 Conference on New Interfaces for Musical Expression*. Singapore: National University of Singapore Press, 74–79.
Oliveros, P. (2010): *Sounding the Margins: Collected Writings 1992-2009*. Kingston: Deep Listening Publications.
Ortiz, M.; Coghlan, N.; Jaimovich, J.; Knapp, R. B. (2011): "Biosignal-Driven Art: Beyond Biofeedback". *Ideas Sonicas/Sonic Ideas* 3:2.
Posner, M. I.; Petersen, S. E. (1990): "The Attention System of the Human Brain". Annual *Review of Neuroscience* 13:1, 25–42.
Pressing, J. (1984): "Cognitive Processes in Improvisation". In: R. Crozier; A. Chapman (eds.): *Cognitive Processes in the Perception of Art*. Amsterdam: North Holland, 345–363.
Pressing, J. (1988): "Improvisation: Methods and Models." In: Sloboda J. A. (ed.): *Generative Processes in Music: The Psychology of Performance, Improvisation, and Composition*. New York: Oxford University Press.
Ramírez-Vizcaya, S.; Froese, T. (2019): "The Enactive Approach to Habits: New Concepts for the Cognitive Science of Bad Habits and Addiction". *Frontiers in Psychology* 10:301.
Rietveld, E.; Kiverstein, J. (2014): "A Rich Landscape of Affordances." *Ecological Psychology* 26:4, 325–352. https://doi.org/10.1080/10407413.2014.958035
Schiavio, A.; De Jaegher, H. (2017): "Participatory Sense-Making in Joint Musical Practices". In: M. Lesaffre; M. Leman; P.J. Maes (eds.): *The Routledge Companion to Embodied Music Interaction*. New York and London: Routledge, 31–39.
Schroeder, F.; Renaud, A.; Rebelo, P.; Gualdas, F. (2007): "Addressing the Network: Performative Strategies for Playing Apart". In: *Proceedings of the 2007 International Computer Music Conference (ICMC)*. Copenhagen, 113–9.
Stewart, J. R.; Gapenne, O.; Di Paolo, E. A. (2010): *Enaction: Toward a New Paradigm for Cognitive Science*. Cambridge, MA: MIT Press.
Thompson, E. (2010): *Mind in Life: Biology, Phenomenology, and the Sciences of Mind*. Cambridge (Mass): Belknap Press.

Turino, T. (2008): *Music as Social life: The Politics of Participation*. Chicago: University of Chicago Press.

van den Herik, J. C. (2018): Attentional Actions—An Ecological-Enactive Account of Utterances of Concrete Words. *Psychology of Language and Communication* 22:1, 90–123.

van der Schyff, D.; Schiavio, A. (2017): "The Future of Musical Emotions". *Frontiers in Psychology* 8, 988.

van der Schyff, D.; Schiavio; A.; Elliott, D. J. (2016): "Critical Ontology for an Enactive Music Pedagogy". Action, Criticism, and Theory for Music Education, 15: 5.

Varela, F.; Thompson, E.; Rosch, E.; (1991): *The Embodied Mind: Cognitive Science and Human Experience*. Cambridge, MA: MIT Press.

Walton, A.; Richardson, M.; Langland-Hassan, P.; Chemero, A.; Washburn, A. (2015): "Musical Improvisation: Multi-Scaled Spatiotemporal Patterns of Coordination". In: D. C. Noelle; R. Dale; A. S. Warlaumont; J. Yoshimi; T. Matlock; C. D. Jennings et al. (eds.): *Proceedings of the 37th Annual Meeting of the Cognitive Science Society*. Austin, TX: Cognitive Science Society, 2595–2600.

Wehrle, M. (2020): "Bodies (that) Matter: the Role of Habit Formation for Identity". *Phenomenology and the Cognitive Sciences* 20, 356–386.

Zahavi, D. (2014): *Self and Other: Exploring Subjectivity, Empathy, and Shame*. Oxford: Oxford University Press.

Short Biography

Lauren Hayes, PhD, is a Scottish musician and sound artist who builds hybrid analogue/digital instruments and unpredictable performance systems. As an improviser, her music has been described as 'voracious' and 'exhilarating'. Her research explores embodied music cognition, enactive approaches to digital instrument design, and haptic technologies. She is currently Assistant Professor of Sound Studies within the School of Arts, Media and Engineering at Arizona State University where she leads PARIESA (Practice and Research in Enactive Sonic Art). She is Director-At-Large of the International Computer Music Association and is a member of the New BBC Radiophonic Workshop.

Juan Loaiza, PhD, is an independent scholar. His interests lie in the crossroads of southamerican traditional music, participatory projects, and the philosophy of cognitive science. His research comprises explanatory, descriptive, and creative levels of engagement, seeking to establish links between post-cognitivist, ecological-enactive, approaches to cognition and mind, and the lived experience of the practice of music within the social groups where Juan situates himself. He has received a PhD from Queen's University Belfast, a master in music composition at the University of Newcastle upon Tyne and a Bachelor of Engineering in Medellín, Colombia.

www.ingramcontent.com/pod-product-compliance
Lightning Source LLC
Chambersburg PA
CBHW060351190426
43201CB00044B/1980